Artificial Intelligence in Action

Business practices are being drastically altered and improved using AI and machine learning (ML), a branch of AI that enables autonomous machines. Big data can assist AI to automatically learn and acquire knowledge to assist managers in achieving their practical and technical objectives. AI brings new methods for conducting business, finding new opportunities in the global market, and executing work.

Artificial Intelligence in Action: Transforming Industries and Society focuses on novel ideas and research for providing a broader outlook on AI in business and society. The book:

- Reflects the diverse applications of AI, from healthcare and education to finance and e-commerce, while emphasizing sustainability.
- Highlights the transformative role of AI in addressing modern challenges across multiple sectors, including public distribution, retail, and human resource management.
- Covers the evolution and prospects of AI in achieving sustainable development goals, improving education, and advancing social and economic outcomes.

The book explores the innovations, disruptions, and changes that AI is causing in industry and business. It looks at the fast-evolving global changes being wrought by the rise of large-language models, real-time tools for decision-making, and AI-powered analytics. Presenting novel ideas for creating robust sustainability strategies, the book explores answers to the moral and ethical questions about using AI in decision-making, cybersecurity, and resource management. Covering a vast swath of industries from healthcare to retail and education, it highlights the meaningful overall impact of AI by addressing key issues, trends, and future perspectives.

Advances in Computational Collective Intelligence

Series Editor: Dr. Subhendu Kumar Pani

Deep Learning Applications in Operations Research
Edited by Aryan Chaudhary, Biswadip Basu Mallik, Gunjan Mukherjee, and Rahul Kar

Computational Intelligence for Analysis of Trends in Industry 4.0 and 5.0
Edited by Joseph Bamidele Awotunde, Kamalakanta Muduli, and Biswajit Brahma

Metaverse and Blockchain Use Cases and Applications
Edited by Dileep Kumar Murala, Sandeep Kumar Panda, and Sujata Priyambada Dash

Leveraging Artificial Intelligence in Cloud, Edge, Fog and Mobile Computing
Edited by Shrikaant Kulkarni, P. William, Vijaya Prakash, and Jaiprakash Narain Dwivedi

Using AI to Develop Sustainability Strategies for a Changing Global Economy
Edited by A.V. Senthil Kumar, Ankita Chaturvedi, Atul Bansal, and Rohaya Latip

Artificial Intelligence, Geographic Information Systems, and Multi-Criteria Decision-Making for Improving Sustainable Development
Edited by Sujoy Kumar Jana, Kamalakanta Muduli, Indrajit Pal, and Purushottam Meena

Advanced AI and Data Science Applications
Edited by Dr. D. Sivabalaselvamani, Dr. G Revathy, and Dr. Ranjit Singh Sarban Singh

Intelligent Business Analytics
Edited by Nitendra Kumar, Lakhwinder Kaur Dhillon, Mridul Dharwal, Elena Korchagina, and Vishal Jain

Augmented Reality and Sustainability: Goals and Challenges
Edited by Sonal Trivedi, Vishal Jain, Balamurugan Balusamy, Subhendu Pani, and Danish Ather

Patient-Centric 6G: A New Era in Smart Healthcare
Edited by Rajeev Kumar, Ankush Joshi, Preeti Bajaj, and Danila Parygin

Generative AI Tools: From Algorithms to Applications
Edited by Priyanka Sharma, Senthil Kumar AV, Monika Jyotiyana, and Adnan Alrabea

IoT Cybersecurity: Trends, Challenges, and Solutions
Edited by Laxmi Shaw, D. Ajitha, Chinmay Chakraborty, and S.M. Prabin

Artificial Intelligence in Action: Transforming Industries and Society
Edited by A.V. Senthil Kumar, Ankita Chaturvedi, Atul Bansal, and Rohaya Latip

https://www.routledge.com/Advances-in-Computational-Collective-Intelligence/book-series/ACCICRC

Artificial Intelligence in Action

Transforming Industries and Society

Edited by

A.V. Senthil Kumar, Ankita Chaturvedi,
Atul Bansal, and Rohaya Latip

CRC Press
Taylor & Francis Group
Boca Raton London New York

CRC Press is an imprint of the
Taylor & Francis Group, an **informa** business
AN AUERBACH BOOK

Designed cover image: Shutterstock

First edition published 2026
by CRC Press
2385 NW Executive Center Drive, Suite 320, Boca Raton FL 33431

and by CRC Press
4 Park Square, Milton Park, Abingdon, Oxon, OX14 4RN

CRC Press is an imprint of Taylor & Francis Group, LLC

Library of Congress Cataloging-in-Publication Data
Names: Kumar, A. V. Senthil, 1966- editor | Chaturvedi, Ankita editor | Bansal, Atul editor | Latip, Rohaya editor
Title: Artificial intelligence in action : transforming industries and society / edited by A.V. Senthil Kumar, Ankita Chaturvedi, Atul Bansal, and Rohaya Latip.
Description: First edition. | Boca Raton, FL : CRC Press, 2026. | Series: Advances in computational collective intelligence | Includes bibliographical references and index.
Identifiers: LCCN 2026002390 (print) | LCCN 2026002391 (ebook) | ISBN 9781041069935 hbk | ISBN 9781041070306 pbk | ISBN 9781003638506 ebk
Subjects: LCSH: Artificial intelligence–Industrial applications
Classification: LCC TA347.A78 A834 2026 (print) | LCC TA347.A78 (ebook)
LC record available at https://lccn.loc.gov/2026002390
LC ebook record available at https://lccn.loc.gov/2026002391

ISBN: 978-1-041-06993-5 (hbk)
ISBN: 978-1-041-07030-6 (pbk)
ISBN: 978-1-003-63850-6 (ebk)

DOI: 10.1201/9781003638506

Typeset in Times
by SPi Technologies India Pvt Ltd (Straive)

Contents

About the Editors

Dr. A.V. Senthil Kumar is working as Principal, Nehru Institute of Information Technology and Management, Coimbatore, India. He has worked as Professor and Director, PG and Research Department of Computer Applications, Hindusthan College of Arts & Science, Coimbatore, Tamilnadu for more than 15 years and as Senior Grade Lecturer in CMS College of Science and Commerce for 14 years. He has five years of industrial experience and 29 years of teaching experience. He has also received his Doctor of Science (D.Sc in Computer Science). He has to his credit 235 papers in national and international journals, 96 book chapters, 92 papers in national and international conferences, 22 edited volumes, and three textbooks. Dr. Kumar is a key member for India, Machine Intelligence Research Lab (MIR Labs). He is involved in many esteemed journals as an Editor-in-Chief, Associate Editor, Editorial Board Member, and reviewer. He is also a committee member for various international conferences and has guided 16 Ph.D scholars.

Dr. Ankita Chaturvedi is a globally recognized expert in Commerce, Accounting, and Business Strategy with over 19 years of experience in higher education. An Associate Professor at IIS (Deemed to be University), Jaipur, she is celebrated for her pioneering research, leadership, and contributions to curriculum innovation and digital pedagogy. Dr. Chaturvedi has authored over 100 research publications in top international journals and edited volumes and has supervised multiple Ph.D. scholars. As a keynote speaker at international platforms across Indonesia, Uzbekistan, Mauritius, Liberia, and India, she shares insights on sustainability, AI in commerce, financial inclusion, and NEP 2020. A recipient of numerous accolades, including "Most Influential Women in Academics," she has chaired global conferences, led academic committees, and edited books on AI, innovation, and digital transformation, exemplifying the intersection of scholarly excellence and visionary leadership.

Dr. Atul Bansal is a dedicated academic and lifelong learner with over twenty-seven years of rich experience in teaching, research, and academic administration across universities in India, Africa, and the Middle East. Known for his self-motivation and deep commitment to academic excellence, he has consistently blended theory with practice to create meaningful learning experiences for students and professionals alike. Currently, Dr. Bansal serves as member of faculty of Accounting in the College of Business Administration at the University of Bahrain, Kingdom of Bahrain. His work reflects a strong focus on accounting education, research rigor, and institutional development, shaped by years of engagement with internationally accredited academic environments. Through his teaching, research, and scholarly writing, he continues to contribute thoughtfully to the advancement of accounting knowledge and professional practice.

Dr. Rohaya Latip is an Associate Professor at the Department of Communication Technology and Network, Faculty of Computer Science and Information Technology, Universiti Putra Malaysia (UPM). She leads the Network, Parallel and Distributed Computing (NPDC) Research Group and previously served as Head of Department (2017–2022). Her international engagements include visiting professor appointments at L.N. Gumilyov Eurasian National University and Monash University, Australia. She also served as Associate Professor at Najran University, Saudi Arabia. Dr. Latip serves several renowned journals in editorial capacities. Her research areas include Big Data, Cloud, Fog and Edge Computing, Network Management, and Distributed Databases, with over 148 publications to her name.

Contributors

Arunima K.V.
T A Pai Management Institute
Manipal Academy of Higher Education
Manipal, India

Deepali Bhatnagar
Amity Business School
Amity University
Jaipur, India

Kartikeya Bolar
T A Pai Management Institute
Manipal Academy of Higher Education
Manipal, India

K.S. Chandrasekar
Institute of Management in Kerala (IMK)
University of Kerala
Thiruvananthapuram, India

Ankita Chaturvedi
IIS University
Jaipur, India

Deeksha S.
Department of Commerce
Bengaluru City University
Bengaluru, India

Khushi Desai
Amity Law School
Amity University
Jaipur, India

V. Ajantha Devi
AP3 Solutions
Chennai, India

Asha Devi J
Institute of Management in Kerala (IMK)
University of Kerala
Thiruvananthapuram, India

Surendra Kumar Gautam
Department of Life Science
Shri Rawatpura Sarkar University
Raipur, India

Vikas Godara
Department of Mechatronic Engineering
School of Mechanical Science and Engineering
Huazhong University of Science and Technology
Wuhan, China

Youmin Hu
Mechatronic Engineering
Huazhong University of Science and Technology
Wuhan, China

Anju Kalluvelil Janardhanan
Southern Cross Institute
Parramatta, Australia

Veena Roshan Jose
Department of Law
Maharashtra National Law University
Nagpur, India

Ruchika Joshi
Faculty of Commerce and Management
Amrapali University
Haldwani, India

Mrinalini Kacker
Humanities and Social Sciences
Symbiosis International University
Pune, India

Sushil Kalyani
School of Business
University of Petroleum and Energy Studies
Dehradun, India

Riyanshi Khandelwal
Amity Business School
Amity University
Jaipur, India

Mukesh Kondala
Department of Operations
GITAM School of Business, GITAM Deemed to be University
Visakhapatnam, India

Yiqun Kou
Mechatronic Engineering
Huazhong University of Science and Technology
Wuhan, China

Nisha Kumari
Department of General Management
Aditya School of Business
Aditya University
Surampalem, India

Mallika Sankar M.
School of Business and Management
Christ University
Bengaluru, India

Manigandan R.
Faculty of Management - MBA
SRM Institute of Science and Technology
Vadapalani, India

Manjushree G. S.
Dayananda Sagar University
Bengaluru, India

Manu K. S.
School of Business and Management
CHRIST University
Bengaluru, India

Nayana N
Dayananda Sagar University
Bengaluru, India

Bindu K. Nambiar
Union Bank of India
Bengaluru, India

Mohd Naved
Department of Business Analytics
Jaipuria Institute of Management
Noida, Uttar Pradesh, India

Durdana Ovais
BSSS Institute of Advanced Studies
Bhopal, Madhya Pradesh, India

Girja Shankar Patel
Department of Library and Information Science
Shri Rawatpura Sarkar University
Raipur, Chhattisgarh, India

Rajani H. Pillai
School of Commerce
Mount Carmel College, Autonomous
Bengaluru, India

Prabhakara S.
Indira Gandhi National Open University
Indian Institute of Banking and Finance
Mumbai, India

Shivender Rahul
Department of Humanities
Maharashtra National Law University
Nagpur, India

Lakshmi Raj
Marketing Department
IIM Bangalore
Bengaluru, India

Kavitha Rajamohan
School of Sciences
CHRIST University
Bangalore, India

Sangeetha Rangasamy
School of Business and Management
CHRIST University
Bangalore, India

Santosh N.
Department of Commerce
Bengaluru City University
Bengaluru, India

V. Sasirekha
School of Business
Galgotias University
Greater Noida (NCR), India

Abhineet Saxena
Amity Buisness School
Amity University
Rajasthan, India

Priti Saxena
Atlas Skilltech University
Mumbai, India

V. Selvalakshmi
Saveetha School of Management
Saveetha Institute of Medical and Technical Sciences (SIMATS)
Chennai, India

Garima Sharma
Welingkar Business School
Mumbai, India

V. Suganya
Faculty of Management - MBA
SRM Institute of Science and Technology
Chennai, India

Ye Yang
Mechatronic Engineering
Huazhong University of Science and Technology
Wuhan, China

Xi Zhang
Mechatronic Engineering
Huazhong University of Science and Technology
Wuhan, China

Hang Zheng
Mechatronic Engineering
Huazhong University of Science and Technology
Wuhan, China

1 Applications of Artificial Intelligence in the Healthcare Industry

A Meta-Analysis

V. Sasirekha, V. Suganya, and Manigandan R.

1.1 INTRODUCTION

Artificial intelligence (AI) has the potential to revolutionize healthcare in numerous ways, from improving patient care to streamlining administrative processes. AI has several applications in medicine, including hospitals, clinical laboratories, and research facilities. Healthcare administration and operation, clinical decision support, healthcare predictions, patient monitoring, and healthcare interventions are the key domains where AI is applied. AI applications in healthcare is immense and are used by doctors, patients, drug industries, health services, insurance agencies, and medical institutions. AI also helps in screening, monitoring, and providing medical and clinical studies for patients. One of the most promising areas of AI and analytics applications is the healthcare sector. With the introduction of AI, India's healthcare industry is about to undergo a significant transformation. Computerized reasoning has a few applications in medicine, including emergency clinics, clinical labs, and medical examination offices, operations and management of the healthcare system, clinical decision support, and the future of healthcare monitoring of patients. AI is also used extensively in healthcare interventions. Simulated intelligence application in medical care is gigantic, which is utilized by specialists as well as patients, drug businesses, well-being administrations, insurance companies, and medical institutions. Additionally, AI aids in screening, monitoring, and providing patients with medical and clinical help.

1.2 APPLICATIONS OF ARTIFICIAL INTELLIGENCE (AI) IN THE HEALTHCARE SECTOR

1.2.1 Medical Imaging and Diagnostics

AI algorithms can analyze medical images like X-rays, MRIs, and CT scans to assist in early disease detection and diagnosis. While AI has shown tremendous potential in medical imaging, it is not meant to replace radiologists and other healthcare

DOI: 10.1201/9781003638506-1

professionals. Instead, AI serves as a valuable tool to assist and enhance their capabilities, leading to more accurate and timely diagnoses, improved patient outcomes, and better resource allocation. These are some common ways in which AI is used in medical imaging and diagnostics.

1.2.2 Image Segmentation and Annotation

AI algorithms, especially convolutional neural networks (CNNs), can accurately segment and annotate structures within medical images. This includes identifying organs, tumors, blood vessels, and other anatomical features. This segmentation assists radiologists in understanding complex structures and measuring various parameters.

1.2.3 Image Classification

AI can classify medical images into different categories, such as normal versus abnormal or benign versus malignant. AI models are trained on large datasets of annotated images to learn patterns and features associated with different conditions.

1.2.4 Object Detection

AI can detect and localize specific objects or anomalies within medical images. For instance, it can identify and highlight regions suspected of containing tumors or lesions.

1.2.5 Early Disease Detection

AI algorithms can analyze subtle changes in medical images that might be difficult for human radiologists to detect. This enables the early detection of diseases like cancer, often before they are clinically apparent.

1.2.6 Quantitative Analysis

AI can provide precise measurements and quantifications of structures within medical images. This is particularly useful for tracking disease progression and treatment response over time.

1.2.7 Radiomics

AI can extract and analyze a large number of quantitative features from medical images, such as texture, shape, and intensity patterns. These features can be used to create predictive models for diagnosis, prognosis, and treatment response assessment.

1.2.8 Image Fusion

AI can combine information from multiple imaging modalities, such as MRI and CT scans, to provide a more comprehensive view of a patient's condition. This aids in accurate diagnosis and treatment planning.

1.2.9 Quality Enhancement

AI algorithms can improve image quality by reducing noise, artifacts, and distortions, thereby enhancing the clarity and visibility of structures.

1.2.10 Automated Report Generation

AI systems can generate preliminary diagnostic reports based on image analysis. These reports can include findings, measurements, and potential diagnoses, which can be reviewed and finalized by radiologists.

1.2.11 Workflow Optimization

AI can prioritize and triage images, flagging cases that require urgent attention. This helps radiologists manage their workload more effectively.

1.2.12 Clinical Decision Support

AI systems provide radiologists with relevant information from medical literature and databases, aiding in making informed diagnoses and treatment decisions.

1.2.13 Customized Treatment Planning

In radiation therapy, AI can help plan the precise delivery of radiation by analyzing medical images and optimizing treatment plans for individual patients.

1.3 DRUG DISCOVERY AND DEVELOPMENT

AI is used in identifying potential drug candidates by analyzing large datasets and predicting their effectiveness and potential side effects. It speeds up the drug discovery process and helps researchers identify novel targets. The application of AI in drug discovery has the potential to significantly reduce costs, improve success rates, and accelerate the overall drug development timeline. However, challenges such as data quality, interpretability, and regulatory considerations need to be addressed to ensure the safe and effective deployment of AI-driven solutions in this critical field. These are some ways in which AI is useful in this domain.

1.3.1 Virtual Screening

AI algorithms can efficiently screen vast chemical libraries to identify molecules that have a high probability of binding to a target protein. This reduces the time and resources needed for experimental screening.

1.3.2 Predicting Drug–Target Interactions

AI models can predict the interactions between potential drug compounds and target proteins. This helps researchers understand how a drug might affect a biological pathway or disease.

1.3.3 Molecular Design and Optimization

AI can design and optimize drug molecules by predicting their physical and chemical properties, such as solubility, bioavailability, and toxicity. This leads to the creation of more effective and safer drug candidates.

1.3.4 Compound Synthesis Planning

AI assists in designing synthetic routes for chemical compounds. It predicts the most efficient and cost-effective ways to synthesize drug candidates.

1.3.5 Quantitative Structure–Activity Relationship (QSAR) Modeling

AI develops QSAR models that predict a molecule's biological activity based on its structural features. These models guide the selection of compounds with desired properties.

1.3.6 Drug Repurposing

AI identifies existing drugs that could potentially be repurposed for new therapeutic indications. This reduces the time and costs associated with traditional drug development.

1.3.7 Clinical Trial Design

AI helps in designing more efficient and informative clinical trials by analyzing large datasets, patient characteristics, and disease progression patterns.

1.3.8 Data Integration and Analysis

AI integrates and analyzes diverse data sources, including omics data (genomics, proteomics, metabolomics), to identify potential drug targets and biomarkers.

1.3.9 Biological Pathway Analysis

AI models analyze complex biological pathways and networks to identify potential drug targets and understand the mechanisms of disease.

1.3.10 High-Throughput Screening Data Analysis

AI handles the analysis of data from high-throughput screening experiments, identifying hits and optimizing compounds for further development.

1.3.11 Chemical Library Design

AI designs focused on compound libraries that are more likely to contain promising drug candidates, optimizing the chances of success in experimental testing.

1.3.12 Optimizing Clinical Trial Recruitment

AI helps in identifying suitable patient populations for clinical trials, improving recruitment and trial success rates.

1.3.13 Genomic Medicine and Pharmacogenomics

AI analyzes genetic data to identify patient populations that are more likely to respond positively to specific drugs, enabling personalized treatment approaches.

1.4 PERSONALIZED TREATMENT

AI can analyze patient data to provide personalized treatment plans. By considering a patient's medical history, genetic information, and other factors, AI can assist doctors in tailoring treatments that are more effective and have fewer adverse effects. However, challenges such as data privacy, interoperability, and ethical considerations must be addressed and must be aligned with patient preferences. Uses of AI in personalized treatment are as follows:

1.4.1 Data Integration and Analysis

AI integrates and analyzes diverse patient data, including medical history, genetic information, lifestyle factors, and treatment outcomes. This comprehensive view helps identify patterns and correlations that guide personalized treatment decisions.

1.4.2 Predictive Analytics

AI models predict patient responses to specific treatments based on historical data. This assists in selecting the most effective interventions and avoiding treatments that are likely to be ineffective or cause adverse effects.

1.4.3 Genomic Analysis

AI analyzes genetic data to identify genetic mutations, variations, and markers that influence disease susceptibility and treatment responses. This information guides targeted therapies and precision medicine.

1.4.4 Treatment Recommendation

AI algorithms consider a patient's individual characteristics, medical history, and genetic makeup to recommend the most appropriate treatment options, including drug selection and dosages.

1.4.5 Adverse Event Prediction

AI identifies patients who might be at a higher risk of experiencing adverse effects from certain treatments, enabling proactive monitoring and adjustments to treatment plans.

1.4.6 Clinical Decision Support

AI systems provide healthcare professionals with evidence-based recommendations for treatment options and interventions, aiding in shared decision-making.

1.4.7 Therapeutic Monitoring

AI continuously monitors patient data, such as vital signs and biomarkers, to track treatment effectiveness and adjust interventions in real time.

1.4.8 Drug Interaction Assessment

AI helps identify potential drug interactions, considering the patient's current medications and medical conditions. This ensures the safety and efficacy of treatment regimens.

1.4.9 Personalized Dosage Optimization

AI models tailor medication dosages based on factors like age, weight, genetics, and metabolism, optimizing therapeutic outcomes while minimizing side effects.

1.4.10 Prognostic Assessment

AI assesses a patient's disease progression risk, helping healthcare providers determine the best treatment strategy to achieve positive outcomes.

1.4.11 Telehealth and Remote Monitoring

AI-driven remote monitoring solutions track patients' health outside traditional healthcare settings. This enables personalized interventions and early detection of issues.

1.4.12 Nutrition and Lifestyle Recommendations

AI analyzes patient data to offer personalized advice on diet, exercise, and lifestyle modifications that can complement medical treatments.

1.4.13 Real-world Evidence and Long-term Outcomes

AI leverages real-world data to evaluate treatment effectiveness over the long term and refine personalized treatment strategies.

1.4.14 Clinical Trial Matching

AI matches patients to relevant clinical trials based on their profiles, enhancing their access to cutting-edge treatments.

1.5 HEALTH MONITORING AND WEARABLES

Wearable devices equipped with AI can monitor vital signs and activity levels. They can provide real-time data to both patients and healthcare providers, enabling timely interventions. AI has revolutionized health monitoring and wearable devices by enabling continuous tracking, analysis, and interpretation of health data. AI's ability to analyze large volumes of health data in real time and providing actionable insights makes wearables a valuable tool for preventive care, early intervention, and overall health management. AI is useful as follows in health monitoring

1.5.1 Continuous Monitoring

AI-powered wearables continuously monitor vital signs such as heart rate, blood pressure, temperature, and oxygen saturation. This allows for early detection of anomalies and trends that might indicate health issues.

1.5.2 Data Collection and Integration

AI aggregates data from wearables and other sources, such as electronic health records (EHRs) and mobile apps, creating a comprehensive view of a person's health status.

1.5.3 Real-time Alerts

AI algorithms can detect critical changes in health metrics and send real-time alerts to users and healthcare professionals, prompting immediate actions if necessary.

1.5.4 Personalized Insights

AI interprets data to provide personalized insights and recommendations related to physical activity, sleep patterns, stress levels, and overall well-being.

1.5.5 Disease Management

AI-driven wearables support individuals with chronic diseases by monitoring their condition and providing guidance on medication adherence, lifestyle modifications, and symptom management.

1.5.6 Early Disease Detection

AI analyzes patterns and trends in health data to detect early signs of diseases such as heart arrhythmias, sleep disorders, and respiratory issues.

1.5.7 Remote Patient Monitoring

AI-enabled wearables allow healthcare providers to remotely monitor patients, especially those with chronic conditions, reducing hospitalizations and improving outcomes.

1.5.8 Fall Detection

AI algorithms can identify sudden falls or changes in movement patterns, sending alerts to caregivers or emergency services.

1.5.9 ECG and Cardiac Monitoring

Wearables with AI-powered electrocardiogram (ECG) capabilities can monitor heart rhythm and detect irregularities, such as atrial fibrillation.

1.5.10 Stress and Mental Health

AI analyzes biometric and behavioral data to assess stress levels and mental well-being. Wearables can provide guidance and interventions for stress management.

1.5.11 Physical Activity Monitoring

AI tracks physical activity levels, providing users with insights into exercise intensity, duration, and goals.

1.5.12 Sleep Analysis

Wearables equipped with AI assess sleep quality, duration, and disruptions. They provide recommendations for improving sleep patterns.

1.5.13 Blood Glucose Monitoring

AI-powered wearables track blood glucose levels in individuals with diabetes, offering insights for managing insulin dosages and dietary choices.

1.5.14 Medication Reminders

AI-enhanced wearables provide reminders for medication schedules, ensuring adherence to treatment plans.

1.5.15 Aging in Place

AI-enabled wearables support older adults to live independently by monitoring health parameters and safety, while providing assistance as needed.

1.5.16 Health Insights for Athletes

Athletes can use AI-powered wearables to track performance metrics, optimize training, and prevent injuries.

1.5.17 Telemedicine

AI-integrated wearables facilitate remote telehealth consultations by sharing real-time health data with healthcare providers.

1.6 PREDICTIVE ANALYTICS

AI can analyze patient data to predict disease outbreaks, patient admissions, and other healthcare trends. This assists in resource allocation and planning for healthcare facilities. AI predictive analytics in the healthcare industry involves using machine learning and data analysis techniques to forecast future outcomes, trends, and events based on historical and current data. Here are some key ways AI predictive analytics is used in healthcare.

1.6.1 Disease Outbreak Prediction

AI analyzes epidemiological data, environmental factors, and population trends to predict disease outbreaks. This helps public health agencies allocate resources and implement preventive measures.

1.6.2 Patient Admissions and Resource Allocation

Predictive models analyze patient data to forecast admission rates, enabling hospitals to allocate beds, staff, and resources more effectively.

1.6.3 Readmission Risk

AI assesses patient data to predict the risk of hospital readmission, enabling early interventions and reducing healthcare costs.

1.6.4 Patient Flow and Wait Times

Predictive analytics optimizes patient flow by forecasting emergency department and clinic wait times. This improves patient satisfaction and resource allocation.

1.6.5 Medication Adherence

AI predicts patients' likelihood of adhering to medication regimens based on historical data, improving treatment outcomes and reducing complications.

1.6.6 Healthcare Fraud Detection

Predictive models analyze billing and claims data to identify patterns of fraudulent activities, saving healthcare organizations money and resources.

1.6.7 Population Health Management

AI analyzes patient data to identify at-risk populations and predict disease prevalence. This informs preventive strategies and targeted interventions.

1.6.8 Chronic Disease Management

Predictive analytics help monitor chronic disease progression and predict exacerbations, enabling timely interventions and personalized treatment plans.

1.6.9 Genomic Data Analysis

AI predicts disease susceptibility, treatment response, and genetic traits based on genomic data, aiding in personalized medicine approaches.

1.6.10 Surgical Complications

AI assesses patient data and surgery parameters to predict the likelihood of post-operative complications, allowing surgeons to make informed decisions.

1.6.11 Radiology Workflow Optimization

Predictive analytics assist in scheduling radiology procedures by forecasting demand, reducing patient wait times, and optimizing resource allocation.

1.6.12 Drug Discovery and Development

Predictive models identify potential drug candidates, predict their efficacy, and assess potential adverse effects, streamlining drug development processes.

1.6.13 Patient Risk Stratification

AI evaluates patient data to stratify individuals based on their risk of developing certain conditions, guiding preventive care and treatment decisions.

1.6.14 Clinical Trial Recruitment

Predictive analytics identify potential candidates for clinical trials based on patient profiles, expediting recruitment and study completion.

1.6.15 Healthcare Financial Planning

AI predicts revenue streams, patient volumes, and cost fluctuations, aiding financial planning and budgeting for healthcare organizations.

1.6.16 Personalized Treatment Planning

Predictive models analyze patient data to suggest personalized treatment options, factoring in patient history, genetics, and treatment responses.

1.7 NATURAL LANGUAGE PROCESSING (NLP) FOR ELECTRONIC HEALTH RECORDS (EHR)

AI-powered Natural Language Processing (NLP) systems can convert unstructured data from EHRs into structured information, making it easier for doctors to access and interpret patient histories. AI-based NLP for EHRs is highly beneficial for the healthcare industry, as it enables efficient analysis and extraction of valuable information from unstructured clinical text. Here's how AI-powered NLP is useful in this context.

1.7.1 Efficient Data Extraction

EHRs often contain a wealth of unstructured text data, including physicians' notes, patient histories, and diagnostic reports. AI-based NLP can extract structured information from this unstructured data, making it accessible for analysis and decision-making.

1.7.2 Clinical Documentation

NLP automates the process of transcribing clinical notes, reducing the administrative burden on healthcare professionals and improving documentation accuracy.

1.7.3 Semantic Understanding

AI-NLP models can understand the context and semantics of clinical text, allowing for more accurate interpretation and extraction of medical terms, conditions, and relationships.

1.7.4 Diagnosis and Treatment

NLP analyzes clinical notes to identify symptoms, diseases, and treatments mentioned, aiding in accurate diagnosis and suggesting appropriate treatment options.

1.7.5 Drug Interaction Detection

AI-NLP can identify mentions of medications and potential drug interactions from clinical text, supporting medication safety and decision-making.

1.7.6 Coding and Billing

NLP assists in automatically assigning appropriate codes for medical procedures and diagnoses, ensuring accurate billing and reimbursement.

1.7.7 Research and Clinical Trials

NLP helps researchers identify eligible patients for clinical trials by extracting relevant criteria from EHRs. It also aids in retrospective research studies by analyzing large volumes of patient data.

1.7.8 Population Health Management

NLP identifies patterns and trends within EHRs, assisting in population health analysis, disease surveillance, and proactive interventions.

1.7.9 Outcome Measurement

NLP analyzes clinical narratives to assess treatment outcomes, complications, and patient responses, supporting quality improvement initiatives.

1.7.10 Predictive Analytics

AI-NLP models analyze historical EHR data to predict patient outcomes, readmission risks, and disease progression, guiding preventive measures.

1.7.11 Clinical Decision Support

NLP provides healthcare professionals with relevant information from EHRs and medical literature, assisting in evidence-based clinical decision-making.

1.7.12 Patient Engagement

NLP translates complex medical jargon into understandable language, facilitating patient communication and comprehension of their medical conditions.

1.7.13 Early Warning Systems

AI-NLP models detect early signs of deteriorating patient conditions by analyzing clinical narratives, triggering timely interventions.

1.7.14 Privacy and Security

NLP can de-identify sensitive patient information in EHRs, ensuring compliance with data privacy regulations while allowing researchers access to anonymized data for analysis.

1.7.15 Telemedicine and Virtual Health Assistants (VHA)

NLP powers VHAs that interact with patients through natural language, collecting relevant medical history and symptoms before a telehealth consultation.

1.7.16 Operational Efficiency

NLP automates administrative tasks, such as appointment scheduling and prescription renewals, enhancing operational efficiency and patient experience.

1.8 ROBOT-ASSISTED SURGERY

AI-assisted robots can aid surgeons in performing complex surgeries with higher precision and accuracy. These robots can reduce the margin of error and enhance patient outcomes. AI-based robot-assisted surgery is a transformative technology in the healthcare industry that enhances surgical procedures by combining the precision of robotics with advanced AI capabilities. Here's how AI-based robot-assisted surgery is useful.

1.8.1 Enhanced Precision

Robots equipped with AI can perform highly precise movements, minimizing the risk of human error and enabling surgeons to carry out complex procedures with greater accuracy.

1.8.2 Minimally Invasive Surgery

Robot-assisted surgery often involves smaller incisions, reducing tissue trauma, pain, and recovery time for patients. AI assists in translating surgeon movements into precise robotic actions.

1.8.3 Visual Enhancement

AI algorithms enhance the visual information from surgical cameras, offering surgeons clearer and magnified views of the surgical site, even in difficult-to-reach areas.

1.8.4 Tremor Reduction

AI-controlled robots can stabilize and eliminate tremors in a surgeon's hand movements, ensuring steady and precise actions during surgery.

1.8.5 Teleoperation and Telesurgery

AI-powered robotic systems enable remote surgery, allowing expert surgeons to perform procedures on patients located far away. This is particularly valuable for providing specialized care to underserved areas.

1.8.6 Image Analysis

AI can analyze real-time surgical images, providing augmented visualization, highlighting important structures, and assisting in decision-making.

1.8.7 Real-time Feedback

AI provides surgeons with real-time feedback on tissue properties, blood flow, and potential complications, enhancing intraoperative decision-making.

1.8.8 Procedural Standardization

AI helps standardize surgical procedures by ensuring that robot-assisted actions follow predefined protocols, reducing variability in outcomes.

1.8.9 Surgical Simulation

AI-driven simulations allow surgeons to practice procedures in a virtual environment, improving their skills and familiarizing them with complex cases.

1.8.10 Complex Procedures

Robot-assisted surgery is particularly valuable for intricate procedures, such as cardiac surgery, neurosurgery, and microsurgery, where precision is crucial.

1.8.11 Reduced Recovery Time

Patients undergoing robot-assisted surgeries often experience shorter hospital stays and faster recovery compared to traditional open surgeries.

1.8.12 Accessibility to Expertise

AI-powered robots enable less-experienced surgeons to benefit from the expertise of skilled surgeons, ensuring high-quality care for patients.

1.8.13 Reduced Fatigue

Robot-assisted systems can reduce surgeon fatigue during long and complex procedures, contributing to better outcomes.

1.8.14 Customized Approaches

AI can adapt to individual patient anatomies, allowing for more tailored and patient-specific surgical approaches.

1.8.15 Data Collection and Learning

AI can learn from vast datasets of surgical procedures, leading to continuous improvement in robotic techniques and patient outcomes.

1.8.16 Reduced Radiation Exposure

In procedures like fluoroscopy-guided surgeries, AI assists in minimizing radiation exposure for both patients and healthcare professionals.

1.9 VIRTUAL HEALTH ASSISTANTS

AI-driven chatbots and virtual assistants can offer medical information, appointment scheduling, and even basic diagnostics to patients, improving accessibility to healthcare information. AI-based VHAs are playing a transformative role in the healthcare industry, offering a wide range of benefits and applications. Here's how AI-based VHAs are useful.

1.9.1 24/7 Accessibility

VHAs provide patients with round-the-clock access to medical information, assistance, and support, improving patient engagement and satisfaction.

1.9.2 Immediate Responses

VHAs offer immediate responses to patient queries, reducing waiting times and providing timely information, especially in non-emergency situations.

1.9.3 Health Information

VHAs deliver accurate and reliable health information, helping patients understand medical conditions, treatment options, medications, and preventive measures.

1.9.4 Symptom Assessment

AI-powered VHAs ask patients about their symptoms and medical history, offering initial assessments and recommendations for seeking medical care.

1.9.5 Appointment Scheduling

VHAs help patients schedule appointments, reducing administrative workload for healthcare staff and enhancing patient convenience.

1.9.6 Medication Reminders

VHAs remind patients to take medications on time, promoting adherence to treatment plans and improving health outcomes.

1.9.7 Post-Discharge Care

VHAs offer post-discharge guidance, ensuring patients understand their recovery instructions, follow-up appointments, and self-care measures.

1.9.8 Wellness Advice

VHAs provide advice on maintaining a healthy lifestyle, including exercise, nutrition, stress management, and sleep hygiene.

1.9.9 Chronic Disease Management

VHAs assist patients with chronic conditions by monitoring symptoms, offering personalized recommendations, and guiding self-management.

1.9.10 Language Translation

VHAs can interact with patients in multiple languages, breaking down language barriers and enhancing accessibility to healthcare information.

1.9.11 Remote Monitoring

VHAs gather and analyze health data from wearable devices, assisting patients with real-time insights and connecting them with healthcare providers as needed.

1.9.12 Preventive Care

VHAs encourage patients to engage in preventive screenings, vaccinations, and health check-ups, promoting early disease detection and intervention.

1.9.13 Mental Health Support

VHAs offer support for mental health concerns by providing coping strategies, mindfulness exercises, and referrals to professionals.

1.9.14 Emergency Response

VHAs can guide users through basic first aid procedures and contact emergency services when necessary.

1.9.15 Healthcare Navigation

VHAs help patients navigate healthcare systems, understand insurance coverage, and find suitable healthcare providers.

1.9.16 Data Collection for Research

VHAs collect anonymous patient data for research purposes, contributing to epidemiological studies and medical advancements.

1.9.17 Reduced Healthcare Workload

VHAs assist healthcare professionals by handling routine inquiries, allowing them to focus on more complex patient needs.

1.9.18 Postoperative Support

VHAs offer postoperative guidance, explaining recovery protocols, wound care, and potential complications.

1.9.19 Cost Savings

VHAs can lead to reduced healthcare costs by minimizing unnecessary office visits, preventing hospital readmissions, and enhancing treatment adherence.

1.10 CLINICAL TRIAL OPTIMIZATION

AI can help in patient recruitment for clinical trials by identifying suitable candidates based on their medical histories and profiles, speeding up the trial process. AI is highly beneficial in optimizing various aspects of clinical trials, making the process more efficient, cost-effective, and informed. Here's how AI is useful in clinical trial optimization.

1.10.1 Patient Recruitment and Eligibility

AI analyzes patient data from EHRs to identify potential candidates who match the trial's criteria, accelerating patient recruitment and reducing trial delays.

1.10.2 Site Selection

AI assists in selecting trial sites by analyzing factors, such as patient demographics, disease prevalence, and healthcare infrastructure, ensuring optimal trial placement.

1.10.3 Protocol Design

AI analyzes historical trial data to suggest optimized trial designs, including sample size calculations and treatment allocation, leading to more informative and successful trials.

1.10.4 Drug Safety and Adverse Event Prediction

AI models analyze patient data to predict potential adverse events and monitor drug safety, aiding in proactive risk management.

1.10.5 Patient Monitoring

AI-powered remote monitoring and wearable devices collect patient data in real time, enhancing the quality and accuracy of data collected during the trial.

1.10.6 Real-time Data Analysis

AI performs real-time analysis of incoming trial data, enabling early identification of trends, treatment responses, and potential issues.

1.10.7 Data Quality and Integrity

AI identifies anomalies and inconsistencies in trial data, ensuring data quality and integrity throughout the trial.

1.10.8 Patient Engagement and Retention

AI-powered tools engage patients through personalized communication, reminders, and education, improving patient retention rates.

1.10.9 Predictive Analytics for Outcomes

AI predicts trial outcomes based on historical data and ongoing patient responses, offering insights into potential results and facilitating decision-making.

1.10.10 Optimized Endpoint Identification

AI identifies optimal trial endpoints that align with regulatory requirements, patient needs, and scientific validity.

1.10.11 Adaptive Trial Designs

AI allows for adaptive trial modifications based on interim results, optimizing trial efficiency and resource allocation.

1.10.12 Optimized Resource Allocation

AI models assess trial resource needs and allocate personnel, supplies, and funds efficiently.

1.10.13 Regulatory Compliance

AI assists in ensuring compliance with regulatory requirements by identifying potential issues in trial design and data collection.

1.10.14 Efficient Data Collection

AI streamlines data collection by automating processes and reducing manual data entry errors.

1.10.15 Decision Support for Investigators

AI provides insights and recommendations to investigators, helping them make informed decisions throughout the trial.

1.10.16 Trial Monitoring

AI algorithms track trial progress, milestones, and potential delays, facilitating timely interventions.

1.10.17 Predictive Enrollment Trends

AI predicts enrollment patterns based on historical data, allowing for proactive adjustments to recruitment strategies.

1.10.18 Post-trial Analysis and Reporting

AI helps analyze trial data, generate reports, and interpret results, expediting the reporting process.

1.11 FRAUD DETECTION AND HEALTHCARE MANAGEMENT

AI can be used to detect fraudulent claims and transactions in the healthcare insurance sector, reducing costs and improving the efficiency of healthcare management. AI plays a vital role in fraud detection and healthcare management by leveraging advanced algorithms and data analysis techniques to identify fraudulent activities, enhance operational efficiency, and improve patient care. Here's how AI is useful in these areas.

1.11.1 Fraud Detection

1.11.1.1 Anomaly Detection

AI algorithms can identify unusual patterns and deviations in healthcare billing data, alerting authorities to potential fraudulent claims.

1.11.1.2 Behavioral Analysis

AI models analyze historical data to understand the typical behavior of healthcare providers, patients, and insurers. Deviations from these patterns can indicate fraudulent activities.

1.11.1.3 Predictive Modeling

AI uses historical fraud data to build predictive models that identify high-risk cases and predict potential fraud before it occurs.

1.11.1.4 Claim Validation

AI verifies claims against medical records and procedures, ensuring that billed services are legitimate and accurate.

1.11.1.5 Network Analysis

AI examines relationships among healthcare providers, patients, and insurers to uncover intricate fraud networks and collusion schemes.

1.11.1.6 Real-time Monitoring

AI monitors transactions and activities in real time, identifying suspicious behavior and transactions as they happen.

1.11.1.7 Text Analysis

AI-NLP techniques analyze text-based data, such as clinical notes and billing descriptions, to detect fraud-related keywords and patterns.

1.11.1.8 Data Integration

AI consolidates data from multiple sources, such as claims data, pharmacy records, and patient histories, for a comprehensive view of potential fraud.

1.11.2 Healthcare Management

1.11.2.1 Predictive Analytics

AI analyzes historical patient data to predict disease outbreaks, patient admission rates, and resource needs, aiding in efficient resource allocation and planning.

1.11.2.2 Patient Risk Stratification

AI assesses patient data to stratify individuals based on their risk of developing specific conditions, guiding personalized interventions.

1.11.2.3 Personalized Treatment

AI uses patient data and medical knowledge to recommend personalized treatment options based on individual characteristics.

1.11.2.4 Population Health Management

AI identifies trends and risk factors within patient populations, enabling proactive interventions and public health initiatives.

1.11.2.5 Optimized Resource Allocation

AI models predict patient demand, enabling healthcare facilities to allocate staff, beds, and resources effectively.

1.11.2.6 Telehealth and Remote Monitoring

AI-powered telehealth platforms and wearable devices allow remote patient monitoring and virtual consultations, enhancing patient care accessibility.

1.11.2.7 Clinical Decision Support

AI provides healthcare professionals with evidence-based recommendations for diagnoses, treatment plans, and interventions.

1.11.2.8 Efficient Workflow

AI optimizes administrative tasks such as appointment scheduling, patient registration, and billing, reducing administrative burden.

1.11.2.9 Disease Surveillance

AI monitors health data sources to identify potential disease outbreaks and public health concerns, allowing for rapid responses.

1.11.2.10 Medication Management

AI assists in medication adherence and management by providing reminders, dosing instructions, and potential interactions.

1.12 BEHAVIORAL ANALYSIS AND MENTAL HEALTH SUPPORT

AI can analyze patient behaviors and speech patterns to identify potential mental health issues. It can also offer therapeutic interventions and support. AI is increasingly valuable for behavioral analysis and mental health support, offering innovative tools to assess, understand, and address various mental health concerns. Here's how AI is useful in these areas.

1.12.1 Behavioral Analysis

1.12.1.1 Speech and Language Analysis

AI analyzes speech patterns, tone, and language to detect subtle changes that might indicate emotional distress or cognitive conditions.

1.12.1.2 Social Media Monitoring

AI assesses social media posts to identify signs of depression, anxiety, or other mental health issues based on language, sentiment, and context.

1.12.1.3 Activity Tracking

AI analyzes data from wearable devices to monitor changes in physical activity, sleep patterns, and daily routines, providing insights into mood and mental well-being.

1.12.1.4 Pattern Recognition

AI identifies patterns and deviations from normal behavior that could indicate potential mental health concerns, alerting caregivers or healthcare professionals.

1.12.1.5 Biometric Monitoring

AI analyzes physiological indicators such as heart rate variability, skin conductance, and eye movement to assess stress levels and emotional responses.

1.12.1.6 Virtual Reality (VR) Environments

AI-powered Virtual Reality (VR) simulations help clinicians observe and analyze patient responses in controlled behavioral situations.

1.12.1.7 Data Fusion

AI integrates data from various sources, including wearables, social media, and EHRs, to provide a comprehensive view of a person's behavioral patterns.

1.12.2 Mental Health Support

1.12.2.1 Chatbots and Virtual Assistants

AI-powered chatbots offer immediate and anonymous support by engaging in conversations, providing coping strategies, and offering resources for self-help.

1.12.2.2 Emotion Recognition

AI analyzes facial expressions and emotional cues to assess a person's emotional state, providing insights for caregivers and mental health professionals.

1.12.2.3 Crisis Intervention

AI-driven tools can recognize signs of crisis or distress in user communication and respond with appropriate interventions or connect them to a human professional.

1.12.2.4 Therapeutic Tools

AI can deliver personalized therapeutic interventions, such as mindfulness exercises, relaxation techniques, and cognitive-behavioral therapy modules.

1.12.2.5 Treatment Monitoring

AI monitors patient responses to treatment interventions, adapting therapy plans based on progress and providing real-time feedback.

1.12.2.6 Predictive Modeling

AI predicts the risk of relapse or worsening symptoms, enabling early interventions and targeted support.

1.12.2.7 Personalized Recommendations

AI tailors mental health resources, educational materials, and support services based on individual needs and preferences.

1.12.2.8 Teletherapy

AI-powered platforms facilitate teletherapy sessions, offering remote access to mental health professionals and support groups.

1.12.2.9 Reducing Stigma

AI-based tools provide an anonymous and non-judgmental platform for users to discuss their mental health concerns, helping reduce stigma.

1.12.2.10 Long-term Monitoring

AI analyzes trends in behavioral data over time, allowing caregivers and mental health professionals to track progress and adjust treatment plans.

1.13 GENOMIC ANALYSIS

AI plays a significant role in analyzing vast amounts of genomic data to identify genetic mutations linked to diseases and potential treatment options. AI-based genomic analysis has significant utility in the healthcare sector, offering a range of benefits that contribute to improved patient care, disease diagnosis, treatment selection, and research advancements. Here are some ways in which AI-based genomic analysis is useful in healthcare.

1.13.1 Personalized Medicine

AI can analyze an individual's genomic data to identify genetic variations that may impact their susceptibility to diseases or response to treatments. This information enables healthcare professionals to tailor treatments to a patient's specific genetic makeup, optimizing therapeutic outcomes and minimizing adverse effects.

1.13.2 Disease Diagnosis

AI algorithms can quickly and accurately analyze large-scale genomic data to aid in the diagnosis of various genetic disorders and diseases. This helps clinicians identify rare and complex conditions, leading to quicker diagnosis and appropriate treatment planning.

1.13.3 Cancer Genomics

AI can analyze genomic data from cancer patients to identify specific genetic mutations and variations associated with different types of cancers. This information helps oncologists choose targeted therapies that have a higher likelihood of success.

1.13.4 Drug Discovery and Development

AI can accelerate the process of drug discovery by analyzing genomic data to identify potential drug targets and predict how certain drugs might interact with specific genetic profiles. This reduces the time and cost associated with traditional drug development methods.

1.13.5 Clinical Trials

AI can aid in patient recruitment for clinical trials by identifying suitable candidates based on their genomic characteristics. This leads to more efficient trials with participants who are more likely to respond positively to the treatment being tested.

1.13.6 Genetic Counseling

AI can assist genetic counselors by providing accurate and up-to-date information about the implications of specific genetic variations. This helps patients and families make informed decisions about their healthcare and reproductive choices.

1.13.7 Population Health Management

AI can analyze large genomic datasets to identify patterns and trends within populations. This can assist public health officials in understanding the genetic factors contributing to certain diseases and designing targeted interventions.

1.13.8 Predictive Medicine

By analyzing an individual's genomic data along with other health information, AI can help predict the risk of developing certain diseases or conditions. This enables proactive preventive measures and early interventions.

1.13.9 Data Interpretation

Genomic data is incredibly complex, and AI can help in interpreting this data by identifying relevant genetic variants, their potential effects, and their significance in the context of a patient's health.

1.13.10 Ethical and Legal Considerations

AI can assist in navigating the ethical and legal challenges associated with genomic data, ensuring patient privacy, data security, and appropriate use of genetic information.

1.14 METHODOLOGY

This chapter attempts meta-analysis by systematically collecting, reviewing, and synthesizing existing research studies to conclude the overall impact, effectiveness, or outcomes of applications of AI in the healthcare sector. Meta-analysis is an investigation of data from a number of independent studies of the same subject to determine overall trends. Meta-analysis is used to intentionally integrate or consolidate the findings and conclusions on the study topic. Performing a meta-analysis in the healthcare sector specifically focusing on the applications of AI involves following a similar process to a standard meta-analysis, but with a specific focus on studies that investigate the applications of AI techniques in healthcare context. Meta-analysis in the healthcare sector with an emphasis on AI involves understanding both healthcare practices and AI methodologies. AI methodologies in the healthcare sector encompass a wide range of techniques and approaches that leverage AI to enhance various aspects of healthcare delivery, diagnosis, treatment, research, and more.

This chapter presents a meta-analytic review of existing literature on AI and its applications in the healthcare sector from the Directory of Open Access Journals (DOAJ). DOAJ is an online directory and database that indexes and provides access to high-quality, peer-reviewed, open-access scholarly journals from various academic disciplines. The DOAJ promotes and increases the visibility and accessibility of open-access research. Researchers and academics often use DOAJ to discover reputable open-access journals in their fields of interest, and it is a valuable resource for researchers looking for freely accessible scholarly content. Additionally, it helps authors and publishers increase the visibility and impact of their open-access research by making it more discoverable to a global audience.

1.15 META-ANALYSIS – INCLUSION AND EXCLUSION CRITERIA

Inclusion and exclusion criteria are the factors that are used to determine the studies that are included in a meta-analysis. The inclusion and exclusion criteria for this chapter focus on the applications of AI in the healthcare sector.

1.15.1 Inclusion Criteria

The following are the inclusion criteria of this research:

- Studies that investigate the application of AI in the healthcare sector
- Studies that use a quantitative or mixed-methods design
- Studies that report at least one outcome of interest (as discussed in Section 1.14 of this chapter)
- Studies published in English
- Studies published between 2020 and October 2023 in the DOAJ.

1.15.2 Exclusion Criteria

The following are the exclusion criteria of this research, and were excluded from this study.

- Studies that focus on the development of AI algorithms or systems, but do not evaluate their application in the healthcare sector
- Studies that do not report any outcomes pertaining to the research topic.
- Studies published in languages other than English
- Studies published before 2020.

1.16 RESULTS AND DISCUSSION

The researchers performed a meta-analysis of the applications of AI in the healthcare sector. The studies published in the DOAJ between 2020 and October 2023 were accessed and analyzed on the topic. DOAJ is an online directory and database that indexes and provides access to high-quality, peer-reviewed, open-access scholarly journals from various academic disciplines. The total number of papers published in the DOAJ during the analysis period is 46 (Calendar year 2020 – six papers, 2021 – 11 papers, 2022 – 18 papers, and 2023 – 11 as on date).

Table 1.1 presents the articles published in the DOAJ based on the inclusion and exclusion criteria.

Table 1.1 presents the meta-analysis of applications of AI in the healthcare sector as published in the DOAJ between January 2020 and October 2023. Meta-analyses of AI in healthcare have shown that AI can be used to improve the diagnosis, treatment, and management of a wide range of diseases and conditions. Based on the analysis of the 46 papers published in the online DOAJ library, applications of AI in the healthcare sector during COVID, using AI techniques such as Image Processing, NLP, Machine learning, Data mining, prediction algorithms and detection and management are the common research topics.

1.17 APPLICATIONS AND EFFECTIVENESS OF AI

1.17.1 Improving the Accuracy of Diagnosis

AI algorithms can be trained on large datasets of medical images and patient data to identify patterns that are associated with different diseases. This can help clinicians to diagnose diseases more accurately and earlier. For example, AI algorithms are more accurate than human radiologists at detecting cancer on medical images.

1.17.2 Developing Personalized Treatment Plans

AI algorithms can be used to analyze patient data and identify the best treatment plan for each patient. This can help to improve the efficacy of treatments and reduce the risk of side effects. For example, AI algorithms have been used to develop personalized treatment plans for patients with cancer.

1.17.3 Monitoring Patients and Predicting Complications

AI algorithms can be used to monitor patients' health data and predict when they are at risk of developing complications. This can help clinicians to intervene early and prevent complications from occurring. For example, AI algorithms have been used to predict which patients with heart disease are at risk of developing a heart attack.

In addition to these specific applications, AI is also being used to improve the efficiency and cost-effectiveness of healthcare delivery. AI algorithms are being used to automate tasks such as medical record keeping and scheduling appointments. This can free up clinicians to spend more time with patients. Overall, the evidence from

TABLE 1.1
Meta-Analysis of Applications of AI in the Healthcare Sector

2020	2021	2022	2023
Artificial Intelligence and technology in COVID Era: A narrative review. – Ahuja, Vanita; Nair, Lekshmi. V	Preferences for Artificial Intelligence Clinicians Before and During the COVID-19 Pandemic: Discrete Choice Experiment and Propensity Score Matching Study – Taoran Liu, Winghei Tsang, Yifei Xie, Kang Tian, *et al.*	Artificial Intelligence–Assisted Diagnosis of Anterior Cruciate Ligament Tears From Magnetic Resonance Images: Algorithm Development and Validation Study – Kun Hui Chen, Chih-Yu Yang, *et al.*	Understanding the impact of an AI-enabled conversational agent mobile app on users' mental health and wellbeing with a self-reported maternal event: a mixed method real-world data mHealth study – Becky Inkster *et al.*
Artificial Intelligence for COVID-19: Rapid Review – Jiayang Chen, Kay Choong See	The next era of Biomedical Research – Prioritizing Health Equity in the Age of Digital Medicine	Introducing JMIR AI – A new journal – Khaled El Emam, Bradley Malin.	Artificial intelligence as a coming revolution in medicine – Rami Ben Rhaiem, Grzegorz Tarsa *et al.*
Telepsychiatry and the Role of Artificial Intelligence in Mental Health in Post-COVID-19 India: A Scoping Review on Opportunities – Thenral M and Arunkumar Annamalai	Ethical Concerns About Reidentification of Individuals from MRI Neuroimages – Elizaveta Garbuzova	Adolescents' well-being While Using a Mobile AI–Powered Acceptance Commitment Therapy Tool: Evidence From a Longitudinal Study – Dana Vertsberger, Navot Naor, Mirene Winsberg	Impact of ChatGPT on Teleconsultants in Healthcare: Perceptions of Healthcare Experts in Saudi Arabia – Alanzi TM
Evaluation of Four Artificial Intelligence-Assisted Self-Diagnosis Apps on Three Diagnoses: Two – Year Follow Up Study	Health Care Professional Association Agency in Preparing for Artificial Intelligence: Protocol for a Multi-Case Study – Caitlin Gillan, Brian Hodges, David Wiljer, Mark Dobrow	Chronic Disease Prediction Using the Common Data Model: Development Study – Chanjung Lee, Brian jo, Hyunki Woo, *et al.*	Prognosis of COVID-19 patients using lab tests: A data mining approach – Fariba Khounraz, Mahmood Khodadoost *et al.*
A New phase of Healthcare: COVID-19 and Medical Advancements – Oshin Puri, Vyas Kumar Rathaur, Nitish Pathania, Monika Pathania	A survey of cardiopulmonary resuscitation in COVID-19 patients – Kapoor, Mukul C; Rao, *et al.*	Visualizing the Interpretation of a Criteria-Driven System That Automatically Evaluates the Quality of Health News: Exploratory Study of 2 Approaches – Xiaoyu Liu, Hiba Alsghaier, *et al.*	Expectations of Anesthesiology and Intensive Care Professionals Toward Artificial Intelligence: Observational Study – Jan Andreas Kloka, Sophie C Holtmann *et al.*

Artificial Intelligence in the Fight Against COVID-19: Scoping Review	A Comprehensive Survey of the Internet of Things (IOT) and AI–Based Smart Healthcare – Fatima Aisheri, Ghulam Muhammad	AI Algorithms In Health Care: Is the Current Food and Drug Administration Regulation Sufficient? – Meghavi Mashar, Shreya Chawla, *et al.*	Blockchain-assisted cyber security in medical things using artificial intelligence – Mohammed Alshehri
	Digital Technology-Based Telemedicine for the COVID-19 Pandemic – Yu Ting Shen, Liang Chen, Wen-Wen Yue, Hui-Xiong Xu.	The Application of AI in Health Care Resource Allocation Before and During the COVID-19 Pandemic: Scoping Review – Hao Wu, Xiaoyu Lu, Hanyu Wang	Artificial Intelligence–Based Consumer Health Informatics Application: Scoping Review – Onur Asan, Euiji Choi, Xiaomei Wang
	Role of Machine Learning Techniques to Tackle the COVID -19 Crisis: Systematic Review – Hafsa Bareen Syeda, Mahanazuddin Syed *et al.*	Strategies to Improve the Impact of AI on Health Equity: Scoping Review – Carl Thomas Berdahl, Lawrence Baker *et al.*	Reliability of Medical Information Provided by ChatGPT: Assessment Against Clinical Guidelines and Patient Information Quality Instrument – Harriet Louise Walker, Shahi Ghani *et al.*
	Application Scenarios for Artificial Intelligence in Nursing Care: Rapid Review – Kathrin Seibert, Dominik Domhoff, Dominik Bruch, Matthias Schulte-Althoff, Daniel Furstenau, Felix Biessman, Karin Wolf-Ostermann.	Enabling Early Health Care Intervention by Detecting Depression in Users of Web-Based Forums using Language Models: Longitudinal Analysis and Evaluation – David Owen, Dimosthenis Antypas *et al.*	A Review on Innovation in Healthcare Sector (Telehealth) through Artificial Intelligence – Ayesha Amjad, Piotr Kordel, Gabriela Fernandes
	Artificial intelligence application versus physical therapist for squat evaluation: a randomized controlled trial – Alessandro Luna, Lorenzo Casertano *et al.*	Developing an Inpatient Electronic Medical Record Phenotype for Hospital-Acquired Pressure Injuries: Case Study Using Natural Language Processing Models – Elvira Nurmambetova, Jie Pan, Zilong Zhang *et al.*	Digital Education for the Deployment of Artificial Intelligence in Health Care – Fernando Korn Malerbi, Luis Filipe Nakayama *et al.*

(Continued)

Table 1.1 (Continued)

2020	2021	2022	2023
	Artificial intelligence performance in detecting tumor metastasis from medical radiology imaging: A systematic review and meta-analysis – Qiuhan Zheng, Le Yang *et al.*	Preparing for an AI-Enabled Future: Patient Perspectives on Engagement and Health Care Professional Training for Adopting AI Technology in Health Care Settings – Tharshini Jeyakumar, Sarah Younus *et al.*	Learning From Experience and Finding the Right Balance in the Governance of Artificial Intelligence and Digital Health Technologies – Stephen Gilbert, Stuart Anderson *et al.*
		Patient Embeddings from Diagnosis Codes for Health Care Prediction Tasks: Pat2Vec Machine Learning Framework – Edgar Steiger, Lars Eric Kroll	Application of explainable artificial intelligence in medical health: A systematic review of interpretability methods – Shahab S Band, Atefeh Yarahmadi *et al.*
		Predicting Patient Mortality for Earlier Palliative Care Identification in Medicare Advantage Plans: Features of a Machine Learning Model – Anne Bowers, Chelsea Drake *et al.*	Future of cardiovascular diagnosis with the support of artificial intelligence – Anna Rymarz, Ewa Grabowska *et al.*
		Forecasting AI Trends in Health Care: Systematic International Patent Analysis – Tan Benjamens, Pranavsingh Dhunnoo, Márton Görög, Bertalan Mesko	ChatGPT Utility in Healthcare Education, Research, and Practice: Systematic Review on the Promising Perspectives and Valid Concerns – Malik Sallam
		Artificial Intelligence Its Uses and Application in Pediatric Dentistry: A Review – Satish Vishwanathaiah, Hytham N. Fageeh *et al.*	

meta-analyses suggests that AI has the potential to revolutionize healthcare. AI can be used to improve the diagnosis, treatment, and management of a wide range of diseases and conditions, and it can also be used to improve the efficiency and cost-effectiveness of healthcare delivery. However, it is important to note that AI is still in its early stages of development in healthcare. More research is needed to validate the effectiveness of AI algorithms in different clinical settings and to develop guidelines for their safe and responsible use.

1.18 CONCLUSIONS

AI is poised to revolutionize healthcare by improving diagnostics, enhancing drug discovery, enabling personalized medicine, and optimizing healthcare delivery. However, as we embrace AI's potential, it is imperative to address ethical concerns and ensure the responsible development and deployment of AI systems in healthcare. The future of medicine is undoubtedly intertwined with AI, and its careful integration has the potential to save lives, improve patient outcomes, and reduce the cost of healthcare delivery. A meta-analysis of AI in healthcare involves systematically reviewing and summarizing the results of multiple studies on the use of AI in the healthcare sector. Such an analysis can help researchers, policymakers, and healthcare professionals understand the overall impact and effectiveness of AI in healthcare, identify trends, and draw evidence-based conclusions. The study revealed that applications of AI in the healthcare sector during COVID, application of AI techniques such as Image Processing, NLP, Machine learning, Data mining, prediction algorithms, and detection and management are the common research topics that have been mostly researched during the period of study.

BIBLIOGRAPHY

Alugubelli, R. (2016). Exploratory study of artificial intelligence in healthcare. *International Journal of Innovations in Engineering Research and Technology*, 3(1), 1–10. https://repo.ijiert.org/index.php/ijiert/article/view/2699

Azzi, S., Gagnon, S., et al. (2020). Healthcare applications of artificial intelligence and analytics: A review and proposed framework. *Applied Sciences*, 10(18), 1–21. https://doi.org/10.3390/app10186486

Beam, A. L., & Kohane, I. S. (2018). Big data and machine learning in health care. *JAMA*, 319(13), 1317–1318. https://doi.org/10.1001/jama.2017.18391

Davenport, T., & Kalakota, R. (2019). The potential for artificial intelligence in healthcare. *Future Healthcare Journal*, 6(2), 94–98. https://doi.org/10.7861/futurehosp.6-2-94

Esteva, A., Kuprel, B., Novoa, R. A., Ko, J., Swetter, S. M., Blau, H. M., & Thrun, S. (2017). Dermatologist-level classification of skin cancer with deep neural networks. *Nature*, 542(7639), 115–118. https://doi.org/10.1038/nature21056

Genomsys. (n.d.). Trends in DNA testing. Retrieved August 28, 2025, from https://genomsys.com/trends-in-dna-testing/

Grant, K., McParland, A., Mehta, S., & Ackery, A. D. (2020). Artificial intelligence in emergency medicine: Surmountable barriers with revolutionary potential. *Annals of Emergency Medicine*, 75(6), 721–726. https://doi.org/10.1016/j.annemergmed.2019.12.024

Jiang, F., Jiang, Y., Zhi, H., Dong, Y., Li, H., Ma, S., … & Wang, Y. (2017). Artificial intelligence in healthcare: Past, present and future. *Stroke and Vascular Neurology*, 2(3), e000101. https://doi.org/10.1136/svn-2017-000101

Liu, J. (2023). Artificial intelligence and data analytics applications in healthcare. *Proceedings of the International Conference on Artificial Intelligence*. https://doi.org/10.1145/3433996.3434006

MDPI. (2023). Personalized medicine: Trends and applications. *MDPI – Journal of Personalized Medicine*, 13(8), 1214. https://www.mdpi.com/2075-4426/13/8/1214

Miotto, R., Wang, F., Wang, S., Jiang, X., & Dudley, J. T. (2018). Deep learning for healthcare: Review, opportunities and challenges. *Briefings in Bioinformatics*, 19(6), 1236–1246. https://doi.org/10.1093/bib/bbx044

Obermeyer, Z., & Emanuel, E. J. (2016). Predicting the future—Big data, machine learning, and clinical medicine. *The New England Journal of Medicine*, 375(13), 1216–1219. https://doi.org/10.1056/NEJMp1606181

Rajkomar, A., Dean, J., & Kohane, I. (2019). Machine learning in medicine. *The New England Journal of Medicine*, 380(14), 1347–1358. https://doi.org/10.1056/NEJMra1814259

Shaheen, M. Y. (2021). Applications of artificial intelligence (AI) in healthcare: A review. *ScienceOpen Research*. https://doi.org/10.14293/S2199-1006.1.SOR-.PPVRY8K.v1

Topol, E. J. (2019). High-performance medicine: The convergence of human and artificial intelligence. *Nature Medicine*, 25(1), 44–56. https://doi.org/10.1038/s41591-018-0300-7

Wang, F., Casalino, L. P., & Khullar, D. (2019). Deep learning in medicine—Promise, progress, and challenges. *JAMA Internal Medicine*, 179(3), 293–294. https://doi.org/10.1001/jamainternmed.2018.7117

Wisetsri, W., & Vijai, C. (2021). Rise of artificial intelligence in healthcare startups in India. *International Journal of Innovative Research in Technology*, 8(2), 45–50. https://www.researchgate.net/publication/349604103_Rise_of_Artificial_Intelligence_in_Healthcare_Startups_in_India

2 Unleashing the Potential

AI's Impact on Sustainable Finance in a Changing Global Economy

Manu K. S., Sangeetha Rangasamy, Anju Kalluvelil Janardhanan, and Kavitha Rajamohan

2.1 INTRODUCTION

Sustainable development is crucial as there is a worldwide need to address environmental and social challenges. In 1987, the Brundtland Report by the World Commission on Environment and Development (WCED) introduced the concept of sustainability as the development that addresses the demands of the current generation without impacting the needs of future generations (WCED, 1987). The United Nations 2030 Agenda for Sustainable Development requires a global partnership of all countries to achieve the 17 Sustainable Development Goals (SDGs). Sustainable finance has a major role in driving SDGs (Ziolo et al., 2021). Concurrently, the growth of Artificial Intelligence (AI) has opened up new opportunities to resolve difficult tasks in various areas. One such area is finance, where combining AI and sustainable finance can revamp how environmental, social, and governance (ESG) factors are considered in decision-making. Nowadays, most of the finance and treasury functions and processes are virtual, which require knowledge workers who make decisions based on analytics related to environmental (local and global) considerations and internal (company needs and trends) considerations, along with corporate financial strategy (hedging, investing, liquidity and capital structure) (Polak et al., 2020). Since a large volume of data is used, it is advisable to apply AI to manage data quickly and accurately.

2.1.1 Sustainability

Sustainability is no longer a buzzword but a call for responsible and equitable global development by taking care of the environment. At its core, sustainability represents the basic understanding that our actions today should not endanger the well-being and prosperity of future generations. It is a concept that surpasses traditional frontiers, echoing not only within environmental frameworks but also within the context of economics, finance, and technology. There is an inherent connection between

DOI: 10.1201/9781003638506-2

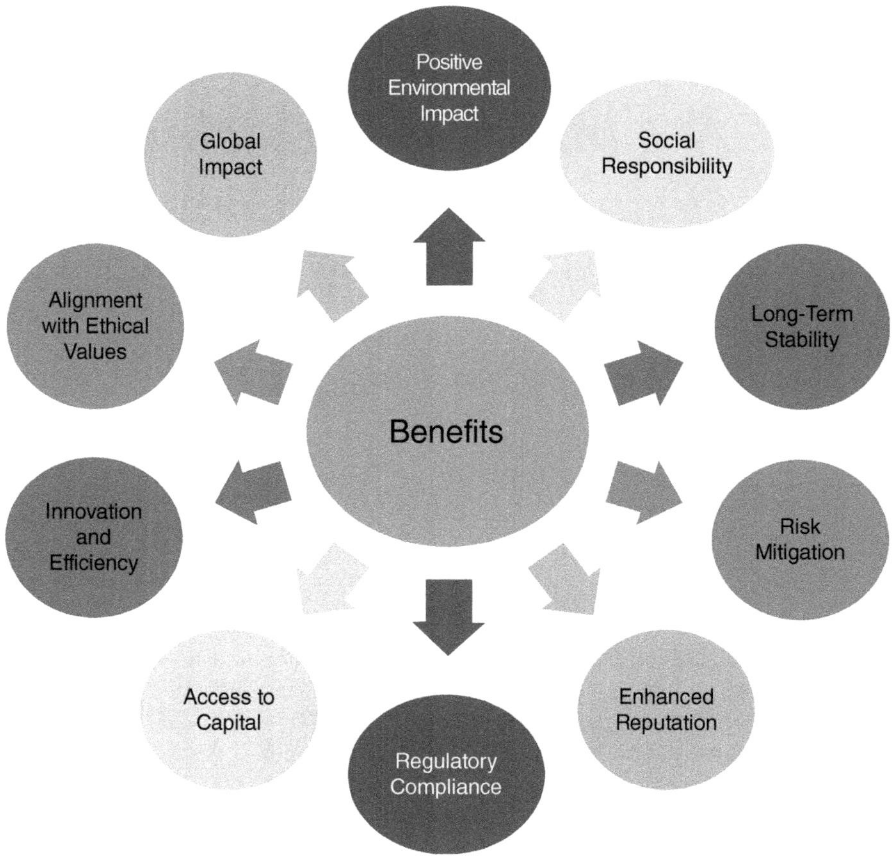

FIGURE 2.1 Benefits offered by sustainable finance.

sustainability and sustainable finance today. Sustainability emphasizes the responsible use of resources and social equality for future generations. The benefits offered by sustainable finance are depicted in Figure 2.1. Sustainable finance provides a practical framework for implementing these principles in the financial sector. Essentially, sustainable finance operation aliases sustainability by incorporating ESG factors into investment decisions, offering a tangible path to a more equitable future. The synergy between sustainability and sustainable finance highlights their collective potential to create positive change and solve some of the most pressing challenges facing our world today (Mathur & Chaturvedi, 2022). Sustainable financing not only helps in preventing but also minimizing and compensating damages made to the environment and to the climate.

2.1.2 ESG Factors

Sustainable finance incorporates ESG factors into financial decision-making processes. It may positively or negatively impact the financial performance of an entity (EBA, 2021). In order to create long-term value and achieve financial returns, investors are keen to assess the environmental and societal impact of their investment in companies. It encompasses various strategies, including responsible investing, impact investing, and green financing. Investors evaluate sustainability risks and opportunities by integrating ESG dimensions into financial decision analysis.

- Environmental – Climate change, pollution and allocation of resources.
- Social – Human rights, community engagement and labor standards.
- Governance – Executive compensation, transparency and board composition.

These factors provide a framework for evaluating companies' and investments' sustainability and ethical performance. Sustainable finance aims to align financial goals with environmental and societal goals by mobilizing capital toward sustainable projects, thus reducing poverty, producing clean energy and fostering social equity.

2.1.3 Emergence of Sustainable Finance

The concept of sustainable finance can be traced back to the sustainable development initiatives that gained popularity in the second half of the 20th century. The integration of sustainability into finance is clearly represented in Figure 2.2.

Nowadays, investing and capital allocation commonly involve sustainable finance. It consists of various financial products such as green finance, impact investing, sustainable bonds and ESG integration. Sustainable finance is becoming increasingly important in addressing global challenges such as climate change and social inequality.

The Morgan Stanley Institute's 'Sustainable Reality' report highlighted the performance of sustainable and traditional funds by different asset classes using Morningstar data during the first half of 2023, as shown in Figures 2.3 and 2.4 (Morgan Stanley Institute for Sustainable Investing, 2023). The study covers more than 96,000 funds globally. Morningstar classifies a fund as 'traditional' if the fund's prospectus does not describe the inclusion of any sustainability or ESG factors and

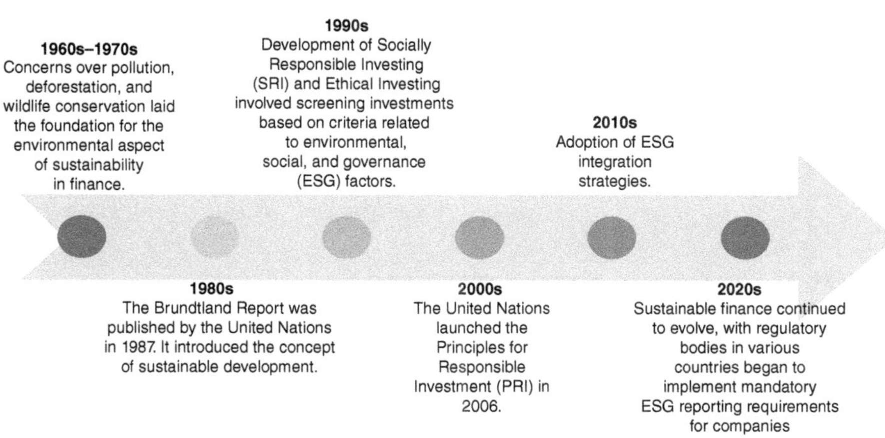

FIGURE 2.2 Integration of sustainability into finance.

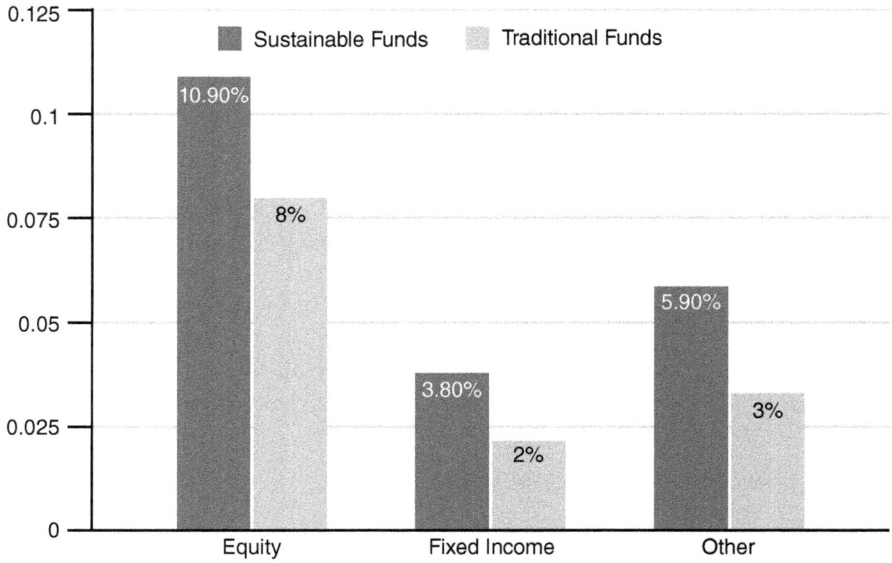

FIGURE 2.3 Sustainable and traditional funds by different asset classes.

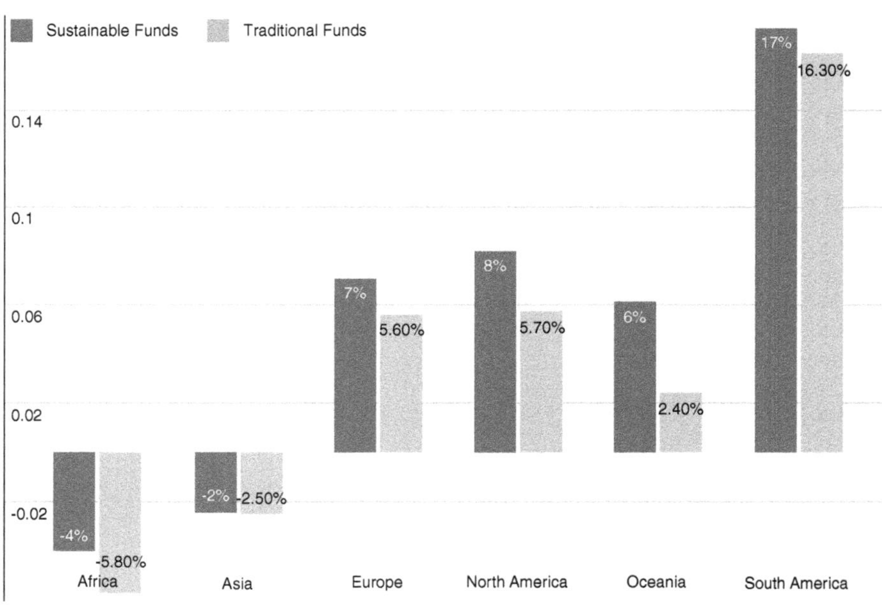

FIGURE 2.4 Region-wise performance of sustainable fund.

'sustainable' if the fund's prospectus or any regulatory disclosure includes ESG or sustainable investing. Figure 2.3 shows the performance of sustainable, traditional and other funds in the first half of 2023. It indicates the sustainable funds' performance outperforms traditional across all three asset classes – Equity, Fixed Income and Others by (1.9%), (1.6%) and (2.6%), respectively. Further, Figure 2.4 depicts region-wise, sustainable funds' performance which outperforms traditional funds across all the regions, Africa, Asia, Europe, North America, Oceania and South America by −1.8%, −0.1%, 1.5%, 2.5%, 3.7% and 1.1%, respectively.

The United Nations Conference on Trade and Development (UNCTAD) (World Investment Report, 2023) highlighted the Top 10 developing Countries by international investment in renewable energy from 2015 to 2022. As shown in Figure 2.5, Brazil (USD 114.8 billion) tops the list in terms of international investment in renewable energy, followed by Vietnam, Chile, and India. With 63%, China (Taiwan Province) tops the list in terms of percentage share of renewable energy investments in total project value followed by Chile (54%), Morocco (34%) and Brazil (32%). The sustainable finance market consists mainly of Voluntary Carbon Markets (VCM) and bonds valued at USD 5.8 trillion and growing more than 10% in 2022. As shown in Figure 2.6, the Sustainable bond market value reached USD 3.3 trillion in 2022. Its issuance has grown significantly from USD 179 billion in 2017 to USD 892 billion in 2022, almost increasing fivefold over the last five years. Further, the significant increase in growth rate of issuance (118%) can be seen from 2019 to 2020. Further, the growth rate decreased by 11.16% in 2022 (UNCTAD, (World Investment Report, 2023)).

The UNCTAD (World Investment Report, 2023) monitored the top 100 sovereign funds, including wealth and pension funds, significantly improving their disclosures on sustainable investments, climate actions and fossil fuel disinvestments. Figure 2.7 shows an increase of green investments to USD 18.7 billion in 2022 from USD 4.2 billion in 2016. Further, the sovereign funds significantly disinvested from fossil fuel securities to renewable energy and other ESG friendly initiatively.

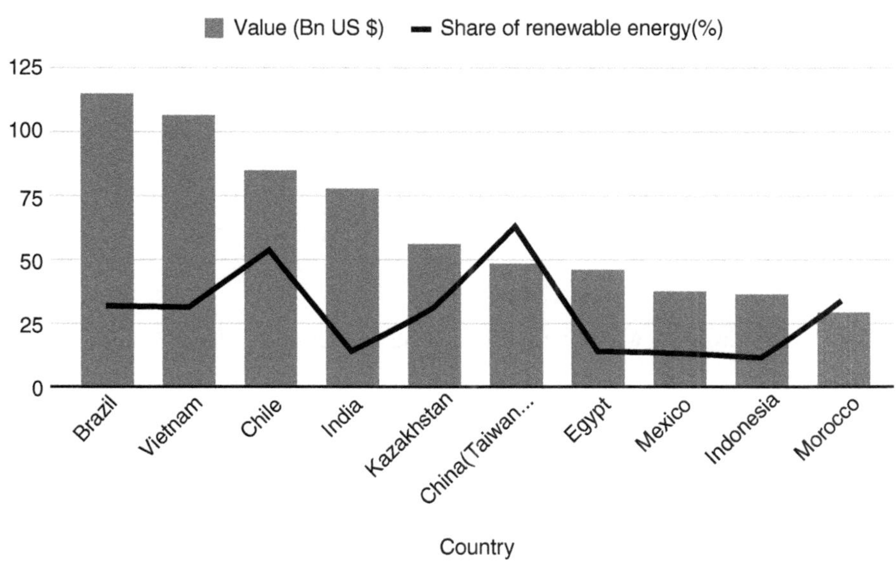

FIGURE 2.5 Top 10 developing countries by international investment in renewable energy (2015–2022).

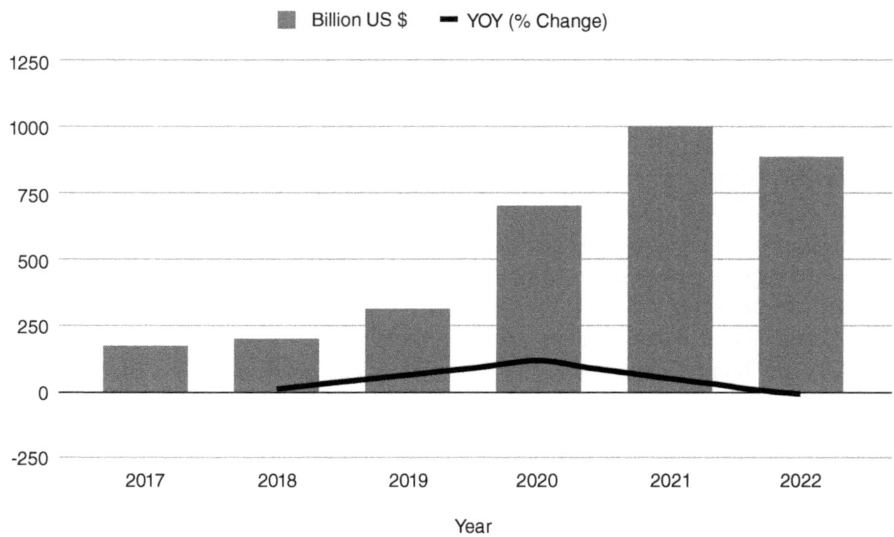

FIGURE 2.6 Global sustainable bond issue size and growth rate (2017–2022).

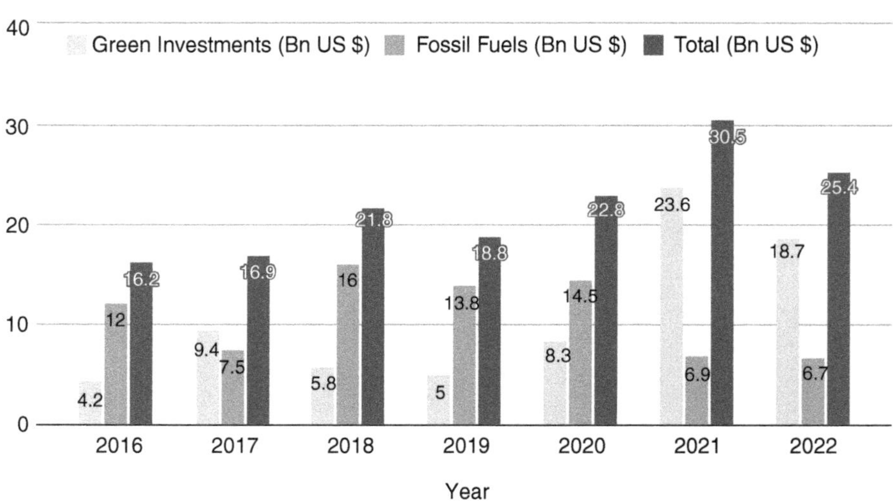

FIGURE 2.7 Sovereign wealth funds increase green investments, divest from fossil fuels (2016–2022).

2.1.4 AI and Sustainable Finance

The ever-evolving landscape of AI and its unlimited potential has transformed the global economy, making it imperative to incorporate sustainability principles into our financial systems. Sustainable finance is the investment based on ESG requirements, which handles huge amounts of complex data like geospatial terrain data, weather data, financial data, and supply chain data (Yang & Broby, 2020). AI has the designated capability to handle this huge data along with other technologies like big data analytics, high-performance computing, etc. AI can facilitate advanced solutions in innovative ways by extracting knowledge from the data and supporting sustainable goals. AI has the potential to address the ESG issues by exploiting AI-based solutions in managing ESG.

In environmental sustainability, AI integrated predictions on climate change, giving meaningful insights into policymakers' design strategies. AI-enabled deforestation monitoring systems recognize the pattern of loss in the forest, which assists officers in taking sensible action. AI-assisted energy management systems (Ahmad et al., 2021) deployed in industries helped to optimize energy conservation and reduce carbon emissions. Related to social responsibilities, AI contributes more by encouraging fairness, extensiveness and ethical code. AI-based recruitment tools help the industry to find a diverse and wide range of workforce. An AI-based drone medicine delivery system provides lifesaving healthcare facilities in remote areas. AI-based FinTech solutions offer services to unbanked people by analyzing their creditworthiness. Related to governance, AI contributes to detecting suspicious activities like anti-money laundering, streamlining the public sector system to provide better services, and tracking ESG performance to identify risks and provide strategies for sustainability. The power of AI-incorporated solutions saves the planet Earth from a red alert by encouraging ESG practices in an innovative way using technologies.

AI-driven disruptive technologies have the potential to lower costs by reducing risks and increasing reliability to offer new solutions to complex issues (Taddeo & Floridi, 2018). Numerous research studies have been conducted to show how AI can be applied to enhance ESG in financial reporting (Leitner-Hanetseder & Lehner, 2023), accounting (Bose & Bhattacharjee, 2022), generating ESG ratings (Crona & Sunsdström, 2023) and business value (Wamba Taguimdje et al., 2020).

Allen et al. (2017) explained various Machine Learning (ML) techniques applied to sustainable finance using ESG Data. Natural Language Processing (NLP) extracts insights from ESG text data. Sentimental analysis is used to classify the public opinions toward the organizations based on ESG metrics like negative, positive and neutral. ESG-related data extracted from audio sources such as interviews, speeches and meetings can be analyzed using ML-based audio analysis. Similarly, ML-based image and video analysis is employed to achieve environmental goals like moving vehicles and pollutant identification. Natalucci (2022) focused on achieving sustainable finance in emerging markets. They discussed the broader classification of ESG investing in debt instruments into activity-based and behavior-based. Activity-based is classified based on the use of proceeds from debt instruments (green bonds or loans) that will be utilized mainly to solve environmental and social problems. Behavior-based includes key performance indicators (KPIs)–based sustainability-linked instruments to achieve pre-specified sustainability goals. The previous research studies showed the use of AI techniques to measure the credibility of companies by analyzing the disclosure of annual reports (Kolbel et al., 2020; Briere et al., 2022). AI can be applied to assess the climate-related risks of European companies (Friederich et al. 2021). Briere et al. (2022) highlighted the major challenges regarding the availability of quality and transparency of financial data, the risk of data manipulation and the high cost associated with collecting and implementing AI / ML techniques.

2.2 ADOPTION OF AI IN SUSTAINABLE FINANCE ACROSS THE GLOBAL MARKET

AI can be used for multiple economic sectors to improve performance. AI-based technologies are embraced by all countries across the globe to achieve the SDG goals. AI has proved that it has enormous potential to improve global sustainability.

The US and EU Green (New) Deals discussed the role of potential innovative technologies, such as AI-based tools and technologies, to achieve SDG goals by balancing economic growth and the environment. They found that AI-based technologies have significant potential to help sustainability. Further, the green deals emphasized the developers and policy-makers to use these advanced technologies responsibly and ethically (Corrigan & Lucaj, 2020).

Francisco and Linner (2023) analyzed the idea of AI-based technologies for sustainability in the policy guidelines and reports of the United Nations (UN), the World Economic Forum (WEF) and the European Union (EU). These organizations are looking for contributions of AI to strengthen the existing sustainability framework and mechanism rather than transformative change. They focus on data accessibility, quantitative data analytics and making the economy more sustainable and energy and cost efficient. They also highlighted the major challenges of correlating sustainable AI with ecological upgradation and green governmentality disclosures, such as cybersecurity, data reliability, ethics and protecting human rights. They also presented solutions like encouraging value-based and ethical AI, flexible ESG frameworks and a stakeholders collaborative approach to make AI sustainable.

AI technology provides multilevel benefits for companies, society and citizens across the globe. The EU is formulating policies and strategies to utilize the opportunities offered by AI-powered tools and technologies. The EU is promoting AI-based innovation capacity to implement ethical AI actors in the EU members' economy. The EU is formulating the 'European strategy for data' to become a global data hub. Further, the EU launched 'a broad consultation of Member States civil society,' including academics and corporate people to formulate the policy to improve investments in new technological research, understand the necessary development skills to support the implementation of AI in SMEs and propose a future regulatory framework. Further, AI offers solutions to some of the challenges faced by the EU, like strengthening EU industries' competitiveness and achieving sustainability by addressing demographic, climate and environmental degradation changes (European Commission, 2020). The Global Partnership on Artificial Intelligence (GPAI), an international stakeholders' collaboration, was started in June 2020, with an objective of guiding responsible use of AI in various countries, organizations and society. It ensures the responsible use of AI-based technologies in international collaborative projects to achieve the SDGs by considering human rights, economic growth and societal benefits. It operates through working groups undertaking AI projects, collecting data, shaping the research and completing the projects. The

United States is important in supporting and guiding GPAI in meeting its mission (Digital Curation Centre, Trilateral Research, & School of Informatics, University of Edinburgh, 2020). The Organisation for Economic Cooperation and Development (OECD) AI Policy Observatory, a platform started in 2020 to provide leadership support and guidance to member countries in AI-based projects. Based on OECD's recommendations on AI, the activities at the GPAI will be executed. Observatory principles encourage safety and security, inclusive growth, accountability, value-based outcomes and transparency. Further, it emphasizes national policies to invest in the digital ecosystem and research and innovation for practicing trustworthy AI in international collaborative projects (Artificial Intelligence (AI) – United States Department of State, 2023).

2.3 BENEFITS OF AI IN SUSTAINABLE FINANCE

2.3.1 Better Risk Management

AI-embedded sustainable finance enables corporates, stock broking companies and portfolio managers to assess the risk involved in the proposed investment. This will facilitate the early identification of risks relevant to ESG factors (Saxena et al., 2022). AI can also handle huge volumes and a variety of data to investigate potential risks and alert decision-makers (Inampudi & Macpherson, 2020).

2.3.2 Portfolio Optimization

AI-driven portfolio optimization techniques will help investors and other stakeholders with enhanced risk management, in which ESG and traditional risk factors will be considered to measure the risk involved in the proposed portfolio (Shanmuganathan, 2020). It can also be evidenced from the work (Serafeim, 2020) that public sentiment is a critical factor in assessing the value of stocks using big ESG data with the help of AI-driven ML models to assess the investors' sentiment on the companies' sustainability activities.

2.3.3 Compliance Monitoring

Regulatory authorities insist that the companies comply with the various regulations, which is made easy by training the existing data with the updated compliance requirements. With little manual work, ESG compliance can be met with the help of an AI-enabled sustainable finance operational policy. Investors are assured that the companies comply with the ESG framework and other regulatory requirements (Lashitew, 2021).

2.3.4 Enhanced Impact Measurement

AI technologies assist in assessing companies' initiatives to reduce carbon emissions, greenhouse gas (GHG), water, electricity and other energy (Imasiku et al., 2019). This will provide investors with the impact investing initiatives of the companies.

2.3.5 Early Detection of ESG Risks

ML techniques help detect the warning signs of ESG risks much earlier before they become uncontrollable (Hao & Demir, 2023). This will help companies overcome not only huge losses but also help in taking corrective measures.

2.3.6 Reduction of Greenwashing

AI can provide a check mechanism to authenticate the companies' claims regarding their sustainable practices (Moodaley & Telukdarie, 2023). This will ensure the genuine practices of the companies in implementing ESG principles in their investment activities.

2.3.7 Behavioral Analysis for Sustainable Investing

AI in sustainable investing will help companies understand the behavior of stakeholders concerning their sustainable practices (Minkkinen et al., 2022). Accordingly, companies can improve their sustainable practices by overcoming greenwashing.

2.3.8 Market Intelligence and Trend Analysis

AI-enabled sustainable finance can help the company track its performance in line with the industry standard and automatically suggest corrective measures to meet the requirements.

2.4 METHODOLOGY

This chapter uses an inductive approach, narrative approach and content analysis to identify the various aspects of ESG metrics and sustainable finance. It also explored the existing viable AI-based models presented in multiple studies.

2.5 SCOPE OF AI IN SUSTAINABLE FINANCE

'Integrating AI into sustainable investing could mark a profound turning point in investors' ability to navigate the complex web of ESG factors,' says Matthew Slovik, Head of Global Sustainable Finance at Morgan Stanley. 'By harnessing AI's analytical capabilities, investors can identify companies with strong ESG performance, mitigate risks and shape portfolios that better align with sustainability objectives.' To increase responsible thinking in investment products' marketing, marketing communications should strongly emphasize the responsible side of the investment products and bring out that ESG issues (in financing) are actively considered.

2.5.1 Investment in Environmental Factors – Climate Change, Pollution and Allocation of Resources

Nowadays, due to the introduction of UN SDG goals, countries across the globe give much importance to the environmental impact of corporate investments.

These projects and investments should address ecological problems such as air and water pollution, accumulation of hazardous and other wastes, and depletion of natural resources (Ziyadin et al., 2019). Nevertheless, in China, it is identified that green investment can positively impact environmental pollution by improving the effect of conservation of energy and reducing carbon emissions through technological innovation through upgradation in the industry structure (Ren et al., 2022). Evidence shows that firms have taken initiatives to implement environmental Management Control Systems packages to monitor their green capabilities to enhance sustainable performance (Rahaman et al., 2023). If so, can it be possible to integrate the responsibility component into the investment products profitably, and can AI be taught to learn to incorporate responsibility in making investment decisions? (Govindan, 2022). An interesting example in which AI in sustainable finance can be elaborated is the research methodology adopted by rating agencies to rate a firm on its ESG performance. Technology innovation has brought the advent of the internet, resulting in real-time data capturing. Big Data, processed through advanced statistical methods and models, AI and NLP, which can incorporate languages, can facilitate better solutions for ESG rating. It also helps provide high-quality data, overcoming biased and self-reported greenwashing.

There are ML-based techniques, namely deep reinforcement learning, to retain neural networks to rebalance the portfolio of investment, which could help the firms achieve better financial performance with improved focus on social impact (Lin & Hsu, 2023). Figure 2.8 developed by Hughes et al. (2021) presents the difference between steps undertaken by traditional analysts compared to the AI-driven model to rate the firms on their ESG performance.

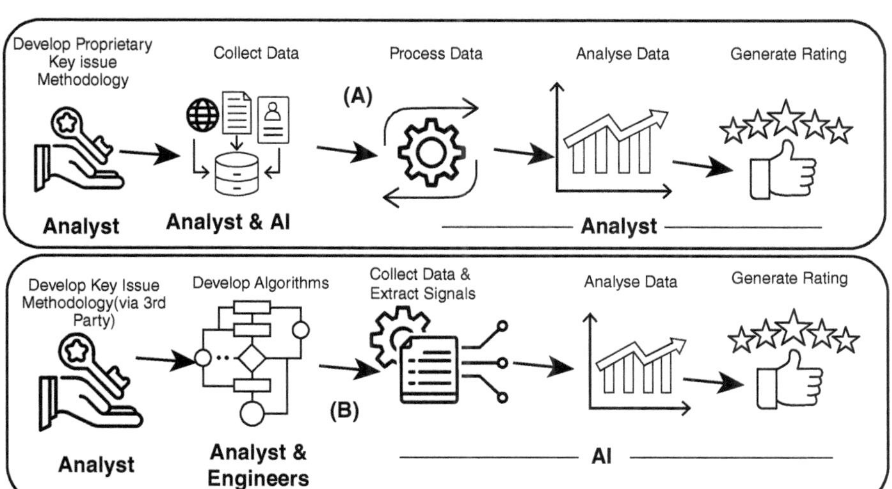

FIGURE 2.8 Top: Analyst-driven ESG research derives rating in a structured data model. Bottom: AI-driven ESG research drives signals from unstructured data.

2.5.2 Investment in Social Factors – Human Rights, Community Engagement and Labor Standards

The firms need to affirm that the resources, mostly technology, are used responsibly to improve the relationships with society to resolve problems like poverty, environmental degradation, climate crisis, pollution and inequality with the use of advanced technology (Musleh Al-Sartawi et al., n.d.). There are ML models and an extended global minimum variance (GMV) model (Double-Screening Socially Responsible Investment – DSSRI) to screen stocks and asset allocation of SRI in constructing portfolios with companies focusing on good ESG performance and high return (Zhang & Chen, 2021). The other dimension to employ social factors is responsible urban innovation, a process involving technology, policy and community. AI can help identify the best model to balance costs, benefits, risks, and positive and negative externalities in society (Yigitcanlar et al., 2021). The concept called smart cities presents the potential of AI technology to ensure the integration of critical dimensions of culture, metabolism, and governance. This will help improve the liveability of urban fabric and economic growth (Allam & Dhunny, 2019). The same concept can be applied to smart factories, embedding AI in the factory's production processes to enhance productivity (Benotsmane et al., 2019). Alternative energy consumption is strongly recommended to preserve the natural resources for future generations. An example of SRI infrastructural investment in a wind energy project requires a complex involving technical, physical, economic, social, environmental and political requirements. Decisions on the optimum utilization of resources need an unbiased model to ensure better community engagement. This can be achieved through fuzzy logic methodology and MATLAB to assess wind sites and decide which sites should be prioritized for their benefits and costs to establish wind energy power stations (Aljicevic et al., 2016). AI technology in investment projects is seen as a threat to the existing job roles. There is evidence to prove that several new categories of human jobs are emerging, requiring skills and training that will take many companies by surprise, and roles are not replacing old ones. They are novel, requiring skills and training that have no precedents. This will bring new labor standards to keep the manpower engaged and productive (James Wilson et al., 2017). AI applications in radiology should aim at well-being, reduce harm and ensure that the benefits and harms are distributed equally among the stakeholders. It should be designed in such a way as to respect human rights, freedom, privacy and maximum transparency (Geis et al., 2019).

2.5.3 Investment in Governance Factors – Executive Compensation, Transparency and Board Composition

AI technology, a 60-year-old invention, has transformed many industries across the globe. Yet another promising area where there is transparency in activities is governance. The performance of the Board of Directors or Management is exhibited through Corporate Governance Performance (CGP) metrics. Institutional investors and other stakeholders need to assess companies' performance and overcome the cost of non-compliance with fiduciary obligations of the board, auditors and management. There are 30 items identified focusing on integrity, frequency, and skills of the board as metrics to assess the CGP to support the initiatives taken by the company to showcase their social responsibility. There are ML algorithms that can be considered to predict the firm that could probably be involved in CGP controversies. Most commonly used ML algorithms are Nearest neighbors (NN), Linear Support Vector Machine (Linear SVM), RBM Support Vector Machine (RBMSVM), Random Forest (RR), Logistic Regression (LR), Artificial Neural Network (ANN), Gradient Boosting (GB), Naive Bayes (NB), and Quadratic Discriminant Analysis (QDA) (Svanberg et al., 2022). AI techniques can improve the policy-making processes with optimization and decision support techniques, data and opinion mining, game theory, and agent-based simulation (Sharma et al., 2020).

2.6 APPLICATION OF AI IN SUSTAINABLE FINANCE – GOVERNMENT, BUSINESS AND CITIZENS – USE CASE

Investors are facing enormous challenges to achieve sustainability across borders. In order to achieve this, a systematic collaboration across the counties and disciplines is required. Adoption of AI technologies helps to solve the complex interdependent problems and scale the world toward a sustainable future.

The world's population is significantly increasing year on year, and it is projected to reach 9.8 billion by 2050. Thus, the demand for resources also increases proportionately. It directly impacts the sustainability of the world's economy. Leveraging AI- and ML-based technologies helps to prevent the world from depletion of resources to preservation (Stanford Institute for Human-Centered Artificial Intelligence (HAI), 2023).

AI-powered NLP techniques are used to analyze the public sentiments in the form of text, which is generated from different social media on a real-time basis. The investors can use these results to identify and understand the companies with allegations and criticism on corruption, human rights and ESG violations. Investors finally assess the impact of companies' public perception on the performance of their share prices. An AI-based satellite sensor system is designed to identify companies that involve negative ESG practices and suggest investors to make informed decisions. Further, the Methane Alert and Response System (MARS) was introduced at the UN Climate Change Conference in 2022 to detect and monitor methane and GHG emissions in the environment.

Jupiter Intelligence, a leading US-based climate risk analytics solution provider, has developed an AI-based system that accurately predicts how climate change affects tangible assets like physical supply chain, infrastructure and buildings by analyzing various climate-related data. This system assists investors in making decisions by assessing the climate change risks on physical assets (Jupiter Intelligence, 2022).

RepRisk, a Switzerland-based data science organization mainly specialized in ESG and business conduct risk research, has built an AI-powered system that examines unstructured text data like blogs and social media posts, company reports and news articles. This system helps investors to understand and identify different ESG risks in the investment process (RepRisk, 2020).

The American company BlackRock is a global leader in investment and asset management solutions that has built an AI-powered system called Aladdin Climate to find out companies that are practicing ESG and analyzing ESG data for the best investment opportunities as per the investment goals.

Further, ML algorithms can analyze an individual's lifestyle choices, behaviors and preferences and their energy consumption, resource usage and purchasing habits to identify meaningful patterns and help them understand how their consumption habits and choices impact the environment. Further, these AI-powered systems provide feedback and suggestions on their sustainable alternatives and practices, energy-efficient alternatives, reducing carbon footprint and usage of eco-friendly products (TakeOff, 2023).

Investors are very keen on the transparency and measurability of sustainability factors and their impact on their investment decisions. AI-powered tools and techniques are helping investors to analyze ESG data such as governance and social factors, carbon emission level and resources required. Finally, AI tools are used to compare different investment alternatives based on ESG benefits and make informed decisions (Day, 2023).

2.6.1 Green Finance – ESG Investing

Green finance is sustainable investment or lending to businesses or projects that meet the ESG standards and has potential to achieve SDG goals. AI-based technologies are highly integrated with green finance to make accurate and informed decisions.

AI-powered models and techniques continue to influence the finance sector and green bond trade. An AI system contributes to fair pricing and accurate valuation of green bonds. Additionally, AI algorithms enhance green bond market efficiency by increasing the trade volume and liquidity of green bonds. AI-based portfolio optimization tools greatly assist in optimizing portfolio returns by integrating AI-powered NLP models that can provide insights by analyzing large amounts of unstructured data. Further, AI systems assist investors to make informed decisions on green bond investments by assessing green bond investment opportunities, market trends and historical data (Green Bond Trade Facilitated by AI, 2023).

AI and big data tools are used in ESG investment to analyze a company's ESG-related data for ESG rating using different data, metrics and frameworks. Further, Natural Programming Language (NPL) is used to assess massive amounts of unstructured data related to ESG to perform sentiment analysis. AI-based ESG rating provides an objective and transparent rating so the investors can make informed decisions about ESG investment. AI-driven ESG rating analysis provides more real-time and standardized analysis compared to traditional analysis (OECD, 2021).

ESG investors are able to make accurate information by integrating the advanced AI-based language model GPT3.5. The classifier module of GPT identifies the required factors of companies to predict accurate responses and help to make improved decisions (Jain et al., 2023).

2.6.2 Sustainable Debt – Green Bonds and Loans

On September 18, 2023, an event took place during the United Nations' 78th Session High-Level Week in New York City. There were representatives from different countries, governments, top corporations, leading AI firms, AI developers, the UN,

TABLE 2.1
Major Financial Commitments Announced to Leverage AI Tools to Achieve SDGs

	Country/Company	Sustainable Investment	Sustainable Commitment
1	USA	USD 15 million	To leverage and responsibly use of AI applications globally to achieve SDGs
2	The Kingdom of Spain	8 billion euros	To support AI and deep learning tech startups projects.
		300 million euros	Establishment of National Agency on AI to support Green AI Programs and global AI governance
3	The UK	£1.07 million	Complex Risk Analytics Fund (CRAF) to utilize AI-driven system to predict humanitarian disaster and conflicts arises in different countries
4	Google	USD 10 million	AI projects to support victims of climate change and natural calamities
5	OpenAI	USD 1 million	To develop a world-class AI-powered system to achieve the UN SDGs.
6	Amazon Web Services	USD 10 million	AI & ML Scholarship program to support students from 85 countries to start careers in ML.

civil society organizations and speakers to discuss and demonstrate solid applications of AI that could be used in a secure, responsible and ethical manner to achieve SDG goals and solve societal challenges. Table 2.1. shows the major financial commitments announced to leverage AI tools to achieve SDGs (United States Department of State, 2023).

2.7 SUSTAINABLE FINANCE LEADING TO GREEN ECONOMY

Green finance is an important financial service that directly influences the transformation, development and upgradation of the financial sector and economy. Many countries started to implement carbon and GHG emissions reduction programs such as carbon peak and carbon neutral programs (CPCNP), renewable energy projects – solar and wind, sustainable finance programs – green bonds and sustainable investments. As per the Leaders' Summit on Climate (2021), The United States announced to reduce GHG by 50% by 2023. The United States and Japan can achieve carbon neutrality by 2050, China and India by 2060 and 2070, respectively. Further, Canada can sell zero-emission vehicles (cars and trucks) from 2025 (Peng et al., 2022).

Green finance has a significant impact on society, economy and sustainability. There are a variety of green and suitable finance services, such as green bonds, green climate funds, green loans, green funds, climate bonds, green credits, green banking and sustainability bonds, that are used to achieve SDG goals. These services mainly help economic development by CO_2 and GHG emission reduction, pollution control, adaptation to change of climate, waste processing and recycling, cleaner production, extraction of renewable resources, climate-resilient infrastructure, water and waste management, green and social projects, green buildings, sustainable transport, wind and solar projects, agriculture and land use and forestry protection (Kumar et al., 2023).

Sustainable finance highlights the significance of sustainability reporting on ESG practices and performance. The sustainable investors and other stakeholders require corporations to disclose their ESG practices, procedures, policies, accountability and transparency mechanisms. This disclosure process allows businesses to display their commitment to achieving sustainability and helps investors to make better decisions. ESG practices of businesses contribute to a sustainable and inclusive economy.

The Singapore Exchange (SGX) has taken the initiative to make it mandatory for all the listed companies to disclose and report their ESG practices. Further, the Monetary Authority of Singapore (MAS) is taking significant steps to encourage sustainable fiancé practices. It has introduced the MAS Green Bond Grand Scheme and issued $2 billion of green bonds.

2.8 ETHICAL CONSIDERATIONS AND CHALLENGES

Ethical considerations for the adoption of AI in complex social contexts are an important prerequisite that is to be focused upon by firms across the globe (Kerr et al., 2020). There are five essential ethical principles: transparency, justice and fairness, non-maleficence, responsibility and privacy (Jobin et al., 2019).

There are possibilities that AI implementation in one aspect of the ESG factor in sustainable finance will provide unintended consequences for others. Hence, it is important to identify a trade-off between the parameters considered for ESG metrics.

Job displacement and inequality are the consequences of AI-driven sustainable finance. This can be avoided if firms and policymakers develop strategies to mitigate negative impacts through reskilling and upskilling to facilitate a successful AI-enabled transition.

Green and ethical washing is another alarming challenge that might occur because of poorly implemented sustainable practices. Greenwashing refers to false representations made by companies to exhibit environment-based CSR initiatives. Ethical washing refers to the false representations made to fulfil ESG requirements toward societal and governance aspects. This can be avoided by a genuine commitment to sustainability and social responsibility.

Data privacy and security are a crucial ethical issue since AI requires personal and sensitive information. Hence, it is the responsibility of the firms to take measures to prevent unauthorized access to the data.

Transparency in the principles and frameworks of ESG metrics to measure the sustainable practices of the firms. A greater level of transparency and explainability in metrics and AI algorithms will help the firms to overcome this issue.

2.9 FUTURE TRENDS AND PROSPECTS

The concept of sustainable investing and finance existed in 2020. Accordingly, the companies started compiling the available data to prove Socially Responsible Investing. The data used to predict the ESG score are unstructured and have noise without a clear regulatory framework. There is much research in sustainable finance and AI applications using ML models. However, a strong understanding of ESG principles, sustainability and responsible practices is necessary for the hour. Ethical awareness and knowledge of potential biases, privacy concerns and ethical considerations in AI and ESG require a multidisciplinary approach. Another crucial question is how AI governance can be integrated into existing ESG metrics or criteria. It can be included either in the social and/or governance pillars, independent pillars to evaluate the organization's ability to incorporate AI and sustainability or in the technology pillar in which technology governance is measured.

2.10 CONCLUSION

The incorporation of AI in sustainable finance marks the start of a brand-new era, complete with opportunities and demanding situations. As this chapter suggests, introducing AI into sustainable finance isn't always an easy task. It's far more important that organizations embark on this journey with a thorough knowledge of ethical problems and are prepared to take a proactive approach to address future trends. Ethical considerations play a vital role in the successful use of AI within the context of sustainable finance. The five essential ethical standards, transparency, fairness and honesty, non-malice, obligation and confidentiality, guide organizations all over the world. Those standards ensure that primarily sustainable AI-based practices are not only profitable but also assist the common good. But it is vital to remember that the introduction of AI in a single factor of ESG indicators might also lead to unintentional outcomes in other factors that need careful consideration.

Challenges like job displacement and inequality in AI-driven sustainable finance are real but can be mitigated through considerate reskilling and upskilling strategies. Organizations and policymakers have to proactively cope with those challenges to ensure an effective and equitable transition to an AI-driven world. Ecological and ethical laundering poses an extreme threat to the credibility of sustainable finance. The most effective manner to solve this hassle is to really commit to sustainability and social responsibility. Organizations must take proactive measures to prevent unauthorized access to confidential information to ensure privacy and security. Transparency of ESG factors and AI algorithms is critical to building trust in an organization's practices. The future prospects of AI in sustainable finance are promising. As this space continues to evolve, a deep understanding of ESG factors, sustainability and responsible practices will provide a critical foundation for success. Ethical awareness, recognition of potential bias, privacy concerns and ethical considerations require a multidisciplinary approach to management and resolution. It is important to integrate AI governance into existing ESG factors. Whether addressing social, governance or technological issues, it will be critical to assess an organization's ability to responsibly integrate AI and sustainability. In summary, the impact of AI on sustainable finance in our growing global economy is profound and diverse. The challenges are enormous, but with a strong commitment to ethics, responsible practices and a forward-looking management approach, AI can truly become a powerful force for good and lead us toward a brighter, better, more sustainable and fairer future. It is the responsibility of all stakeholders involved to seize this opportunity to create a more responsible and ethical financial landscape that benefits not just profits, but society and the entire planet.

REFERENCES

Ahmad, T., Zhang, D., Huang, C., Zhang, H., Dai, N., Song, Y., & Chen, H. (2021). Artificial intelligence in sustainable energy industry: Status Quo, challenges and opportunities. *Journal of Cleaner Production*, 289(289), 125834. https://doi.org/10.1016/j.jclepro.2021.125834

Aljicevic, Z., Kostic, A., Dautbasic, N., & Karli, G. (2016). Model of fuzzy logic for selection infrastructural investment project of wind farm locations. In *Proceedings of the 27th International DAAAM Symposium 2016* (pp. 0743–0748). DAAAM International Vienna. https://doi.org/10.2507/27th.daaam.proceedings.107

Allam, Z., & Dhunny, Z. A. (2019). On big data, artificial intelligence and smart cities. *Cities*, 89, 80–91. https://doi.org/10.1016/j.cities.2019.01.032

Allen, E., Lyons, K., & Tavares, R. (2017). The application of machine learning to sustainable finance. *Journal of Environmental Investing*, 8(1), 115–128.

Artificial Intelligence (AI) – United States Department of State. (2023). United States Department of State. https://www.state.gov/artificial-intelligence/

Artificial intelligence for accelerating progress on the sustainable development goals: Addressing society's greatest Challenges – United States Department of State. (2023). United States Department of State. https://www.state.gov/artificial-intelligence-for-accelerating-progress-on-the-sustainable-development-goals-addressing-societys-greatest-challenges/

Benotsmane, R., Kovács, G., & Dudás, L. (2019). Economic, social impacts and operation of smart factories in Industry 4.0 focusing on simulation and artificial intelligence of collaborating robots. *Social Sciences*, 8(5), 143. https://doi.org/10.3390/socsci8050143

Bose, D., & Bhattacharjee, S. (2022). Big data, data analytics and artificial intelligence in accounting: An overview. In S. Akter & S. F. Wamba (Eds.), *Handbook of big data methods*. Edward Elgar Publishing. https://doi.org/10.4337/9781800888555.00007

Briere, M., Keip, M., & Berthe, T. L. (2022). Artificial intelligence for sustainable finance: Why it may help. SSRN. https://doi.org/10.2139/ssrn.4252329

Caitlin Corrigan and Laura Lucaj (2020). The potential for AI in implementing the green deal and ethical implications. Retrieved from https://ieai.sot.tum.de/wp-content/uploads/2022/03/Research-Brief_GreenDeal_Final_Update.pdf

Crona, B., & Sunsdström, E. (2023). Sweet spots or dark corners? An environmental sustainability examination of big data and AI in ESG. In T. Rana, A. Lowe, & J. Svanberg (Eds.), *Handbook of big data and analytics in accounting and auditing*. Springer Singapore. https://dx.doi.org/10.2139/ssrn.4037299

Day, R. (2023, June 5). *The AI revolution in sustainable infrastructure investing*. Forbes. https://www.forbes.com/sites/robday/2023/06/05/the-ai-revolution-in-sustainable-infrastructure-investing/?sh=5e047be838bc

Digital Curation Centre, Trilateral Research, & School of Informatics, University of Edinburgh. (2020, November). *The role of data in AI: Report for the Data Governance*

Working Group of the Global Partnership of AI (Final report). Global Partnership on AI. https://wp.oecd.ai/app/uploads/2025/05/role-of-data-in-ai.pdf

EBA. (2021). EBA report on management and supervision of ESG risks for credit institutions and investment firms. https://www.eba.europa.eu/sites/default/documents/files/document_library/Publications/Reports/2021/1015656/EBA%20Report%20on%20ESG%20risks%20management%20and%20supervision.pdf

European Commission (2020), On artificial intelligence – A European approach to excellence and trust. https://commission.europa.eu/system/files/2020-02/commission-white-paper-artificial-intelligence-feb2020_en.pdf

Francisco, M., & Linner, B. (2023). AI and the governance of sustainable development. An idea analysis of the European Union, the United Nations, and the World Economic Forum. *Environmental Science & Policy*, 150, 103590. https://doi.org/10.1016/j.envsci.2023.103590

Friederich, D., Kaack, L. H., Luccioni, A., & Schmidt, T. S. (2021). Automated Identification of Climate Risk Disclosures in Annual Corporate Reports. arXiv (Cornell University). https://doi.org/10.48550/arxiv.2108.01415

Geis, J. R., Brady, A. P., Wu, C. C., Spencer, J., Ranschaert, E., Jaremko, J. L., Langer, S. G., Borondy Kitts, A., Birch, J., Shields, W. F., van den Hoven van Genderen, R., Kotter, E., Wawira Gichoya, J., Cook, T. S., Morgan, M. B., Tang, A., Safdar, N. M., & Kohli, M. (2019). Ethics of artificial intelligence in radiology: Summary of the joint european and North American Multisociety Statement. *Radiology*, 293(2), 436–440. https://doi.org/10.1016/j.jacr.2019.07.028

Govindan, K. (2022). How artificial intelligence drives sustainable frugal innovation: A multitheoretical perspective. *IEEE Transactions on Engineering Management*, 1–18. https://doi.org/10.1109/TEM.2021.3116187

Green bond Trade: Facilitated by AI. (2023). ADVFN newspaper. https://uk.advfn.com/newspaper/advfnnews/71005/green-bond-trade-facilitated-by-ai#:~:text=Portfolio%20management%20is%20a%20critical,opportunities%20and%20diversify%20portfolios%20effectively

Hao, X., & Demir, E. (2023). Artificial intelligence in supply chain decision-making: An environmental, social, and governance triggering and technological inhibiting protocol. *Journal of Modelling in Management*, Ahead-of-print(ahead-of-print). https://doi.org/10.1108/JM2-01-2023-0009

Hughes, A., Urban, M. A., & Wójcik, D. (2021). Alternative ESG ratings: How technological innovation is reshaping sustainable investment. *Sustainability: Science Practice and Policy*, 13(6), 3551. https://doi.org/10.3390/su13063551

Imasiku, K., Thomas, V. M., & Ntagwirumugara, E. (2019). Unraveling green information technology systems as a global Greenhouse gas emission Game-Changer. *Administrative Sciences*, 9(2), 43. https://doi.org/10.3390/admsci9020043

Inampudi, K., & Macpherson, M. (2020). The impact of AI on environmental, social and governance (ESG) investing. In *The AI book* (pp. 129–131). Wiley. https://doi.org/10.1002/9781119551966.ch35

Jain, Y., Gupta, S., Yalciner, S., Joglekar, Y. N., Khetan, P., & Zhang, T. (2023). Overcoming complexity in ESG Investing: The role of generative AI integration in identifying contextual ESG factors. *Social Science Research Network*. https://doi.org/10.2139/ssrn.4495647

James Wilson, H., Daugherty, P. R., & Morini-Bianzino, N. (2017). The jobs that artificial intelligence will create: A global study finds several new categories of human jobs emerging, requiring skills and training that will take many companies by surprise. MIT Sloan Management Review. https://sloanreview.mit.edu/article/will-ai-create-as-many-jobs-as-it-eliminates/

Jobin, A., Ienca, M., & Vayena, E. The global landscape of AI ethics guidelines. *Nature Machine Intelligence* **1**, 389–399 (2019). https://doi.org/10.1038/s42256-019-0088-2

Jupiter Intelligence. (2022). The new standard in investment-grade climate risk analytics. Retrieved August 31, 2025. https://www.jupiterintel.com/

Kerr, A., Barry, M., & Kelleher, J. D. (2020). Expectations of artificial intelligence and the performativity of ethics: Implications for communication governance. *Big Data & Society*, 7(1), 2053951720915939. https://doi.org/10.1177/2053951720915939

Kolbel, J. F., Leippold, M., Rillaerts, J., & Wang, Q. (2020). Does the CDS market reflect regulatory climate risk disclosures? *Social Science Research Network.* https://doi.org/10.2139/ssrn.3616324

Kumar, B., Kumar, L., Kumar, A., Kumari, R., Tagar, U., & Sassanelli, C. (2023). Green finance in circular economy: A literature review. *Environment, Development and Sustainability*. https://doi.org/10.1007/s10668-023-03361-3

Lashitew, A. (2021). Corporate uptake of the sustainable development goals: Mere Greenwashing or an advent of institutional change? In *Corporate Uptake of the sustainable development goals: Mere greenwashing or an advent of institutional change?*. Lashitew, Addisu. [S.l.]: SSRN. https://dx.doi.org/10.2139/ssrn.3789574

Leitner-Hanetseder, S., & Lehner, O. M. (2023). AI-powered information and big data: Current regulations and ways forward in IFRS reporting. *Journal of Applied Accounting Research*, 24(2), 282–298. https://doi.org/10.1108/JAAR-01-2022-0022

Lin, H.-Y., & Hsu, B.-W. (2023). Empirical study of ESG score prediction through machine learning—A case of non-financial companies in Taiwan. *Sustainability: Science Practice and Policy*, 15(19), 14106. https://doi.org/10.3390/su151914106

Mathur, S., & Chaturvedi, A. (2022). Future of green financing-banking customer's perspective. Available at SSRN. https://doi.org/10.2139/ssrn.4247169

Minkkinen, M., Niukkanen, A., & Mäntymäki, M. (2022). What about investors? *ESG analyses as tools for ethics-based AI auditing*. AI & Society. https://doi.org/10.1007/s00146-022-01415-0

Moodaley, W., & Telukdarie, A. (2023). Greenwashing, sustainability reporting, and artificial intelligence: A systematic literature review. *Sustainability: Science Practice and Policy*, 15(2), 1481. https://doi.org/10.3390/su15021481

Morgan Stanley Institute for Sustainable Investing. (2023). Sustainable reality: Despite challenging market conditions in 2022, investor demand for sustainable funds remains strong (Report No. CRC-5440003). Morgan Stanley. https://www.morganstanley.com/content/dam/msdotcom/en/assets/pdfs/Sustainable-Reality_2022_Final_CRC-5440003.pdf

Musleh Al-Sartawi, A. M. A., Hussainey, K., & Razzaque, A. (n.d.). The role of artificial intelligence in sustainable finance. *Journal of Sustainable Finance & Investment*, 1–6. https://doi.org/10.1080/20430795.2022.2057405

Natalucci, F. M., Gautam, D., & Goel, R. (2022). Sustainable finance in emerging markets: Evolution, challenges, and policy priorities. *IMF Working Papers*, 2022(182), 1. https://doi.org/10.5089/9798400218101.001

OECD (2021), Artificial intelligence, machine learning and big data in finance: Opportunities, challenges, and implications for policy makers. https://www.oecd.org/finance/artificial-intelligence-machine-learningbig-data-in-finance.htm

Peng, W., Lu, S., & Lü, W. (2022). Green financing for the establishment of renewable resources under carbon emission regulation. *Renewable Energy*, 199, 1210 1225. https://doi.org/10.1016/j.renene.2022.08.140

Polak, P., Nelischer, C., Guo, H., & Robertson, D. C. (2020). "Intelligent" finance and treasury management: What we can expect. *AI & Society*, *35*(3), 715–726. https://doi.org/10.1007/s00146-019-00919-6

Rahaman, M. S., Ansari, K. M. N., Kumar, H., & Shah, K. (2023). MCS package and entrepreneurial competency influence on business performance: The moderating role of business strategy. *European Journal of Management and Business Economics*, *32*(1), 1–23. https://www.emerald.com/ejmbe/article-pdf/32/1/1/653455/ejmbe-04-2020-0088.pdf

Ren, S., Hao, Y., & Wu, H. (2022). How does green investment affect environmental pollution? Evidence from China. *Environmental & Resource Economics*, 81(1), 25–51. https://doi.org/10.1007/s10640-021-00615-4

RepRisk methodology overview. (2020). https://www.reprisk.com/news-research/resources/methodology

Saxena, A., Singh, R., Gehlot, A., Akram, S. V., Twala, B., Singh, A., Montero, E. C., & Priyadarshi, N. (2022). Technologies Empowered Environmental, Social, and Governance (ESG): An Industry 4.0 landscape. *Sustainability: Science Practice and Policy*, 15(1), 309. https://doi.org/10.3390/su15010309

Serafeim, G. (2020). Public sentiment and the price of corporate sustainability. *Financial Analysts Journal*, 76(2), 26–46. https://doi.org/10.2139/ssrn.3265502

Shanmuganathan, M. (2020). Behavioural finance in an era of artificial intelligence: Longitudinal case study of robo-advisors in investment decisions. *Journal of Behavioral and Experimental Finance*, 27, 100297. https://doi.org/10.1016/j.jbef.2020.100297

Sharma, G. D., Yadav, A., & Chopra, R. (2020). Artificial intelligence and effective governance: A review, critique and research agenda. *Sustainable Futures*, 2, 100004. https://doi.org/10.1016/j.sftr.2019.100004

Stanford Institute for Human-Centered Artificial Intelligence (HAI). (2023, January). *Sustainability and AI [Industry Brief 6]*. Stanford University. https://hai.stanford.edu/sites/default/files/2023-01/HAI_IndustryBrief6_v8.pdf

Svanberg, J., Ardeshiri, T., Samsten, I., Öhman, P., Neidermeyer, P. E., Rana, T., Semenova, N., & Danielson, M. (2022). Corporate governance performance ratings with machine learning. *Intelligent Systems in Accounting Finance & Management*, 29(1), 50–68. https://doi.org/10.1002/isaf.1505

Taddeo, M., & Floridi, L. (2018). How AI can be a force for good. *Science*, 361(6404), 751–752. https://doi.org/10.1126/science.aat5991

TakeOff. (2023, July 1). How sustainable finance can benefit from AI – TakeOff – Medium. *Medium*. https://medium.com/@takeoff.financial/how-sustainable-finance-can-benefit-from-ai-1af9178c3f46

Wamba-Taguimdje, S. L., Fosso Wamba, S., Kala Kamdjoug, J. R., & Tchatchouang Wanko, C. E. (2020). Influence of artificial intelligence (AI) on firm performance: The business value of AI-based transformation projects. *Business Process Management Journal*, 26(7), 1893–1924. https://doi.org/10.1108/BPMJ-10-2019-0411

WCED. (1987). *Report of the World Commission on Environment and Development: Our Common Future*. United Nations Documents.https://www.are.admin.ch/are/en/home/media/publications/sustainable-development/brundtland-report.html

World Investment Report 2023. (2023, July 5). UNCTAD. https://unctad.org/publication/world-investment-report-2023

Yang, Z., & Broby, D. (2020). Sustainable finance: AI applications in satellite imagery and data. Strathprints.strath.ac.uk https://strathprints.strath.ac.uk/73828/

Yigitcanlar, T., Corchado, J. M., Mehmood, R., Li, R. Y. M., Mossberger, K., & Desouza, K. (2021). Responsible urban innovation with local government Artificial Intelligence (AI): A conceptual framework and research agenda. *Journal of Open Innovation: Technology, Market, and Complexity*, 7(1), 71. https://doi.org/10.3390/joitmc7010071

Zhang, J., & Chen, X. (2021). Socially responsible investment portfolio construction with a double-screening mechanism considering machine learning prediction. Discrete dynamics in nature and society. *An International Multidisciplinary Research and Review Journal*, 2021. https://doi.org/10.1155/2021/7390887

Ziolo, M., Bak, I., & Cheba, K. (2021). The role of sustainable finance in achieving sustainable development goals: Does it work. *Technological and Economic Development of Economy*, 27(1), 45–70. https://doi.org/10.3846/tede.2020.13863

Ziyadin, S., Streltsova, E., Borodin, A., Kiseleva, N., Yakovenko, I., & Baimukhanbetova, E. (2019). Assessment of investment attractiveness of projects on the basis of environmental factors. *Sustainability: Science Practice and Policy*, 11(9), 2544. https://doi.org/10.3390/su11092544

3 Exploring the Role of Artificial Intelligence in Educational Technology

Lakshmi Raj, Mallika Sankar M., Durdana Ovais, and V. Selvalakshmi

3.1 INTRODUCTION

The rapid advancements in technology over the past decade, coupled with the increased computing power, have led to significant transformations in various aspects of our daily lives, with a particular focus on understanding human intelligence in all its forms. Artificial intelligence (AI) is a set of computer programs and technologies that imitate and replicate the functioning and intelligence of the human brain. AI is "an area of study in the field of computer science. It is concerned with the development of computers able to engage in human-like thought processes such as learning, reasoning and self-correction" (Kok et al., 2009). Russell and Norvig (2010) in their book "Artificial Intelligence: A Modern Approach" assert that the birth of AI occurred in 1956 with the work of John McCarthy. AI encompasses diverse fields like information, logic, cognition, and biology, utilized in knowledge processing and pattern recognition (Zhang and Lu, 2021). The rapid integration of AI into our daily lives can be attributed to three key technological advancements: improved algorithms, abundant data resources, and cost-effective computational power and storage, which have all reached significant development (Ergen, 2019). AI is advancing toward emotional capabilities, potentially surpassing human intelligence (Zhang and Lu, 2021). AI's wide-ranging applications encompass autonomous driving, financial risk management, healthcare diagnosis, retail automation, media content generation, smart payment systems, and voice-controlled smart homes (Zhang and Lu, 2021). In business, AI offers a multitude of benefits, including enhancing product features and functions, improving decision-making, creating new products, optimizing internal operations, reducing labor costs through automation, streamlining external processes like marketing and sales, harnessing scarce knowledge, exploring new markets, and liberating workers to be more creative by automating tasks (Davenport and Ronanki, 2018). In summary, AI has made significant contributions across various business departments, including HR, marketing, finance, and operations (Guenole and Feinzig, 2018; Dhamija and Bag, 2020; Dogru and Keskin, 2020; Verma et al., 2021; Weber et al., 2023). As per data from Statista, the AI market is anticipated to experience substantial growth, increasing from

DOI: 10.1201/9781003638506-3

$244 billion in 2025 to nearly $800 billion by 2030, with a notable compounded annual growth rate (CAGR) of 17.3% (Statista, 2025).

In today's rapidly evolving technological landscape, numerous tech-related companies harness the power of AI to transform their operations and services. Among these pioneering companies, educational technology (EdTech) stands out as a promising player in the field, leveraging AI to enhance its student offerings. However, it is crucial to emphasize that the effective utilization of AI is key to unlocking the full potential of EdTech's services and delivering even greater value to the student community. In the education sector, AI is believed to make personalized learning, spread education everywhere, and bring new and creative teaching and learning methods (Bulathwelaet al., 2021). EdTech refers to

> the use of tools, technologies, processes, procedures, resources, and strategies to improve learning experiences in a variety of settings, such as formal learning, informal learning, non-formal learning, lifelong learning, learning on demand, workplace learning, and just-in-time learning.
>
> (Huang et al., 2019a)

According to the report by Statista, Revenue in the Online Education market is expected to reach US$166.60 billion in 2023, with a projected annual growth rate (CAGR 2023–2027) of 9.48%, leading to a market volume estimated at US$239.30 billion by 2027 (Statista, 2023). Over the past decade, the introduction of learning applications, STEM kits, interactive videos, and other technological advancements, such as personalized educational content, internet-based live discussions, and online classroom activities, has been the driving force behind the evolution of EdTech (Datalabs, 2020). According to the PGA Labs report, EdTech companies like Byjus, Cuemath, Toppr, Mindler, and others are employing AI to automate operational tasks for teachers, enabling them to dedicate more time to higher-order activities like teaching and guiding students, while also implementing "intelligent teaching systems" (ITSs) that adapt to each learner's knowledge level and goals (PGALabs, 2023). AI in EdTech presents a paradigm shift in how we approach learning and teaching. AI tools have already been integrated into various aspects of the educational process, encompassing content development, teaching methodologies, student assessment, and teacher–student communication (Chassignola et al., 2018). Leveraging AI's capabilities, personalized learning, intelligent tutoring, enhanced teaching and assessment, and intelligent content creation can reshape the educational landscape, making education more accessible, engaging, and effective. The bibliometric analysis titled "Identification and Assessment of Technology Trends in K-12 Education from 2011 to 2021" anticipates that emerging technologies such as virtual reality (VR) and AI are poised to become prominent trends in the forthcoming years (Dubé and Wen, 2022). Integrating AI in EdTech can unlock the full potential of each learner, empower educators with data-driven insights, and foster a dynamic, learner-centric educational ecosystem. While there has been considerable research on the use of AI in education (Chen et al., 2022), there exists substantial untapped potential for deeper investigations into AI's role within the domain of EdTech. Building upon the existing body of knowledge, it is evident that further exploration and examination of the multifaceted applications and implications of AI within EdTech is both relevant and promising, offering a fertile ground for scholarly inquiry and innovation. The potential of AI to transform and optimize teaching and learning experiences, improve educational outcomes, and cater to individual student needs warrants continued research and development (Khandelwal et al., 2021). Therefore, the purpose of this study is to investigate the role of AI in EdTech, with a particular focus on how AI is used in EdTech services, including personalized learning, experiential learning, automating administrative tasks, and data privacy, aiming to understand their roles in detail.

3.2 SIGNIFICANCE OF THE STUDY

The study exploring the role of AI in EdTech holds significant importance for various stakeholders in the education sector. This study delves into the transformative potential of AI in revolutionizing teaching and learning practices and promoting adaptive and personalized educational experiences. By understanding the capabilities of AI in EdTech, educators can enhance instructional delivery, create interactive learning environments, and increase student engagement. Integrating AI-based systems enables the improvement of educational outcomes through personalized learning pathways, adaptive assessments, and real-time feedback, catering to individual student needs. Additionally, the study recognizes the importance of addressing ethical considerations, such as data privacy and algorithmic bias, and provides insights into responsible implementation. The significance of the AI factors under consideration in our study is delineated as follows: the incorporation of AI significantly enhances EdTech services, particularly in terms of personalized learning. *Personalized learning*, an essential component of modern education, benefits substantially from AI's capability to customize educational content and learning pathways to cater to the diverse needs and learning preferences of students. This not only empowers educators to provide more effective instruction but also allows EdTech companies to actively leverage AI to enhance their platforms, ensuring more effective and engaging learning experiences for students. Furthermore, for academics, this research offers invaluable insights for optimizing their teaching methods to adapt to the evolving EdTech landscape. AI's role in *experiential learning* is pivotal. It enables immersive, hands-on experiences that are invaluable for skill development and real-world preparation. AI-driven simulations, VR, and interactive scenarios are reshaping education. This research sheds light on the future of experiential learning, impacting how educators design engaging content and prepare students for practical challenges. The integration of AI for *automating administrative functions* within the EdTech sector brings about a transformation of operations. By streamlining tasks, it effectively lessens the administrative workload for educators and educational institutions. This shift toward AI-driven automation allows educators to allocate more of their time to the core aspects of teaching and mentoring. Consequently, this study offers valuable insights to the academic community, highlighting the substantial efficiency improvements AI brings to educational processes. Simultaneously, for EdTech companies, it emphasizes the paramount importance of AI integration for enhancing operational efficiency and, subsequently, the overall quality of educational services. In the digital era, safeguarding *data privacy* is of utmost importance.

AI assumes a pivotal role in securing sensitive educational information. Grasping the present and future implications of AI for data privacy is essential. Academics acquire valuable insights to educate both students and professionals in this critical domain. Simultaneously, EdTech enterprises stand to gain from this research by fortifying their data protection protocols, nurturing trust, and ensuring alignment with the ever-evolving privacy regulations. In a time marked by heightened data sharing and technological integration, these measures become indispensable. By shedding light on the future of AI in EdTech, including emerging trends and potential growth areas, this study guides policymakers, researchers, and technology developers in making informed decisions to advance the field and optimize educational experiences.

3.3 RESEARCH OBJECTIVES

1. To study the role of AI in EdTech.
2. To investigate the diverse factors through which AI supports EdTech firms.

3.4 RESEARCH QUESTIONS

1. What is the role of AI in EdTech?
2. What are the various factors that contribute to AI's support for EdTech firms?

3.5 RESEARCH METHODOLOGY

This study adopts an exploratory research design and utilizes a research synthesis approach to systematically review and consolidate the extensive body of literature on the role of AI in EdTech. The research aims to integrate diverse theories and insights from experts in the field, offering a comprehensive perspective on AI's contributions to EdTech. To ensure the study's credibility and comprehensiveness, reputable sources, including highly cited documents from the Google Scholar database, are leveraged as the theoretical foundation. This is particularly important since the study explores a relatively new area that lacks extensive prior research linking AI and EdTech. The study's selection criteria focus on the past decade, considering the relatively recent emergence of AI and EdTech as significant educational paradigms. By adhering to this recency criterion, the research aims to provide timely insights into the contemporary landscape of AI in EdTech, offering a deeper understanding of its potential benefits, challenges, and ethical implications. For source selection, the study includes peer-reviewed journal articles, book chapters, conference proceedings, technical reports, and white papers, regardless of the specific population, intervention, or outcome in the context of AI in EdTech. The studies are limited to those published in English or languages translatable to English, ensuring accessibility to a wider audience. This methodological approach ensures that the study's findings are grounded in reliable and recent sources, establishing a strong foundation for understanding the intricate relationship between AI and EdTech. It also paves the way for a valuable resource that can aid educators, policymakers, and researchers in harnessing the full potential of AI to enhance modern educational experiences and outcomes.

3.6 ROLE OF ARTIFICIAL INTELLIGENCE IN EDUCATIONAL TECHNOLOGY

Chen et al. (2022) conducted a bibliometric study titled "Two Decades of Artificial Intelligence in Education: Contributors, Collaborations, Research Topics, Challenges, and Future Directions," which identifies the existing gap in the research conducted on AI applications in the education sector between 2000 and 2019, emphasizing the necessity for additional exploration of AI in EdTech. The introduction of learning apps, steam kits, interactive videos, and other technological advancements, such as personalizing education content, internet-based live discussion, and internet-based classroom activities, has been the dominant force in the progress of information and evolution in EdTech (Datalabs, 2020). EdTech is a phrase for "Educational technology," where IT tools and education practices are utilized to facilitate and enhance learning. Companies that provide these facilities are called EdTech start-ups or EdTech companies. EdTech solutions have found their presence in various domains, including B2B, B2C, B2B–B2C, and C2C (Datalabs, 2020). In the B2B model, firms either create their own learning modules or collaborate with third-party content curators, a model particularly popular in higher education institutions and corporate settings (Datalabs, 2020). In the B2C model, EdTech startups serve as content curators and marketplaces. In the B2B–B2C model, both ends of the target audience's needs are met, and in the C2C domain, platforms like YouTube connect potential content curators with consumers (Datalabs, 2020). The subsector-wise classification of Ed Firms is Test preparation, K-12, Certification, and Enterprise solution, where K-12 and Test Preparation has larger customer segment compared to other subsectors (EdTechreport, 2020). The report "The Great 'Un-Lockdown': Indian EdTech" states that the Indian education market is $117 billion, with about 360 million learners in 2020 (PGALabs, 2020). Around 4,450 start-ups are operating in the EdTech space in India (EdTechreport, 2020). The prominent features or trends that are major influencers in the EdTech landscape are AI, blockchain technology, VR, online proctoring or mentorship, personalized learning, and gamification (Chen et al., 2018; Pedro et al., 2019; Ameer and Vineeth, 2021). "Artificial Intelligence is a set of computer programs and technologies that mimic the human's brain functioning and intelligence" (Huang et al., 2019b). AI has the potential to assist in identifying and addressing students' difficulties, promoting collective imagination, and revolutionizing the educational experience (Chassignola et al., 2018). Special education has also benefited from applying AI techniques to solve problems (Drigas and Ioannidou, 2013). Implementing intelligent systems can facilitate the assessment of numerous assignments and identify gaps in learning and teaching (Chassignola et al., 2018). In the field of EdTech, AI-driven personalized learning enables students to explore their strengths, interests, learning styles, and areas for improvement while providing targeted support and reinforcing concepts to optimize educational outcomes (Chassignola et al., 2018; Jaiswal and Arun, 2021). Moreover, the data-driven nature of education facilitated by AI allows teachers to customize their teaching goals based on students' needs and comprehension, leading to an improved curriculum (Jaiswal and Arun, 2021). With AI-powered platforms, teachers can allocate more time to nurturing students' soft skills, such as emotional intelligence, leadership, and interpersonal abilities, thus ensuring holistic learning (Jaiswal and Arun, 2021).

3.7 CONCEPTUAL FRAMEWORK

According to Anderson et al. (1985), Baker (2016), Roll et al. (2018), Seo et al. (2020), and VanLehn (2011), there are many opportunities for AI in online learning and teaching. These opportunities range from customized instruction for students and automated execution of instructors' routine tasks to AI-powered examinations. For instance, AI tutoring systems can offer individualized advice, assistance, or feedback by customizing learning materials based on unique student learning styles or skill sets (Hwang et al., 2020). AI teaching assistants enable teachers to free up time from responding to students' straightforward, repetitive inquiries in online discussion forums so that they can focus on more important tasks (Goel and Polepeddi, 2016). By decrypting clickstream data, AI analytics enables teachers to comprehend students' performance, progress, and potential (Roll and Winne, 2015; Fong et al., 2019). The study's conceptual framework elucidates the AI factors that underpin EdTech services, and they are detailed as follows.

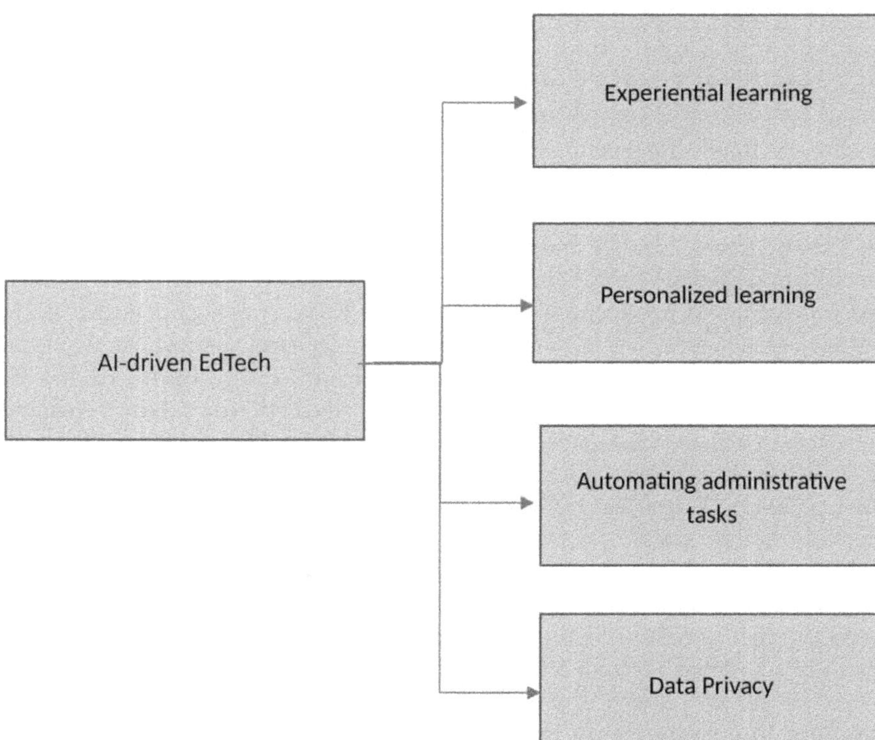

FIGURE 3.1 Conceptual framework.

3.7.1 Experiential Learning

Kolb introduced experiential learning in 1984, emphasizing active involvement and knowledge development (Kolb, 1984). This approach assumes that active involvement in real experiences and deliberate reflection are the best ways to learn (Jarmon et al., 2009). Researchers are increasingly studying experiential learning and AI (Chittaro and Ranon, 2007). AI can produce realistic and immersive experiences that would be impossible without AI, which is changing experiential learning (Goldberg and Robson, 2023). The rise of virtual environments in higher education has led researchers to investigate their potential for project-based learning and online collaboration (Junior, 2020). Integrated support in the virtual platform, including various communication tools, allows for social interaction, collaboration, a sense of shared presence, reduced social barriers, and reduced social anxiety (Jarmon et al., 2009). Performance-based activities, including storytelling, role play, improvisation, and scaffolded action-oriented activities, promote experiential learning (Bateson, 1993; Taussig, 1993; Wertsch, 1985). AI is said to foster creativity in a multimedia-rich setting, social interaction and community formation, collaboration, a sense of shared presence, the dissolution of social barriers, the reduction of social anxiety, student motivation and engagement, and learning preferences. AI-enabled instructional technology may improve certain performance attributes. VR/AR simulations are a potential use of AI in experiential learning. VR simulations may transfer students to various time periods and locales, allowing them to experience other cultures, historical events, and scientific studies. Augmented reality (AR) simulations allow learners to engage with virtual objects and their surroundings (Chiang et al., 2014). VR simulations have been employed in medical, pilot, and military training (Howard and Gutworth, 2020; Rogers et al., 2021; Xie, 2021). AR simulations have been used to teach anatomy, chemistry, and engineering (Viglialoro et al., 2021). VR and AR simulations may make language instruction more entertaining and effective (Parmaxi, 2023). Lamb (2015) examined how experiential learning affects student engagement and active participation in his project and found that it allowed students to apply classroom information in real life, making learning more dynamic and engaging. AI may be used to create customized learning environments to improve experiential learning. Mikropoulos and Natsis (2011) suggest that AI-based methods might make experiential learning more fun and engaging. According to Google AI education and Kim et al. (2021), this strategy may improve learners' knowledge intake and retention, which is crucial to learning. Mikropoulos and Natsis (2011) say VR and simulation technology may enhance students' learning experiences using AI. ITSs may ask and prod students until they can fully express their ideas, including their rationales. This improves information acquisition and retention, according to Kochmar et al. (2022). The influence of AI on experiential learning is still developing, but it has great potential (Chen et al., 2020).

3.7.2 Personalized Learning

Personalized learning is an educational technique that tailors teaching to each learner's unique requirements. This may be accomplished by taking into account the learner's past knowledge, learning style, interests, and objectives. AI has the ability to transform personalized learning by providing instructors and students with tools and resources to facilitate individualized teaching. AI is being utilized to increase student learning by customizing and personalizing the curriculum and material to each learner's requirements, talents, and capacities (Mikropoulos and Natsis, 2011). AI may improve learning by generating more realistic and immersive settings, personalizing education to specific learners, and offering timely feedback and assistance. Nazaretsky et al. (2022) state that AI systems can analyze student data to determine their strengths and weaknesses. These algorithms may then customize learning experiences for each learner. This intervention may improve kids' study skills. AI-driven educational platforms like tutoring systems may provide pupils with personalized academic feedback. According to Nazaretsky et al. (2022), AI-powered learning applications can adjust to each student's speed and learning style, giving the right amount of difficulty. Knewton, for example, employs machine learning algorithms to identify students' learning styles and provide real-time suggestions for course materials and content (Chen et al., 2020; Tang et al., 2023). Cerego, Immersive Reader, and CALL are some more systems with comparable features. Pokrivcakova also noted that integrating AI and using chatbots improves student learning experiences since they employ machine learning algorithms and give material tailored to students' learning requirements and capabilities (Pokrivcakova, 2019). Learners in personalized learning environments may access a range of learning tools, including as interactive classes, simulations, and exams (Google AI education, 2023). Based on their specific requirements and interests, the student may pick which resources to utilize and how to use them. Personalized learning environments have been shown to be beneficial in a wide range of contexts, including schools, residences, and libraries. The use and usage of simulation-based learning and ITSs were proved to support deep learning, which improves the learning experience of students (Sharma et al., 2019). In terms of learning materials, AI provides flexible learning digital interfaces that are applicable to students of various ages and grades. Furthermore, AI allows instructors to acquire student knowledge based on the full ecosystem of learning resources throughout the learning process. AI systems instruct students depending on their struggles with class topics. Previously, students had a limited window of time to contact their teachers, which meant attending office hours or hoping they responded to emails (Chen et al., 2020; Chaib et al., 2023; Görer and Aydemir, 2023; Larsson and Eriksson, 2023). AI will soon be able to function as a full-fledged assistant capable of adapting to a broad range of learning methods to support teachers and pupils (Alnajjar et al., 2021). An introduction to high-tech social agents, intelligent instructors, and curricular tools. It specifically assists teachers and students with their educational demands in almost any field (Alam, 2021). AI-powered tutoring systems can give personalized feedback and help with students' work while also adapting to their speed and learning style (Nazaretsky et al., 2022). AI-powered tutoring systems, for example, have been proven to be beneficial in helping kids learn math, science, and language arts. Adaptive learning technology can monitor a learner's progress and change the materials' difficulty levels correspondingly (Nazaretsky et al., 2022). This might assist in guaranteeing that the student is constantly challenged at a suitable level. The use of simulation, VR, and other parts of AI in learning has been proven to better prepare students for future trends, with a progressive shift toward keeping up with the use of AI in business (Wartman and Combs, 2019). Furthermore, AI algorithms forecast each student's career path by collecting study data, which in turn customizes university course choices for pupils. Students may improve their grades and gain skills that will be useful in the real world by taking into account their specific abilities and career paths.

3.7.3 Data Privacy

In the Industry 4.0 landscape, the critical importance of safeguarding sensitive information within data privacy issues underscores the necessity for an extensive examination (Onik et al., 2019). This emphasis on data privacy is a significant concern for the future of industrial revolutions. The high demand for edge AI services preserving data privacy has led to research on novel machine learning paradigms like federated learning and differential privacy, supported by the Sherpa.ai Federated Learning framework, offering a holistic approach and methodological guidelines for AI services development (Rodríguez-Barroso et al., 2020). Using machine learning requires a lot of data, which can pose privacy risks, and removing personal information to protect privacy is challenging, impacting fields like ophthalmology and other areas where data can be linked to identify individuals (Tom et al., 2020). AI in the healthcare industry, from its inception, not only enhances safety, accuracy, and cost-efficiency but also ensures data security through advanced privacy features, emphasizing the importance of an ethical balance between data access and privacy (Bak et al., 2022). In the case of businesses, the enduring importance of data across their developmental stages, in conjunction with technological progress, highlights the role of federated learning as a privacy-preserving approach that decentralizes training, empowering mobile devices to securely collect and train data (Li et al., 2020). This approach effectively tackles significant data privacy challenges and explores future applications. In the context of Mobile Edge Crowdsensing (MECS) utilizing Internet of Things (IoT) devices, an AI-enabled Three-Party Game (ATG) framework combines machine learning and game theory to ensure data privacy, using a classification-anonymity model based on the random forest classifier and k-anonymity algorithm, thus addressing the challenge of privacy disclosure in MECS of IoT (Xiong et al., 2019). To protect personal data, the proliferation of AI has prompted a reassessment of current threat modeling methods (Azam et al., 2022). In EdTech, amidst increasing educational costs and the expanding influence of online software, particularly in the realm of personalized learning programs driven by big data, the central ethical focus centers on data privacy, encapsulating six distinct dimensions—information privacy, anonymity, surveillance, autonomy, non-discrimination, and data ownership (Regan and Jesse, 2019).The adoption of Learning Analytics (LA) is increasingly vital for next-generation education technology (EdTech), utilizing AI and Analytics to gain insights into large-scale student learning, aligning with the "Ethical AI" movement, and addressing technical and procedural challenges to create a privacy-preserving analytics framework for the ethical application of LA (Marshall et al., 2022). The provided examples and technical insights play a vital role in elucidating the ways in which AI can serve as a valuable tool for the EdTech sector in offering effective data privacy services. AI's capabilities, such as robust data encryption, user authentication, and intelligent data monitoring, have the potential to significantly bolster data protection within EdTech platforms, ensuring the privacy and security of user information.

3.7.4 Automating Administrative Tasks

AI enhances administrative efficiency in EdTech through automated grading and feedback, leveraging machine learning to ensure objective evaluation, allowing educators to allocate more time to personalized instruction (Palav, 2023). Predictive analytics driven by AI enable early intervention and tailored support for at-risk students, ultimately boosting student retention and academic success (Palav, 2023). Additionally, AI-driven chatbots and virtual assistants streamline administrative tasks, offering prompt responses and simplifying class scheduling, resource allocation, and enrollment processes, reducing administrative workloads (Palav, 2023). Educators frequently invest substantial time in grading, but AI offers swift grading and learning guidance, including evaluating written responses. AI's administrative automation enhances teacher–student interaction and streamlines enrollment processes, promising more efficient education management (Marr, 2018). Conversely, some educators believed that AI's capacity to automate administrative duties and deliver data-driven insights could liberate teachers to concentrate on instructional leadership, student mentoring, and fostering innovation in the classroom, allowing them to become more influential in shaping students' educational journeys (Ghamrawi et al., 2023). AI can simplify student record management, streamline grading and assessments, enhance course scheduling, optimize financial aid, and support tutoring and mentorship (Zhai, 2022). However, challenges include ethical concerns, technological limitations, teacher buy-in, student engagement, and integration with existing systems (Zhai, 2022).

3.8 DISCUSSION

AI-powered EdTech presents several key factors that support its implementation. Firstly, AI enables personalized learning experiences by tailoring educational content to students' individual needs and preferences. This adaptability fosters increased engagement and improved academic performance. Secondly, AI streamlines administrative tasks such as grading and data analysis, allowing educators to dedicate more time to instruction and student support. Additionally, AI-driven educational platforms offer interactive and immersive learning opportunities, leveraging technologies like VR and AR to create dynamic and captivating educational content. Moreover, AI assists in identifying students' learning gaps, enabling targeted interventions and personalized remediation. These findings underscore the potential benefits of AI in EdTech, including personalized learning, administrative efficiency, interactive experiences, and targeted support, ultimately leading to a more effective and inclusive education system. AI is reshaping the landscape of EdTech in profound ways. With its capabilities to analyze vast amounts of data, personalize learning experiences, and streamline administrative tasks, AI is rapidly becoming a central component in modern education. One of the primary roles of AI in EdTech is personalized learning. AI has the remarkable ability to tailor educational content and support to the unique needs, learning styles, and progress of individual students. By analyzing student data, AI can identify strengths and weaknesses, allowing for the customization of learning materials and pacing. This personalization not only enhances student engagement but also leads to improved learning outcomes. The advent of AI-powered chatbots and virtual assistants further contributes to personalized learning by offering prompt responses to students' queries and guiding them through their learning journey. Another significant aspect of AI in EdTech is its role in experiential learning. AI-driven virtual and AR simulations create immersive educational experiences that would be otherwise unattainable. These technologies transport students to different time periods, cultures, or scientific experiments, allowing them to engage with the subject matter in a profound and interactive way. Additionally, the integration of AI in online communication tools enhances collaboration, reduces social barriers, and fosters a sense of shared presence. The performance-based activities that experiential learning often entails, such as storytelling, role-playing, and scaffolded action-oriented tasks, benefit from AI's capacity to foster creativity in multimedia-rich environments. AI enriches experiential learning by making it more engaging, dynamic, and interactive. However, the increasing role of AI in EdTech also raises significant concerns, primarily regarding data privacy. With the collection and analysis of vast amounts of student data, the potential for privacy breaches and data misuse is a pressing issue. Ethical questions surrounding the use of this data and its security are paramount. In response to these concerns, the field is evolving to address ethical considerations, incorporating privacy-preserving analytics frameworks and data encryption, user authentication, and intelligent data monitoring. Balancing the advantages of AI in EdTech with the need to protect students' privacy is a critical challenge for educational institutions and policymakers. Administrative tasks in EdTech are also being streamlined through AI. Automated grading and feedback systems not only save educators time but also provide objective evaluations. Predictive analytics, driven by AI, enable early intervention and tailored support for at-risk students, ultimately boosting student retention and academic success. AI-driven chatbots and virtual assistants further reduce administrative workloads by offering prompt responses and simplifying tasks such as class scheduling, resource allocation, and enrollment processes. By automating these administrative tasks, educators can focus more on personalized instruction and mentorship, fostering innovation in the classroom. In conclusion, AI's role in EdTech is expanding and transformative. It facilitates personalized learning, enhances experiential learning, streamlines administrative tasks, and promises improved educational outcomes. However, this transformation also presents challenges, primarily in the realm of data privacy and ethical considerations. Striking the right balance between the benefits of AI in EdTech and the ethical use of data is an ongoing concern. As AI continues to shape the future of education, educators, institutions, and policymakers must harness its potential while safeguarding students' privacy and ensuring that the educational experience remains equitable and accessible to all.

3.9 IMPLICATIONS

Incorporating AI into EdTech carries significant practical, social, and managerial implications.

3.9.1 Practical Implications

From a practical perspective, AI can enhance teaching and learning experiences by offering personalized and adaptive learning pathways, improving educational outcomes, and automating administrative tasks. It can also enable the creation of ITSs and virtual mentors, expanding access to high-quality education.

3.9.2 Social Implications

On a societal level, AI in EdTech has the ability to reduce disparities in education by offering personalized assistance for students with a wide range of learning requirements and styles. AI-powered EdTech has the potential to democratize education by making it more accessible and cheap to students of all socioeconomic levels. AI can tailor learning experiences to each student's unique requirements, regardless of their ability or learning style. AI in EdTech will allow Howard Gardeners' Multiple intelligence kinds to be honored. It will allow for customization based on personality types and learning methods. AI may aid in the early identification and remediation of learning gaps, preventing pupils from falling behind. This will result in the formation of a more knowledgeable society. AI can help students remain motivated and on track by making learning more entertaining and interactive. AI can offer real-time feedback on student achievement to instructors, allowing them to better their education. However, the incorporation of AI in EdTech poses serious ethical questions about data privacy, algorithmic bias, and the changing role of instructors in AI-powered classrooms. A big concern will also be the centralization of information and the politicization of learning, which may restrict the sort of knowledge provided through technology. There is a risk of abusing technology in such circumstances.

3.9.3 Managerial Implications

EdTech companies may employ AI to develop innovative and novel learning products and services. AI can assist academic institutions, including schools, in improving their teaching and learning procedures. In the future, even more technological advancements will be necessary to clarify complex and difficult topics. Such as the extinction of species in recent years, the influence of human activities on climate, the consequences of climate change for human beings, and so forth. EdTech companies may use AI to develop individualized learning programs for each pupil based on their specific needs and objectives. Institutions may employ AI to identify students who are at risk of falling behind and provide them with individualized assistance. Teachers may shift their focus from information dissemination to knowledge creation. Educators may utilize AI to obtain immediate feedback on their lessons and implement necessary modifications. Additionally, educators would be required to undertake additional responsibilities as a consequence of this. Policymakers may establish regulations to ensure the equitable and moral utilization of EdTech powered by AI.

3.10 FUTURE TRENDS AND CONCLUSIONS

The future of AI in EdTech promises to be dynamic and transformative, driven by several key trends. One prominent trend is the further evolution of personalized learning at scale. AI's capacity to gather and analyze vast amounts of data will enable even more tailored and effective learning experiences for individual students, accommodating their unique preferences and needs. As AI continues to advance, chatbots and virtual assistants will become more integral to the educational landscape, streamlining administrative tasks, providing instant support to students, and freeing up educators to concentrate on teaching and mentorship. Additionally, the integration of virtual and AR into EdTech will expand, offering immersive, experiential learning opportunities where students can explore historical events, scientific phenomena, and various cultures. Furthermore, AI-driven data analytics will play an increasingly vital role, offering educators invaluable insights into student performance, enabling early interventions for at-risk students, and ultimately boosting overall student success and retention. However, alongside these promising trends, the importance of ethical AI and robust data privacy measures cannot be overstated. As AI's role in EdTech grows, it is crucial to maintain a focus on responsible and secure data handling to ensure the privacy and security of user information in EdTech platforms. The integration of AI in EdTech represents a revolutionary shift in the way we approach and deliver education. This synthesis of AI and EdTech has the potential to make learning more accessible, engaging, and effective for students of all ages. It enables personalized learning experiences, administrative task automation, and data-driven insights that can significantly enhance education quality. However, this journey is not without its challenges, particularly in the realm of data privacy and ethical considerations. As AI in EdTech continues to advance, it is paramount that educators, policymakers, and developers prioritize ethical AI frameworks and robust data protection measures to ensure that the educational experience remains safe and secure for all. The future of AI in EdTech is exciting, promising more tailored and immersive learning experiences, and with the right ethical foundations, it has the potential to revolutionize education for generations to come.

REFERENCES

Alam, A. (2021, December). Should robots replace teachers? Mobilisation of AI and learning analytics in education. In *2021 International Conference on Advances in Computing, Communication, and Control (ICAC3)* (pp. 1–12). IEEE. https://doi.org/10.1109/ICAC353642.2021.9697300

Alnajjar, F., Bartneck, C., Baxter, P., Belpaeme, T., Cappuccio, M., Di Dio, C., . . . & Reich-Stiebert, N. (2021). *Robots in education: An introduction to high-tech social agents, intelligent tutors, and curricular tools*. Routledge. https://doi.org/10.4324/9781003142706

Ameer, P. A., & Vineeth, K. (2021). Impact of Covid-19 Pandemic on Ed Tech Industry in India: A Multidimensional Analysis. In *Virtual National Conference on The Changing Economic Models and Trends in the Post Covid-19 World (5).*

Anderson, J. R., Boyle, C. F., & Reiser, B. J. (1985). Intelligent tutoring systems. *Science*, 228(4698), 456–462.

Azam, N., Michala, L., Ansari, S., & Truong, N. B. (2022). Data privacy threat modelling for autonomous systems: A survey from the GDPR's perspective. *IEEE Transactions on Big Data*, 9(2), 388–414. https://doi.org/10.1109/TBDATA.2022.3227336

Bak, M., Madai, V. I., Fritzsche, M. C., Mayrhofer, M. T., & McLennan, S. (2022). You can't have AI both ways: Balancing health data privacy and access fairly. *Frontiers in Genetics*, 13, 1490.

Baker, R. S. (2016). Stupid tutoring systems, intelligent humans. *International Journal of Artificial Intelligence in Education*, 26(2), 600–614.

Bateson, M. C. (1993). Joint performance across cultures: Improvisation in a Persian garden. *Text and Performance Quarterly*, 13, 113–121. https://doi.org/10.1080/10462939309366037

Bulathwela, S., Pérez-Ortiz, M., Holloway, C., & Shawe-Taylor, J. (2021). Could AI democratise education? Socio-technical imaginaries of an edtech revolution. arXiv preprint arXiv:2112.02034.

Chaib, S. O., Joti, I., & Khoulji, S. (2023). Evaluation of a computer-assisted language learning system based on adaptive learning designed for self-training in scientific French language. *International Journal of Information and Education Technology*, 13(8). https://doi.org/10.18178/ijiet.2023.13.8.1930

Chassignol, M., Khoroshavin, A., Klimova, A., & Bilyatdinova, A. (2018). Artificial Intelligence trends in education: A narrative overview. *Procedia Computer Science*, 136, 16–24. https://doi.org/10.1016/j.procs.2018.08.233

Chen, G., Xu, B., Lu, M., & Chen, N. S. (2018). Exploring blockchain technology and its potential applications for education. *Smart Learning Environments*, 5(1), 1–10. https://doi.org/10.1186/s40561-017-0050-x

Chen, L., Chen, P., & Lin, Z. (2020). Artificial intelligence in education: A review. *IEEE Access*, 8, 75264–75278. https://doi.org/10.1109/ACCESS.2020.2988510

Chen, X., Zou, D., Xie, H., Cheng, G., & Liu, C. (2022). Two decades of artificial intelligence in education. *Educational Technology & Society*, 25(1), 28–47.

Chiang, T. H., Yang, S. J., & Hwang, G. J. (2014). An augmented reality-based mobile learning system to improve students' learning achievements and motivations in natural science inquiry activities. *Journal of Educational Technology & Society*, 17(4), 352–365.

Chittaro, L., & Ranon, R. (2007). Web3D technologies in learning, education and training: Motivations, issues, opportunities. *Computers & Education*, 49(1), 3–18. https://doi.org/10.1016/j.compedu.2005.06.002

Davenport, T. H., & Ronanki, R. (2018). Artificial intelligence for the real world. *Harvard Business Review*, 96(1), 108–116.

Dhamija, P., & Bag, S. (2020). Role of artificial intelligence in operations environment: A review and bibliometric analysis. *The TQM Journal*, 32(4), 869–896. https://doi.org/10.1108/TQM-10-2019-0243

Dogru, A. K., & Keskin, B. B. (2020). AI in operations management: Applications, challenges and opportunities. *Journal of Data, Information and Management*, 2, 67–74.

Drigas, A. S., & Ioannidou, R. E. (2013). A review on artificial intelligence in special education. Information Systems, E-learning, and Knowledge Management Research: 4th World Summit on the Knowledge Society, WSKS 2011, Mykonos, Greece, September 21–23, 2011. *Revised Selected Papers*, 4, 385–391.

Dubé, A. K., & Wen, R. (2022). Identification and evaluation of technology trends in K-12 education from 2011 to 2021. *Education and Information Technologies*, 27(2), 1929–1958.

Ergen, M. (2019). What is artificial intelligence? Technical considerations and future perception. *The Anatolian Journal of Cardiology*, 22(2), 5–7.

Fong, M., Dodson, S., Harandi, N. M., Seo, K., Yoon, D., Roll, I., & Fels, S. (2019). Instructors desire student activity, literacy, and video quality analytics to improve video-based blended courses. In Proceedings of the Sixth (2019) ACM Conference on Learning@ Scale (pp. 1–10). https://doi.org/10.1145/3330430.3333618

Ghamrawi, N., Shal, T., & Ghamrawi, N. A. (2023). Exploring the impact of AI on teacher leadership: Regressing or expanding?. *Education and Information Technologies*, *29*(7), 1–19.

Goel, A. K., & Polepeddi, L. (2016). *Jill Watson: A virtual teaching assistant for online education*. Georgia Institute of Technology.

Goldberg, D. E., & Robson, A. J. (2023). *Handbook of applied artificial intelligence*. CRC Press. [Google Scholar]

Google AI education. (2023). Artificial intelligence for education. Retrieved from https://ai.google/education/

Görer, B., & Aydemir, F. B. (2023). RoboREIT: An interactive robotic tutor with instructive feedback component for requirements elicitation interview training. arXiv preprint arXiv:2304.07538. https://doi.org/10.1002/smr.2608

Guenole, N., & Feinzig, S. (2018). The business case for AI in HR. *With Insights and Tips on Getting Started*. Armonk: IBM Smarter Workforce Institute, IBM Corporation.

Howard, M. C., & Gutworth, M. B. (2020). A meta-analysis of virtual reality training programs for social skill development. *Computers & Education*, 144, 103707. https://doi.org/10.1016/j.compedu.2019.103707

Huang, M.-H., Rust, R., & Maksimovic, V. (2019b). The feeling economy: Managing in the next generation of Artificial Intelligence. *California Management Review*, 61(4), 43–65. https://doi.org/10.1177/0008125619863436

Huang, R., Spector, J. M., & Yang, J. (2019a). Educational technology. *Lecture Notes in Educational Technology*. https://doi.org/10.1007/978-981-13-6643-7

Hwang, G. J., Xie, H., Wah, B. W., & Gašević, D. (2020). Vision, challenges, roles and research issues of Artificial Intelligence in Education. *Computers and Education: Artificial Intelligence*, 1, 100001. https://doi.org/10.1016/j.caeai.2020.100001

Jaiswal, A., & Arun, C. J. (2021). Potential of artificial intelligence for transformation of the education system in India. *International Journal of Education and Development using Information and Communication Technology*, 17(1), 142–158.

Jarmon, L., Traphagan, T., Mayrath, M., & Trivedi, A. (2009). Virtual world teaching, experiential learning, and assessment: An interdisciplinary communication course in Second Life. *Computers & Education*, 53(1), 169–182. https://doi.org/10.1016/j.compedu.2009.01.010

Junior, J. B. B. (2020). Assessment for learning with mobile apps: Exploring the potential of Quizizz in the educational context. *International Journal of Development Research*, 10 (01), 33366–33371.

Khandelwal, A. R., Chaturvedi, A., & Kanwar, M. V. (2021). Effectiveness of virtual classroom: A comparative study of theoretical and practical papers. *Webology*, 1735–188X, 18(4).

Kim, S. W., Kim, M., & Shin, H. S. (2021). Affective empathy and prosocial behavior in rodents. *Current Opinion in Neurobiology*, 68, 181–189. https://doi.org/10.1016/j.conb.2021.05.002

Kochmar, E., Vu, D. D., Belfer, R., Gupta, V., Serban, I. V., & Pineau, J. (2022). Automated data-driven generation of personalized pedagogical interventions in intelligent tutoring systems. *International Journal of Artificial Intelligence in Education*, 32(2), 323–349.

Kok, J. N., Boers, E. J., Kosters, W. A., Van der Putten, P., & Poel, M. (2009). Artificial intelligence: Definition, trends, techniques, and cases. *Artificial Intelligence*, 1, 270–299.

Kolb, D. A. (1984). *Experiential learning: Experience as the source of learning and development*. Prentice-Hall.

Lamb, D. (2015). Learning about events through involvement and participation: The use of experiential and authentic learning experiences. *International Journal of Event and Festival Management*, 6(1), 73–91. https://doi.org/10.1108/IJEFM-12-2013-0043

Larsson, N., & Eriksson, H. (2023). Chatting up the grade: An Exploration on the impact of ChatGPT on self-study experience in higher education. (Master's thesis, Umeå University).DIVA portal. https://www.diva-portal.org/smash/get/diva2:1777754/FULLTEXT01.pdf

Li, Z., Sharma, V., & Mohanty, S. P. (2020). Preserving data privacy via federated learning: Challenges and solutions. *IEEE Consumer Electronics Magazine*, 9(3), 8–16. https://doi.org/10.1109/MCE.2019.2959108

Marr, B. (2018). *How is AI used in education–Real world examples of today and a peek into the future*. Forbes, Forbes Magazine, 25.

Marshall, R., Pardo, A., Smith, D., & Watson, T. (2022). Implementing next generation privacy and ethics research in education technology. *British Journal of Educational Technology*, 53(4), 737–755. https://doi.org/10.1111/bjet.13224

Mikropoulos, T. A., & Natsis, A. A. (2011). Educational virtual reality applications: A state-of-the-art review. *British Journal of Educational Technology*, 42(3), 572–601. [Google Scholar]

Nazaretsky, T., Bar, C., Walter, M., & Alexandron, G. (2022, March). Empowering teachers with AI: Co-designing a learning analytics tool for personalized instruction in the science classroom. In *LAK22: 12th International Learning Analytics and Knowledge Conference* (pp. 1–12). https://doi.org/10.1145/3506860.3506861

Onik, M. M. H., Chul-Soo, K. I. M., & Jinhong, Y. A. N. G. (2019, February). Personal data privacy challenges of the fourth industrial revolution. In *2019 21st International Conference on Advanced Communication Technology (ICACT)* (pp. 635–638). IEEE. https://doi.org/10.23919/ICACT.2019.8701932

Palav A, (2023, July).The impact of artificial intelligence in the edtech industry | The Financial Express. https://www.financialexpress.com/business/blockchain-the-impact-of-artificial-intelligence-in-the-edtech-industry-3161299/

Parmaxi, A. (2023). Virtual reality in language learning: A systematic review and implications for research and practice. *Interactive Learning Environments*, 31(1), 172–184. https://doi.org/10.1080/10494820.2020.1765392

Pedro, F., Subosa, M., Rivas, A., & Valverde, P. (2019). Artificial intelligence in education: Challenges and opportunities for sustainable development. (UNESCO Working Paper). UNESCO. https://repositorio.minedu.gob.pe/bitstream/handle/20.500.12799/6533/Artificial%20intelligence%20in%20education%20challenges%20and%20opportunities%20for%20sustainable%20development.pdf

Pokrivcakova, S. (2019). Preparing teachers for the application of AI-powered technologies in foreign language education. *Journal of Language and Cultural Education*, 7(3), 135–153. https://doi.org/10.2478/jolace-2019-0025

Regan, P. M., & Jesse, J. (2019). Ethical challenges of edtech, big data and personalized learning: Twenty-first century student sorting and tracking. *Ethics and Information Technology*, 21, 167–179.

Rogers, M. P., DeSantis, A. J., Janjua, H., Barry, T. M., & Kuo, P. C. (2021). The future surgical training paradigm: Virtual reality and machine learning in surgical education. *Surgery*, 169(5), 1250–1252. https://doi.org/10.1016/j.surg.2020.09.040

Roll, I., & Winne, P. H. (2015). Understanding, evaluating, and supporting self-regulated learning using learning analytics. *Journal of Learning Analytics*, 2(1), 7–12. https://doi.org/10.18608/jla.2015.21.2

Roll, I., Russell, D. M., & Gašević, D. (2018). Learning at scale. *International Journal of Artificial Intelligence in Education*, 28(4), 471–477.

Rodríguez-Barroso, N., Stipcich, G., Jiménez-López, D., Ruiz-Millán, J. A., Martínez-Cámara, E., González-Seco, G., ... & Herrera, F. (2020). Federated learning and differential privacy: Software tools analysis, the Sherpa. ai FL framework and methodological guidelines for preserving data privacy. *Information Fusion*, 64, 270–292. https://doi.org/10.1016/j.inffus.2020.07.009

Russell, S. J., & Norvig, P. (2010). *Artificial intelligence a modern approach.* Pearson Education, Inc.

Sharma, R. C., Kawachi, P., & Bozkurt, A. (2019). The landscape of artificial intelligence in open, online and distance education: Promises and concerns. *Asian Journal of Distance Education*, 14(2), 1–2.

Seo, K., Fels, S., Yoon, D., Roll, I., Dodson, S., & Fong, M. (2020). Artificial intelligence for video-based learning at scale. In *Proceedings of the Seventh ACM Conference on Learning@ Scale* (pp. 215–217). https://doi.org/10.1145/3386527.3405937

Tang, K. Y., Chang, C. Y., & Hwang, G. J. (2023). Trends in artificial intelligence-supported e-learning: A systematic review and co-citation network analysis (1998–2019). *Interactive Learning Environments*, 31(4), 2134–2152. https://doi.org/10.1080/10494820.2021.1875001

Taussig, M. (1993). *Mimesis and alterity: A particular history of the senses*. Routledge. The New Media Consortium and the EDUCAUSE Learning. https://doi.org/10.4324/9781315227689

Tom, E., Keane, P. A., Blazes, M., Pasquale, L. R., Chiang, M. F., Lee, A. Y., ... & Force, A. A. I. T. (2020). Protecting data privacy in the age of AI-enabled ophthalmology. *Translational Vision Science & Technology*, 9(2), 36–36. https://doi.org/10.1167/tvst.9.2.36

VanLehn, K. (2011). The relative effectiveness of human tutoring, intelligent tutoring systems, and other tutoring systems. *Educational Psychologist*, 46(4), 197–221. https://doi.org/10.1080/00461520.2011.611369

Verma, S., Sharma, R., Deb, S., & Maitra, D. (2021). Artificial intelligence in marketing: Systematic review and future research direction. *International Journal of Information Management Data Insights*, 1(1), 100002. https://doi.org/10.1016/j.jjimei.2020.100002

Viglialoro, R. M., Condino, S., Turini, G., Carbone, M., Ferrari, V., & Gesi, M. (2021). Augmented reality, mixed reality, and hybrid approach in healthcare simulation: A systematic review. *Applied Sciences*, 11(5), 2338. https://doi.org/10.3390/app11052338

Wartman, S. A., & Combs, C. D. (2019). Reimagining medical education in the age of AI. *AMA Journal of Ethics*, 21(2), 146–152.

Weber, P., Carl, K. V., & Hinz, O. (2023). Applications of explainable artificial intelligence in finance—A systematic review of Finance, information systems, and computer science literature. *Management Review Quarterly*, 1–41. https://doi.org/10.1007/s11301-023-00320-0

Wertsch, J. V. (1985). *Vygotsky and the social formation of mind.* Harvard University Press.

Xie, B., Liu, H., Alghofaili, R., Zhang, Y., Jiang, Y., Lobo, F. D., ... & Yu, L. F. (2021). A review on virtual reality skill training applications. *Frontiers in Virtual Reality*, 2, 645153. https://doi.org/10.3389/frvir.2021.645153

Xiong, J., Zhao, M., Bhuiyan, M. Z. A., Chen, L., & Tian, Y. (2019). An AI-enabled three-party game framework for guaranteed data privacy in mobile edge crowdsensing of IoT. *IEEE Transactions on Industrial Informatics*, 17(2), 922–933. https://doi.org/10.1109/TII.2019.2957130

Zhai, X. (2022). ChatGPT user experience: Implications for education. Available at SSRN: https://ssrn.com/abstract=4312418 or http://dx.doi.org/10.2139/ssrn.4312418

Zhang, C., & Lu, Y. (2021). Study on artificial intelligence: The state of the art and future prospects. *Journal of Industrial Information Integration*, 23, 100224. https://doi.org/10.1016/j.jii.2021.100224

WEBSITE

DataLabs. (2020). *The future of EdTech in India: Decoding the $10 Bn market opportunity* (Report). https://inc42.com/reports/the-future-of-edtech-in-india-decoding-the-10-bn-market-opportunity-report-2020/

EdTechreport. (2020). *EdTech in India: The $10 Bn market opportunity* (Report). https://aspirecircle.org/wp-content/uploads/2021/04/Inc42_Edtech_Report_updated.pdf

PGA Labs. (2020). *The great 'Un-Lockdown': Indian EdTech* (Report). Indian Private Equity and Venture Capital Association. https://www.pgalabs.in/PGALabsImages/ReportPdf/pga-labs-ivca-report-the-great-un-lockdown-indian-edtech-pgalabs.pdf

PGA Labs. (2023). *Edtech 2023: Road ahead* (Report). Lattice Technologies Pvt. Ltd. https://storage.googleapis.com/marketreports/Reports/edtech-2023-road-ahead-pgalabs.pdf

Statista. (2023). *Online education - worldwide*. Statista. https://www.statista.com/outlook/dmo/eservices/online-education/worldwide

Statista. (2025). Artificial intelligence (AI) worldwide – statistics & facts. *Statista*. https://www.statista.com/topics/3104/artificial-intelligence-ai-worldwide/

4 Obstacles of Artificial Intelligence Applications in the Public Distribution System in India

Asha Devi J and Dr. K.S. Chandrasekar

4.1 INTRODUCTION

Artificial Intelligence (AI) is the theory and development of computer systems able to perform tasks usually requiring human intelligence, such as visual perception, speech recognition, decision-making, and translation between languages. The Public Distribution System (PDS) is a government-sponsored program commissioned to circulate food grains and non-food commodities to the underprivileged sections of society at a meager price. The Food Corporation of India (FCI) is an apex organization that manages and provides a PDS, and it distributes rice, wheat, sugar, kerosene, and other food grains through ration and fair-price shops (FPS). The ability of machines to think, perceive, learn, solve problems, and make decisions is known as AI. AI was initially intended to be a technology that could imitate human intelligence, but its development has gone far beyond that. With tremendous advances in information assortment, handling, and calculation power, savvy frameworks can now be used to assume control over various assignments, empower networks, and improve efficiency. The applications of AI have increased in a growing number of fields as its capabilities have significantly increased. India has a chance to define its AI leadership due to the technology's truly transformative nature and its nascent stage in global adoption. India's proposed brand, "Artificial Intelligence for All," implies inclusive technology leadership in which AI's full potential is realized following the country's particular requirements and aspirations. The strategy should use AI for economic, social, and inclusive growth and as a "garage" for developing and emerging economies. Even though AI has the potential to add a lot of value to many different industries, adoption has so far been driven mainly by business reasons. Large-scale adoption strategies, particularly national ones, must balance biased descriptions of monetary influence and the greater good because technology disruptions like AI are once-in-a-generation events. NITI Aayog has chosen to zero in on five areas that are imagined to benefit the most from computer-based intelligence in settling cultural necessities:

DOI: 10.1201/9781003638506-4

1. Medical Care: AI will improve healthcare operations and delivery processes. It also helps in cost savings. AI in the healthcare system is essential for implementing AI applications.
2. Agriculture: Increased availability and affordability of high-quality healthcare lowered wastage, increased farm productivity, and raised farmers' incomes.
3. Education: Further developed admittance and nature of schooling.
4. Savvy Urban Communities and Foundations: Efficient and connected for the expanding urban population; and
5. Smart Transportation and Mobility: More brilliant and more secure transportation methods and better traffic; furthermore, clog issues.

The following obstacles must be overcome to achieve the objectives of AI for all to truly reap the benefits of deploying AI at scale:

1. Absence of wide-based mastery in exploration and utilization of AI.
2. Nonappearance of empowering information environments – access to astute information.
3. High asset cost and low mindfulness for the reception of AI.
4. Protection and security, including the absence of formal guidelines around the anonymization of information, and
5. A non-attendance of cooperative ways to deal with reception and utilization of AI.

Predominant examination capacities have been the foundation of administration goals in arising advances, and understanding the development potential requires skill in both center and applied research. AI research in India is still starting and requires large-scale coordinated and collaborative interventions, despite signs of recent positive efforts in this field. The two-layered construction to address India's AI research desires is as follows:

1. A Center of Excellence in Research (CORE) whose mission is to improve comprehension of existing core research and push technological boundaries through the creation of new information;
2. The International Centers of Transformational AI (ICTAI), which are tasked with the creation and implementation of research that is application based. Collaboration with the private sector is anticipated to be a crucial component of ICTAIs.

The National Strategy for Artificial Intelligence is an umbrella organization that aims to direct research efforts by studying global advancements, analyzing socio-economic indicators, and encouraging international collaboration is proposed to supplement the research capabilities. Chasing after "moonshot research projects" through particular groups, improvement of a devoted supranational organization to divert research in settling enormous, nervy issues of AI – "CERN for artificial intelligence," and creating normal computing and other related foundations for computer-based intelligence are further essential parts of research proposed. Skilling and reskilling the workforce are integral parts of our approach to adopting AI, as technology increasingly changes the nature of jobs and shifts the standards for technological aptitude. There is an urgent need to reskill the current labor force and create future ability per the gig market's changing requirements. This should be possible through the reception of decentralized instruction components working as a team with the confidential area and instructive organizations to endorse certificates with esteem. In addition, it is necessary to identify and promote job creation in new fields like data annotation, which can potentially absorb a significant portion of the workforce that may become redundant as automation increases. The reception of simulated intelligence across the value chain, viz., new companies, confidential areas, PSUs, and government elements, will open the potential by creating an upright pattern of market interest. The obstructions to AI advancement and arrangement can be avoided by embracing the commercial center model, which empowers the market revelation of the cost and various methodologies ideal for accomplishing excellent outcomes. Establishing a formal market with three pillars is possible: data collection and aggregation, data annotation, and deployable models. The National AI Marketplace (NAIM) could serve as a unified platform.

In addition, the government needs to play the crucial role of a catalyst in fostering innovation through research, supporting partnerships, providing access to infrastructure, fostering innovation through partnerships, and creating demand by looking for solutions to various governmental needs if AI is to be accelerated in its adoption. Questions regarding security, privacy, and ethics will also arise as AI-based solutions permeate our daily lives. The FAT framework, "Fairness, Accountability, and Transparency," is the basis of most discussions regarding AI's ethical implications. It is possible to establish a consortium of ethics councils at each CORE, and it is expected that all COREs adhere to standard procedures when creating AI products and technologies. The former Union Minister of Consumer Affairs, Food and Public Distribution Piyush Goyal says that we require a Brilliant Public Dissemination System (PDS) drive "to advance the utilization of mechanization" in the food conveyance framework. Goyal stated at a conference attended by food ministers from all states and union territories that all regions should work toward such a system. Individuals in these networks are frequently eliminated from mechanical turns of events. They may need help to acquire food grains if human mediation is removed from this technique or even decreased. Plans for food distribution automation: Goyal said that the auto grain analyzer is based on AI and certified by ICAR-CIPHET, Ludhiana. According to a press release, it can accurately process results for paddy, rice, wheat, pulses, oil seeds, and coarse grain. Goyal said that by providing digitally verifiable results, such a mechanism would reduce human "intervention/error/biases" and save time. He described it as a "transparent and accountable" system that would aid the PDS's free grain supply chain. Furthermore, he added that a direct procurement would be seamless.

Technology is updated from time to time, so our nation also needs updated technology in our PDS system. GPS technology, digitization of ration cards, computerized allocation to FPS, SMS-based monitoring, issuing of smart cards for replacing ration cards, and a web-based citizens portal are important for PDS to ensure the security of the food grains supplied through PDS. Because the food security of our nation is the most important thing, GovTech solutions are often strengthened with Aadhaar integration to filter fake beneficiaries. Moreover, Point of Sale (POS) devices can be handy for last-mile delivery to the targeted beneficiaries. The success of PDS would still need groundwork if they successfully tackle the "identification challenge." Technology online and fieldwork offline have to go in tandem for a more efficient food distribution system.

AI has a major role in optimizing and improving the efficiency of PDS. Here are some potential applications of AI in PDS:

1. Demand Forecasting: AI algorithms can analyze historical data and external factors such as population trends, economic indicators, and weather conditions to forecast demand for essential commodities. This helps in better planning and allocation of resources, ensuring sufficient supply and avoiding wastage.
2. Inventory Management: AI can assist in optimizing inventory levels by analyzing consumption patterns, identifying trends, and predicting future demand. By maintaining optimal stock levels, PDS agencies can reduce storage costs, minimize wastage, and ensure the timely availability of essential goods.
3. Fraud Detection: AI-powered algorithms can analyze the considerable importance of transactional data to identify fraudulent activities in the PDS. By detecting anomalies and patterns indicative of fraud, AI can help minimize leakages and ensure that benefits reach the intended beneficiaries.
4. Targeted Delivery and Subsidy Management: AI can identify and target beneficiaries based on various parameters such as income, demographics, and social indicators. This ensures that subsidies and benefits reach the right individuals or households, reducing leakages and improving the effectiveness of the PDS.
5. Route Optimization: AI algorithms can optimize the delivery routes for transporting essential commodities to various distribution centers. By considering factors like traffic conditions, distance, and vehicle capacity, AI can minimize transportation costs, reduce delivery time, and enhance the efficiency of the distribution system.
6. Customer Service and Support: AI-powered chatbots and virtual assistants can handle customer queries, provide information about available commodities, address complaints, and assist with PDS-related matters. This can enhance the user background and facilitate the burden on human customer support staff.
7. Analytics and Decision Support: AI can generate valuable insights from large volumes of data collected from the PDS. These insights can aid policymakers and administrators in making informed decisions, identifying areas of improvement, and implementing targeted interventions to enhance the efficiency and effectiveness of the system.

The booming implementation of AI in the PDS requires robust data collection mechanisms, data privacy considerations, and stakeholder engagement. Additionally,

AI should augment human decision-making rather than replace human judgment entirely.

Several obstacles and challenges are associated with undertaking AI applications in India's PDS. The blocks are as follows:

1. Data Quality and Availability: AI systems rely heavily on high-quality data for training and decision-making. However, in the case of PDS in India, there are challenges related to the quality and availability of data. The data related to beneficiaries, ration cardholders, inventory, and supply chain may need to be completed, updated, or unreliable, making it difficult to build accurate and effective AI models.
2. Infrastructure and Connectivity: The successful implementation of AI applications requires robust technological infrastructure and reliable internet connectivity. In many regions of India, especially in rural areas, the infrastructure and connectivity could be improved, making it challenging to deploy AI systems that require real-time data processing and communication.
3. Language and Cultural Diversity: India is a linguistically and culturally diverse country with multiple languages and dialects. Developing AI systems that effectively handle this diversity and provide localized solutions can be challenging. Language barriers and cultural nuances must be considered to ensure the inclusivity and effectiveness of AI applications in the PDS.
4. Adoption and Acceptance: The PDS serves millions of beneficiaries across India, and any significant technological change may face resistance and skepticism from various stakeholders. Ensuring the adoption and acceptance of AI applications among government officials, ration shop owners, and beneficiaries can be a hurdle. Education and awareness campaigns may be required to manage situations and build trust in the technology.
5. Ethical and Fair Distribution: AI systems must be developed to ensure fairness and avoid biases. In the context of PDS, it is crucial to prevent discrimination, ensure equitable distribution of food grains, and avoid excluding eligible beneficiaries. Developing AI models that are unbiased and transparent can be a complex task, requiring careful algorithm design and continuous monitoring.
6. Security and Privacy: AI applications in the PDS involve handling sensitive personal information of beneficiaries, such as Aadhaar numbers, addresses, and financial data. Ensuring data security, privacy, and protection against cyber threats is paramount. Robust mechanisms for data encryption, access control, and secure storage need to be implemented to safeguard the confidentiality of beneficiaries.
7. Human Intervention and Support: AI can automate specific tasks and enhance efficiency, but it should differ from human intervention and support. The PDS involves complex social and administrative processes that require human judgment and decision-making in certain situations. Striking the right balance between automation and human involvement is necessary to integrate AI in the PDS effectively.

Managing these obstacles requires a comprehensive approach involving collaboration between government agencies, technology experts, and stakeholders. It consists of improving data collection and management systems, investing in infrastructure development, promoting digital literacy, and ensuring design transparency, fairness, and inclusivity.

4.1.1 Implementation of Artificial Intelligence (AI) Applications in the Public Distribution System (PDS)

The governments of developed countries use information and communication technology to facilitate their people with quicker and more open services. These facilities are inaccessible to developing economies due to insufficient infrastructure, ineffective policies, and unskilled labor. This section summarizes the current literature on technology adoption in PDS, which is extremely limited. The intervention of IT is to transform the transaction process and bring flexibility to beneficiaries' access to PDS. The study's outcome explained that lower incentives, less IT competency, lack of infrastructure, lack of trust in technology, and poor understanding among beneficiaries could be significant challenges for government bodies. Gowd addressed that using technology in PDS would enable the capabilities to fight corruption and diversion of commodities and help the distributors distribute the goods to the proper beneficiaries. It can transform the non-functional PDS into a more secure and transparent functional PDS with the active involvement of communities. Digitalizing data could improve PDS performance by increasing transparency and reducing leakages and corruption. Masiero suggested data application of population and biometric infrastructure to enhance beneficiaries' identification and reduce fake transactions. Recently, Biswal et al. investigated the adoption of radio-frequency identification (RFID) for improving operational performance and resource utilization. It was observed that using RFID reduces supply shrinkage and misplacement of products, but it comes at an expense that requires infrastructure development and a skilled workforce. The study suggested that whenever the loss reaches a threshold value, the analysis between the device with and without RFID indicates that, for a specified tag cost and deterioration conversion efficiency, RFID deployment is advantageous. Kumar recommended the adoption of more transparent and secure platforms in PDS, like Blockchain technology (BT), to make the supply chain more reliable and robust. Chaturvedi et al. discuss leveraging banking customer experience through AI, which is relevant to understanding the challenges faced in implementing AI in the PDS in India. Singh et al. proposed that by using BT, government agencies can remove and decentralize the PDS's manual transaction and record-keeping process and strengthen it by integrating it with AI. By year-end, the government-owned FCI will have a chemical-based testing system for determining the age of both rice and wheat during procurement. Similarly, it would finalize the testing kits that can use AI to test the quality of food grains and moisture meters, which can measure the moisture content automatically. Speaking at a conference on the PDS reforms in India to ensure food security, then FCI Chief Atish Chandra said they had used the chemical-based technique for testing the age of rice during procurement, and a hackathon has been awarded for testing the age of wheat. He said that by doing this, they are trying to restrict rice recycling. Currently, a massive scope of subsidized rice is coming back to the procurement cycle. Chandra also said an expression of interest (tender) has been floated for an AI-based grain quality testing system for automated testing and also an automated moisture meter. He added that trials are being conducted, and they will finalize the equipment by December 2021 and then deploy them. This will eliminate the manual systems that are currently being used. All these tests are critical, considering that the FCI stores procured wheat and rice for up to two years for disposal. Chandra said ensuring food grain quality and reducing loss during transport and storage are the most critical tasks. Nearly 600 lakh tonnes of wheat and rice are allocated from the central pool annually to meet the requirements of public distribution based on the National Food Security Act and other welfare schemes. The FCI will install 50 AI-based automatic grain analyzers at its food grain depots for assessing the quality of rice, thus bringing automation to the procurement system. This aims to minimize the discretionary decision-making power of the corporation's ground-level officials in grain handling. Sources told FE that performance evaluation of such machines at a few locations has been encouraging, and the corporation will install such analyzers in all its 2,000 depots across the country. The corporation is also initiating the digitization of quality control laboratories from depot levels for real-time data on grain testing reports by March this year. This comes after a recent nationwide crackdown against a few FCI officials by the Central Bureau of Investigation (CBI) after apparent malpractices in the procurement, storage, and distribution of food grains. Rice millers say that because of the testing of samples in manual laboratories, the local-level officials have a crucial say in the procurement process. The FCI aims to install grain analyzers across depots to eliminate the human interface in the rice procurement process. "Gradually roll out across procurement centers after observing its efficacy."

Meanwhile, the FCI has decided to set up a quick response team to see whether the spot where the millers bring in rice meets the fair and average (FAQ) standard. After the FCI and state agencies procure paddy from the farmers, it is handed over to millers for conversion into rice. The government agencies procured nearly 60 million tonnes of rice in the 2021–2022 season (October–September). The FCI supplied rice for distribution to more than 800 million beneficiaries under the National Food Security Act and other welfare schemes. The rice procured from grain-surplus states is also used to keep a buffer stock with the FCI. To prevent malpractice and inappropriate business activities and to retain good quality and norms of rice being supplied under the Targeted Public Distribution System (TPDS), the FCI has introduced a new technology, the Mixed Indicator Method, which determines the age of the milled rice stocks during acceptance. It mainly controls the millers from combining old rice with fresh supplies.

4.2 OBJECTIVES OF THE STUDY

1. To identify the present applications of AI in India's PDS.
2. To find out the obstacles to AI applications in India's PDS.

4.3 REVIEW OF LITERATURE

Siau & Wang (2020) state that the cost of implementing innovations, experimentation, and trials is prohibitive. The governments of developed nations have taken initiatives toward investing in strengthening research and development, while developing countries like India are limited in implementing such projects. There are also issues around trust with AI in India.

Chatterjee (2020) states that AI-powered systems have many potential benefits. Yet, there are still concerns related to data privacy and data sharing. Users are hesitant to share their data. There is a need to define standard practices to preserve privacy. This would attract international organizations to facilitate AI implementation in India. The government must enhance research facilities to address privacy and ethical concerns to accelerate successful AI implementation in India's PDS (Keding 2021). Investing in technology can help an organization stay ahead of the competition. With AI, the cost may decrease to build the next generation of decision-support systems.

Berman and Dalzell-Payne (2018) state that many organizations need help implementing AI due to its high initial cost. Recent literature suggests that AI implementation by public manufacturing organizations is increasing but involves high upfront costs with long-term, unquantifiable benefits.

Dwivedi et al. (2021) state that in the context of an AI-led economy, India, with the strategy "AI for All," has a unique opportunity to contribute to AI-based solutions that may revolutionize healthcare, agriculture, manufacturing, education, and other sectors. The "AI for All" strategy has a vast pool of AI-trained workers and an emerging start-up system. AI-enabled systems in India have been implemented in several areas, including healthcare, to detect and prevent diseases. AI-based systems are projected to have a ripple effect on economic growth and wealth. It will add $957 billion to the Indian economy by 2035. Based on current literature, it is accepted that the Government of India is focusing on AI. Setting out standardized policies is crucial at this time. Thus, achieving "AI for All" in developing nations like India is a long-term target. Engagement and institutional collaboration among all stakeholders, including the general public, is needed.

McKinsey (2018) analyzed 400 use cases spanning nine business functions and 19 sectors and found that AI outperformed conventional analysis techniques for 69% of cases. The wide scope of tasks that can be widely applied in business is identified as the competitive advantage of AI over primary technological principles (modeling, statistics, heuristics, and other classical optimizations) because objective intelligence solves uncertain, complex situations and generates new solutions. AI in the food supply chain alone will not provide a competitive advantage due to its growing use. In the future, creative applications of AI that positively impact competition will need to be investigated. Technologies based on AI will unavoidably alter the agricultural sector as well, affecting every stage of the food supply chain.

4.4 METHODOLOGY

A systematic literature review is the empirical approach taken in this study. This is a well-known academic method for finding and evaluating relevant literature on a particular subject. Peer-assessed research articles were gathered from the web through

Google Scholar, Scopus, Web of Science, Elsevier, Shodhganga, and Academia, utilizing search terms connected with AI and their applications in India's PDS.

4.5 RESULTS AND DISCUSSIONS

In the 2009 report, the Wadhwa Committee identified that some states implemented computerization and other technology-based reforms to TPDS; technology-based reforms help to stop leakages of food grains during TPDS. The committee discovered that the current manual recording of beneficiaries' eligibility and transactions needs to be corrected and tampering prevented. Also, there was pilferage through the distribution network and no central monitoring system to ensure end-to-end delivery. The Wadhwa Committee observed that end-to-end computerization controls large-scale diversion and helps track food grain delivery from state depots to beneficiaries.

AI has benefited various sectors, including the PDS, which involves the distribution of essential commodities to the general public. However, there are also potential negative impacts that need to be considered. Here are some of the adverse effects of AI in the PDS:

1. Job Displacement: AI technologies, such as automated systems and robotics, can replace human labor in various aspects of the PDS. This can lead to job losses for workers involved in inventory management, packaging, and transportation tasks. Automating these processes may increase efficiency, but it can also lead to unemployment and socioeconomic challenges for those whose jobs are automated.
2. Inequality and Accessibility: Adopting AI in the PDS requires advanced technology and infrastructure access. This may create a digital divide, where individuals or regions needing access to such technology are left behind. Using AI systems to distribute essential commodities could further exacerbate existing inequalities in access to goods and services.
3. Privacy Concerns: AI systems in the PDS often rely on collecting and analyzing large amounts of data, including confidential information. There is a risk of privacy breaches and unauthorized use of personal data, which can compromise individuals' privacy rights. Safeguarding sensitive data and ensuring transparency and accountability in data handling have become critical challenges.
4. Algorithmic Bias: AI systems are trained using historical data, which can contain biases and reflect societal inequalities. When these biased datasets are used to make decisions in the PDS, they can perpetuate and amplify existing disparities. For example, if an AI system's algorithms are biased against specific demographics or regions, it could result in inequitable distribution of resources.
5. Lack of Human Oversight: Overreliance on AI systems without adequate human oversight can be problematic. AI can automate operations and increase efficiency, but it may lack the capability to comprehend complex social contexts, cultural nuances, or unforeseen circumstances. This can lead to errors or unfair outcomes, as the system may need to be more capable of adapting to changing situations or addressing individual needs.

6. Dependency and System Vulnerability: Relying heavily on AI systems in the PDS can create a dependency that may be vulnerable to system failures, hacking, or malicious attacks. Technical issues or disruptions can lead to significant problems, such as disruptions in the supply chain or the inability to predict demand and distribution patterns.

For these negative impacts, it is crucial to ensure ethical AI practices, promote AI systems' transparency, address data set biases, prioritize human oversight, and invest in education and reskilling programs to support workers affected by automation. Additionally, policy frameworks and regulations can play a vital role in minimizing the negative impacts of AI in the PDS and fostering a more equitable and inclusive use of technology.

India is a developing country with a massive transition in technology and infrastructure, yet there is a greater scope for advancement in the times to come. The government prioritizes expanding production through various schemes, such as the upcoming Production Linked Incentive Scheme, where technological influence can significantly change. INR 10,900 crore has been allotted to the food processing industry through this scheme, as it is a priority sector. AI is an upcoming field and has played a pivotal role in revolutionizing the internet as we see it, along with many other industries, ranging from letter delivery systems to Netflix recommendation algorithms. According to a report by the Central Institute of Post-Harvest Engineering and Technology (CIPHET), around 16% of India's total agricultural produce is wasted. Alternatively, they could have been processed, or maybe more demanded crops could have been grown in place of them, which could be further exported, leading to an increase in the farmer's income and the nation's GDP at the same time. Most crop selection in agriculture in the country is based on region and visible options, whereas the soil may be more optimized to grow other crops that are not known to the farmers. This is where Machine Learning can come to the rescue. We can feed data from various states, even different nations, regarding several parameters which may be diverse, ranging from soil type, crop yield, cost of growing, selling price expectations, previous data, wastage percentage, and various other parameters that can be used to fit a prediction model which can analyze the parameters through neural networks which may predict the best crop based on your parameters and also an expected profit after growing that particular crop. As the success of this model is based on the data input into the neural network, the prediction model will get more efficient with time and give better predictions in the years to come. Thus, if there is a central agency monitoring the demand for particular crops in the nation and also the export requirements, then feeding this demand into the system can make the distribution of all the crops demanded a remarkably optimized process.

AI can play a role in crop management in various innovative ways. Recently, a food-tech start-up in Switzerland, Gayama, developed a surveillance system using drones to map the growth of crops and report changes in water, fertilizer, pests, and yield using Computer Vision. Computer vision is widely used in the analysis of physical entities, and there is further scope for research into this revolutionary technology and its application in the agriculture and food industries. Supply chain management can also be optimized to prevent food wastage, as AI can be used to tackle the inefficiency of primitive supply chains. There have been various start-ups, such as LogiNext and Locus, working on improving the efficiency of the supply chains in the food processing industry, from the farms where crops are grown to the manufacturing units where they are converted into finished goods ready for retail markets. Exploring unobserved domains in the food processing industry and finding opportunities to enhance the efficiency of food production in the nation can be a revolutionary step. The industry will have rapid technological advancements, leading to a sustainable shift.

India is a developing nation, and there have been rapid strides in the development of the infrastructure and industry-specific facilities in the country over the last few years. Cold storage capacity is an essential facility for the food processing industries and other similar units. According to a report by the National Bank for Agriculture and Rural Development Consultancy Services (NABCONS), the current cold storage capacity of India stands at 374.25 Lakh MT as of September 2020, and there is a requirement of at least 350 Lakh MT more, which is somewhere around 193% of the current capacity. The government is infusing vast amounts of capital into building such facilities and infrastructure to support the food processing industries. The Pradhan Mantri Kisan Sampada Yojana (PMKSY) has specific provisions for financial support as a grant aid in the integrated cold chain and value-addition infrastructure scheme. Machine Learning can also come in handy for this step. Based on the production statistics of various states and the amount of crop wastage due to a lack of infrastructure, models can be implemented that can define a loss function proportional to the wastage and optimize the location of upcoming cold storage plants to enhance resource utilization.

4.6 CONCLUSION

From this study, it is identified that AI is very much required for the expansion of the world of computers and has brought processes to the world through different innovations. AI, namely Machine Learning, can be used to select and save crops, optimize operations, and many more possibilities in the field. McKinsey reported that the efficient use of AI to control food wastage is a monetary opportunity worth $127 billion by 2030. Increasing investor sentiment toward the food-tech startup ecosystem and the phenomenal success of various agri-tech startups is the commencement of the expansion of a bigger economy in the inventive food processing sector, and we hope to see further capital infusions in the industry in the coming time. Every state needs to follow technology-based reforms. Implementing technology gradation in all forms will reduce the technological issues in the PDS; implementing computerized allocation to FPS in all states will reduce mistakes and errors. The GPS tracking system has to be implemented in all states to identify and reduce the loss of food grains during transportation. It is concluded that many of the states in India still need to adopt technological upgrades in their PDS, which will affect the smooth functioning of the PDS. Leakage of food grains at the time of transportation, issues related to the storage of food grains, environmental issues, identification of beneficiaries, inclusion and exclusion errors, and issues with the procurement of food grains are the problems faced by the PDS. Reinventing the PDS in India is essential for its technological development. Technological developments such as the digitization of ration cards, computerized allocation to FPS, issuance of smart cards despite ration cards, use of GPS technology, SMS-based monitoring, and use of web-based citizens' portals are some of the adequate changes that should be implemented in the PDS in all states of India. The PDS is not fully upgraded with AI as a public sector. Due to the high expenses and features of the geographical locations, there are some limitations to the applications of AI.

4.7 FUTURE SCOPE OF THE STUDY

There are very few research papers and news articles on "Barriers or Obstacles of Artificial Intelligence Applications in the Public Distribution System in India." As a result, there is scope for future research into the topic, as well as the challenges and barriers to the applications of AI in the PDS in India.

BIBLIOGRAPHY

Aly, H. (2022). Digital transformation, development and productivity in developing countries: Is artificial intelligence a curse or a blessing?. *Review of Economics and Political Science*, 7(4), 238–256.

Ambekar, S., Kapoor, R., & Mehta, P. (2015). Structural mapping of public distribution system using multi-aystems. *Business Process Management Journal*, 21(5), 1066–1090.

Bačiulienė, V., & Petrokė, I. (2020). The impact of artificial intelligence on growth in the agri-food industry: Lithuanian case. *European Science*, 2(4), 34–40.

Berman, S., & Dalzell-Payne, P. (2018). The interaction of strategy and technology in an era of business re-invention. *Strategy & Leadership*., 46(1), 10–15.

Bhattacharya, S., Falcao, V. L., & Puri, R. (Eds.). (2017). *The public distribution system in India: Policy evolution and program delivery trends*. Oxford University Press.

Bohtan, A., Mathiyazhagan, K., & Vrat, P. (2019). Modeling the public distribution system: A PO-P approach. *Opsearch*, 56, 1024–1066.

Bughin, J., & Van Zeebroeck, N. (2018). Artificial intelligence: Why a digital base is critical. *The McKinsey Quarterly*.

Chakraborty, S., & Sarmah, S. P. (2019). India 2025: The public distribution system and national food security act 2013. *Development in Practice*, 29(2), 230–249.

Chatterjee, S. (2020). AI strategy of India: Policy framework, adoption challenges and actions for government. *Transforming Government: People, Process and Policy*, 14(5), 757–775.

Chaturvedi, A., Mathur, S., & Porwal, A. (2024). *Leveraging Banking Customer Experience Through Artificial Intelligence: An Application of Structural Equation Modelling, Artificial Intelligence and Machine Learning for Business*, CRC Press, Taylor & Francis group. https://appleacademicpress.com/artificial-intelligence-and-machine-learning-for-business-/9781774917237

Chaudhuri, B. (2022). Programmed welfare: An ethnographic account of algorithmic practices in the public distribution system in India. *New Media & Society*, 24(4), 887–902.

Davenport, T., Guha, A., Grewal, D., & Bressgott, T. (2020). How artificial intelligence will change the future of marketing. *Journal of the Academy of Marketing Science, 48*, 24–42.

Dwivedi, Y. K., Hughes, L., Ismagilova, E., Aarts, G., Coombs, C., Crick, T., & Williams, M. D. (2021). Artificial intelligence (AI): Multidisciplinary perspectives on emerging challenges, opportunities, and agenda for research, practice and policy. *International Journal of Information Management*, 57, 101994.

Haenlein, M., & Kaplan, A. (2019). A brief history of artificial intelligence: On the past, present, and future of artificial intelligence. *California Management Review*, 61(4), 5–14.

Keding, C. (2021). Understanding the interplay of artificial intelligence and strategic management: Four decades of research in review. *Management Review Quarterly*, 71, 91–134.

Kumar, R., Ganapathy, L., Gokhale, R., & Tiwari, M. K. (2022). Scaling up the digitalization of the public distribution system: A COVID-19 pandemic impetus. *IFAC-Papersonline*, 55(10), 55–60.

McKinsey Global Institute. (2018). *Notes from the AI frontier: Applications and value of deep learning*. McKinsey & Company. https://www.mckinsey.com/featured-insights/artificial-intelligence/notes-from-the-ai-frontier-applications-and-value-of-deep-learning

Mogale, D. G., Cheikhrouhou, N., & Tiwari, M. K. (2020). Modelling of sustainable food grain supply chain distribution system: A bi-objective approach. *International Journal of Production Research*, *58*(18), 5521–5544.

Pillai, S. V., Vipin, H., Mohan, A., & Shruthi, D. A. (2020). Infusing blockchain in supply chain operations of a public distribution system. *Journal of Supply Chain Management Systems*, 9(2–3), 1–7.

Sharma, N., & Gupta, S. (2019). An investigation of IT-intervention adoption in public distribution system: A stakeholder and agency theory perspective. *Information Development*, *35*(2), 203–219.

Siau, K., & Wang, W. (2020). Artificial intelligence (AI) ethics: Ethics of AI and ethical AI. *Journal of Database Management (JDM)*, 31(2), 74–87.

Singh, S. K., Jenamani, M., Dasgupta, D., & Das, S. (2021). A conceptual model for Indian public distribution system using consortium blockchain with on-chain and off-chain trusted data. *Information Technology for Development*, 27(3), 499–523.

WEB RESOURCES

https://medium.com/frontier-technologies-hub/pds-survey-for-ai-and-blockchain-based-digital-platform-for-offering-public-distribution-system-b1008c3dd21e

https://niti.gov.in/sites/default/files/2019-01/NationalStrategy-for-AI-Discussion-Paper.pdf

https://www.medianama.com/2023/03/223-indian-govt-smart-public-distribution-system/

5 Is the Insurance Sector Ready to Adopt Blockchain Technology? *A Detailed SLR on Critical Success Factors*

Sushil Kalyani and Ankita Chaturvedi

5.1 INTRODUCTION

Blockchain works on a decentralized ledger system that all stakeholders interested in a set of transactions can access. The technology is the most feasible choice for all the transactions required to be transparent and cost-effective. Also, many challenges are to be analyzed, and a solution is needed before adopting it, and these include more clarity on its functionality, governance issues, legal standing, privacy issues, and security threats (Upadhyay, 2020).

Blockchain, the technology behind the cryptocurrency Bitcoin, also has the characteristics of making data shareable on a network that many computers called nodes can access from anywhere. This feature makes the recording and traceability of transactions very easy and accurate in all such industries, where the amounts involved in data, whether related to assets or property, are humongous (Barlyn, 2017). Blockchain has a very distinctive set of characteristics, such as an algorithm-based agreement between the parties involved, the distributed nodes, the decentralized structure, smart contracts, security-enabled encryption to avoid network security breaches, and it provides a transparent system of recording, which is accessible to all the people interested (Dutta et al., 2020). Blockchain is said to be the most disruptive discovery of the twenty-first century (Li et al., 2021). Though it is mainly recognized to have contributed to the development of Bitcoin, the cryptocurrency, the world can now see its applications in various sectors, and not only are financial services using it, but it is finding its place in many other sectors, such as Logistics and Supply Chain, Governance, Business Operations, and Taxation (Kimani et al., 2020).

Blockchain has recently gained attention as it has made new value creation methods and has decentralized ways to manage it. Banks, companies working in the ICT domain, manufacturers, and, in some countries, governments have also started working on the applications of the nascent technology where they need the

DOI: 10.1201/9781003638506-5

transactions to be more secure, efficient, and able to handle a large number of transactions with the most efficient costs (Chen et al., 2019).

5.2 WHAT ARE WE TRYING TO ADDRESS – THE RESEARCH OBJECTIVES

Though some work has been done on the applicability of blockchain in insurance, its acceptability and applicability in financial services per se intrigued us to explore how blockchain technology can be implemented in the insurance business. We hypothesize that the adoption of blockchain may reduce the losses of insurance companies due to fraudulent claims filed by the insured and make the insurance business more effective in terms of time and money.

In this study, since we focus on exploring the applicability of Blockchain in insurance, our goal was to analyze the adoption of blockchain technology with its perceived positive and negative impacts on the functioning of the insurance sector. We planned to identify the factors affecting the adoption of technology in the industry. We focused on identifying the critical success factors (CSFs) influencing the adoption of Blockchain technology in insurance, their weighted impact perceived by the industry experts, and evaluated by authors using the most effective technique available.

While the main objective is to find the suitability of Blockchain in insurance, we observed that identifying the CSFs and their relevance in adopting the technology would be of utmost importance. While the literature suggests many CSF frameworks, we found that no framework consists of all the factors we were to take up in our research. To fill this gap, we started our research with the following objectives:

1. To find a suitable framework or, if required, to design a framework of CSFs for analyzing the implementation of blockchain technology in insurance.
2. To analyze the impact and magnitude of all CSFs identified in the framework with the industry experts' inputs from the blockchain industry and insurance.
3. To provide a ranking of CSFs based on their weightage in implementing Blockchain technology in the insurance sector's decision-making process.

5.3 HOW DOES BLOCKCHAIN WORK?

Blockchain, being one of the technologies called Distributed Ledger Technology (DLT), as the name suggests, works on distributed ledgers. Blockchain is a technology that works with various components, including the distributed ledger and a database that is available to all the people in the network containing the information required, without any intermediary and with no central storage required for the data depository. Another component is an algorithm, which makes this technology a true innovation that keeps changing and needs to be agreed upon by all with a digital agreement whenever there is a change in the data. This algorithm is a set of decision rules coded and encrypted in the system, which accepts an entry into the system and then shares it with everyone to whom it is relevant (Bris & Cordon, 2018).

A blockchain contains several blocks in a chain that comprises information. This technique was initially termed in the 1990s by a group of researchers, and it was discovered as a technique to assign timestamps to digital documents so that they are tamper-proof. It is a set of distributed ledgers that is openly accessible by anyone. These ledgers are encrypted so that once the data is recorded in a block, it is nearly impossible to alter the information unless some specific people permit the same. All the blocks in a chain have data and two hashes calculated when a block is created as a unique identity, which will change if the data changes, making all the subsequent blocks invalid. Another technique that blockchain uses is the "Proof of Work" mechanism. Blockchain uses a peer-to-peer network open for anyone to make it more secure, unlike a central place or organization that manages data. Whosoever is part of the Blockchain can access the full copy of the transaction, called nodes. Once a block is made, it is sent to everyone in the network, and each node validates the block, like putting a signature, to make sure that nothing has been altered. Thus, to change the data, one needs to alter the data of all the blocks, create new hashes, recreate the proof of work of all the blocks, and then take the approval of more than half of the peer-to-peer nodes, making it quite impossible and tamper-proof.

Blockchain has fetched much attention from the media, and organizations from the financial services sector have started finding it interesting because of the potential usage it may provide, with security and cost-effectiveness being on the list of priorities for this sector. As blockchain-based systems are still developing, the application of the technology needs to be analyzed before it can fit in within the financial service sector (Walsh et al., 2020). The scientific community is also interested in adopting the blockchain to solve the contractual dimension of natural disaster risk insurance problems, which are a kind of smart contracting (Pagano et al., 2019). Since the adoption of blockchain is gradually increasing, a systematic literature review, exploring the possibility of how the technology is adoptable, and that also with the complementary sources of knowledge, may allow an analysis of the level of diffusion in the industry, making it inevitably relevant (Kar & Navin, 2021).

In this field, another innovation, the Smart Contracts – A newer technology associated with Blockchain has emerged, and these digitally encrypted contracts are like the earlier existing contracts but are automated. The study discusses how Blockchain and smart contracts together are helping to make the operations of an insurance organization more effective and secure, and also how Blockchain and smart contracts are used to design the solutions that help mitigate the risk of fraud in the vehicle insurance domain (Roriz & Pereira, 2019).

Looking at the fast-changing scenario and the importance of digitization across the financial sector, we were intrigued to research how Blockchain is adopted. While going through the literature, we studied that the adoption of Blockchain is being considered in various fields due to its perceived advantages, but there is a long way to go. We also found that the Insurance sector is one where the application of Blockchain is multifaceted.

5.4 WHY INSURANCE?

Insurance business, one of the most ancient, recorded businesses on Earth and accepted by the world at large to be considered as a growing business, is a \$6.3 trillion business as per the data available for the year 2019, and was expected to have a growth of 3%, had COVID-19 not broken out.

Insurance companies offer risk management services to insurance buyers, also called policyholders. The insurer, being the service provider, guarantees protection against events that result in loss, whereas the consumer of this service, the policyholder, pays a premium for the compensation of the loss (Kar & Navin, 2021).

Insurance is transferring the risk of one person to another, specifically from the insured to the insurer, who promises to make good the financial losses of the insured. The insurer, insurance companies, now earn profits in the scenarios where the total insurance premium collected is more than the expenses of running the business and the amounts of claims raised by and settled for the policyholders who have incurred losses due to the occurrence of the events mentioned in the policy agreement. Whether life or non-life, insurance has a particular characteristic in common: "The Trust" of the insured in the insurer. There are many regulations convergent to the philosophy of saving the insured's trust, and many of the organizations, country to country, are established to protect the interest and confidence of the insured – the client side of this business. There are stringent regulations that all companies working in the insurance domain must adhere to, and all these regulations are designed around the focal point, which is protecting the rights of the insured or the policyholders. While most of the regulations concentrate on the client-side, the insurers, who are ignored most of the time, are also at risk of losing out due to the fraudulent acts by the policyholders who file forged claims using manipulative ways to get the reimbursement for cooked-up losses. The data shows that 5 to 10% of the claims are forged or fraudulent and cost the insurance providers in the non-health category a glaringly massive amount of $ 40 billion a year. Also, reports state that if auto insurers adopt blockchain technology, they will lower their operating ratio by 10 to 13 points (Singer, 2019).

These facts bring us to think about this grey side of the insurance business, the risk of loss to the insurers due to these fraudulent claims, which requires more of the regulators' attention, and also, the insurance companies that are to find solutions to control these undesired losses, besides the genuine ones. Secondly, the business of insurance is growing, but at a slow pace. Finding more products based on the contemporary scenarios and requirements of the changing clients may require some latest technology that is changing the business game because digitization and ease have become more critical influencers for any product now.

Further, while doing a review of the literature to understand the possible areas of insurance where the adoption of Blockchain can be beneficial, we came across that smart contracts can automate the authentication and computation process and may enormously change the insurance industry (Hans et al., 2017).

Further, some studies have proposed the usage of Blockchain in Insurance to address some of these issues:

1. Preventing the insurance companies from frauds as the database can be shared and various entities can record the transactions. In the case of insurance, these entities can be medical officers or police officers who may enter the data. After entering, no one can modify the data once entered, and it is already linked to the client's identity created on digital space.

2. It allows entering the data on a shared basis and is linked to the client's digital identity by verifying the proof of ownership. Blockchain may provide a better solution to make the information shareable and writable by people involved and authorized for a transaction.
3. It may be advantageous to provide a new business opportunity like Pay-Per-Use insurance, where the technology can verify that a customer has activated a policy.

In this case, the technology would require guaranteeing all the parties who can be part of the transaction and its entry on the system, whether the insurance company or the customers (Gatteschi et al., 2018). The innovation, Blockchain, has disrupted the flow of information in all industries. As technology changes substantially, it is interesting to see how things can be done. Also, it needs to be analyzed whether the innovation is disruptive or not because it is evident from the literature review that all the technologies classified as disruptive have changed, as far as classification is concerned, since 1997 (Brophy, 2019). There are studies taken up for suggesting ways to improve efficiency, reduce transaction costs, detect fraud, and prevent risks (Tarr, 2018).

In their research study, the authors develop a platform that is distributed and works with Blockchain for supporting recording a transaction and its execution in the process of insurance. The insurance sector requires many processes to take place to take care of various kinds of policies between the insurance companies and the insured to write the policy, to maintain it, and when it comes to a closing, to deal with it. The primary concerns of the sector are the total time taken in the process, the settlement of payments, and the protection of the process by looking at the security threats. Blockchain technology, which is based on distributed and immutable ledgers, is now being seen in the world of FinTech companies to ensure security and productivity are not being compromised. Before the world goes for applying Blockchain to any financial service, it will require a very in-depth understanding of the processes. Smart contracts can be helpful in this domain (Raikwar et al., 2018).

5.5 LITERATURE REVIEW – CSF IDENTIFICATION TO FILL THE RESEARCH GAP

The CSF method can support MIS planning and requirement analysis. CSF are those few critical aspects that need to be taken utmost care of, ensuring success for an organization, and therefore, they represent those aspects of management and enterprise that are essential to be given undivided attention to maintain high performance (Boynton & Zmud, 1984; Li et al., 2012; Peng & Huang, 2020).

While doing the literature review and shortlisting the CSFs for adopting Blockchain technology in the Insurance industry, many models were reviewed. We found that models developed and used by previous studies serve the purposes partially for our study and were suitable for the issues being addressed in the research conducted for a specific field.

Various models, including Unified Theory of Acceptance and Use of Technology (UTAUT), Technology Acceptance Model (TAM), Technology-Organization-

Environment (TOE), Human-Organization-Technology (HOT Fit), and Technology-Organization-Environment-Human (TOEH) frameworks, were analyzed to understand the process of identifying the CSFs. The models contain various factors essential for the adoption of various technologies in businesses.

The UTAUT framework was developed by Venkatesh et al. in 2003. This theoretical framework is quite famous among researchers and is used extensively to predict the intention to adopt new technology. Venkatesh and many other researchers have worked on various models. They have come out with a conclusion that customers' purchase and usage intentions can be measured by effort expectancy (EE), performance expectancy (PE), social influence (SI), and facilitating conditions (FCs). There are many modifications done to the model, and in one such research, the researchers included risk perception and trust factors. Before the UTAUT model, we had the Theory of Reasoned Action (TRA), TAM, Theory of Planned Behavior (TPB), and the Innovation Diffusion Theory (IDT) (Sarfaraz, 2017).

Seven factors were most affected and were treated as determinants of intention in at least one of the models. From these, the authors conceptualize that four will be significant as these will directly affect the acceptance and behavior of users, namely PE, SI, EE, and conditions facilitating. The attitudes toward the usage of the technology, such as self-efficacy and anxiety, are not considered to be directly affecting factors. Different tags used to describe the factors also describe the essence and present that the factors are independent of another theoretical perspective (Venkatesh et al., 2003).

Researchers have worked on and developed vetted scales for variables, usefulness, and ease of use, which are perceived by the users and are considered to be the primary contributors toward the acceptance by users (Davis, 1989).

In one of the pioneer studies, the authors developed a framework to evaluate the quality of information systems and named it "DeLone and McLean IS Success Model (ISSM)" or the "D&M model." As per the authors, six dimensions are to be looked at to determine the quality of the information system. These are the quality of the system, information, use, users' satisfaction, impact on an individual, and the impact of the firm. In 2003, the model was improved by incorporating service quality and replacing individual impact and organization with the net benefits (Delone & Mclean, 1992; Delone & Mclean, 2003).

Another study helps build on the hands-on practices of the objectives of social sustainability using social media, particularly in logistics. The critical success is theorized by stating that the use of social media is affected by factors that help increase the social sustainability of the logistics supply chain and may also help enhance the competitive advantage. By analyzing the HOT and TOE framework's theoretical understanding and finding the usefulness of the framework called TOEH – the Technology, Organization, Environment, and Human, authors found that the framework's factors influence the use of social media for supply chain social sustainability. Since the competition is increasing significantly due to pressures from the demand side and the policy requirement of sustainability goals, considering all the factors per the TOEH framework becomes essential for logistics (Orji et al., 2020).

The authors in one research aim to investigate an enterprise system's CSFs through the life cycle of a smart factory. They provide an extensive literature analysis based on the conscientious criteria of authors in all the research done in the past 22 years (Rashid et al., 2018).

Large firms are more interested in taking advantage of opportunities provided by the Blockchain, but the clarity on the working of the technology and the laws applicable to it and the trust and privacy concerns affect the preparedness of all the involved parties (Porwal et al., 2022). The study shows that important factors affect the application of Blockchain, including motivational readiness, which pertains to the motivation of all the entities involved to recognize the need for change while an existing service is already working. The other one is Engagement Readiness, which relates to the parties' preparedness to recognize the benefits versus challenges of new technology and are now ready to engage in adoption. Engagement readiness also includes the knowledge of the functioning of Blockchain, future implications on the current systems, the impact of developmental costs, the possibility of project failure, and the expected benefits of the new technology, like smart contracts (Li et al., 2012). One more factor can be Technology Readiness, which includes the organizational or individual predilections to deal with the upcoming technologies. These factors include the compatibility of existing systems, including the software, hardware, and all ICT resources, which may make it convenient to adopt the new technology.

Further, the security infrastructure, the physical or cloud-based server, and the bandwidth of the network connection are a few factors under this readiness. Structural readiness includes all other non-technical factors that are an essential part of an organization, like resources taking care of functions related to finance and human resources, who see the adoption of technology positively and have the proper structure and policies supporting the adoption of technology. Implementation of Blockchain would require all the organization's resources to work in sync if the implementation is to succeed, be it time, human resources, or money of the organization (Balasubramanian et al., 2021). Organizations that have better resources available, like the experts in the ICT industry, will be more comfortable using Blockchain, and the firms having scarcity of resources will always be skeptical about experimenting with new technologies. From the customers' perspective, structural readiness also includes the ability to work on computers or smart gazettes with connectivity to complete the daily tasks traditionally done manually (Rezai-Rad et al., 2012).

The researchers analyzed concurrent problems in the domain of travel insurance (Liu et al., 2019). They further studied underlying technologies of Blockchain, such as equality network, Byzantine consensus mechanism, asymmetric encryption, and intelligent contract. Their model of travel insurance system based on Blockchain was construed, and accordingly, the architecture was designed. After that, the system was analyzed to see its application on one travel insurance product and the insurance sector.

Factors include those related to processes, humans, and technology. Human factors include value for the client, consumer satisfaction, retention of customers, and loyalty. The organizational aspects include the change of the culture within the organization, the leadership role, and the fit of the technology in the firm. Technological factors encompass the adoption of technology, resource readiness, impact, and handling the new change with the same efficiency (Mendoza et al., 2007).

The HOT Fit model is used to evaluate the successful implementation of the information systems, along with their acceptance. Then the fitment between the human, organization, and the technology in the system was evaluated. The human factors comprise users of the system and the satisfaction of the users. The factors under organization consist of structure and the environment, while the technology factors include the quality of the system, information, and services (Ayuni et al., 2019).

Another framework called the TOE framework is one of the most frequently followed in research taken up to evaluate technology adoption. It describes three types of factors that affect innovation or technology adoption. One factor is the technological context, like what features are there and whether it is readily available. The second factor is the organizational context, like the business's size, how communication processes take place, hierarchy, and available resources. The third factor is the environmental context, like the characteristics of the industry, market, government regulation, and IT infrastructure (Sila, 2013).

The ever-changing and volatile demand requires organizations to opt for flexible services to fulfill the requirements of customers. Transportation, which is the key for business, requires effective technology to take care of fair and customized services, and for that, the technology has to be robust to take care of the information. Blockchain technology is seen as having the characteristics matching the requirements of the freight transport business. In their research, they identified factors that are instrumental in affecting the process of decision-making of companies. In the technological factor, data security, the speed of the system to process and share the information, and transparency and trust were identified as few. Some other factors like service quality and traceability were also recognized to be relevant. In another category, factors related to cost include the cost related to human resources and the cost of adopting the technology. Finally, environmental issues were identified, such as reducing the usage of paper in the processes and the consumption of power. The development aspects refer to globalization trends, government support, and organizational culture (Shardeo et al., 2020).

Another study was conducted in which a model was used combining the UTAUT framework and the expectation confirmation model (ECM). Two more factors were added to the model: trust and perceived security. The ECM suggests that a user continues with any system if the perceived usefulness and the satisfaction are at par with the level of the user. The earlier use of systems and the usefulness, which is perceived for a new system, both play a substantial role in defining a user's satisfaction. The model used had expectations toward the performance, efforts, SI, and FCs, which all affect the intention of a user to use technology (Singh, 2020).

In one study, the authors studied that the CSFs also depend on some subsets that need to be considered for evaluation. One such subset can be a critical information set, defined as a group of information significant to watch and measure the CSF. For improving the delivery time as one of the CSF, the CIS will be the date and time of the order placed, the arrival, and the target. The other set is the critical decisions set, defined as the participants of critical decisions, a business needs to act on a crucial factor. As per the study, the third set is the critical assumptions set consisting of such conditions on which a CSF is dependent (Flynn & Arce, 1997).

The combined literary research and empirical studies to grasp how the quality of information, service, system, along with the perceived usefulness and reputation, resulting in the consumer's satisfaction, affects the customers' intentions to use bank Social Network Services (Nam & Seong, 2020).

Success factors are the most important phenomena and are likely to be effective in any condition and situation. Some are historically applicable in a set context and are efficient. Some other factors related to differences that exist due to cultural differences include consumer behavior, customer attitude toward services, language, and companies' cultures. These do not change and might make technology adoption difficult (Preissl et al., 1999).

5.6 METHODOLOGY

After the thorough literature review and going through various models explaining the CSFs suggested by the previous researchers, we found that the TOE and HOT fit were not fully fitting for this study and we needed to have a model which includes Technology, Organizational, Human, Environmental and Economic Factors, and due to the unavailability of such model developed or used by other researchers, we decided to name it the ***HEETO*** model.

Still to improve upon our understanding of the model which can appropriately describe the factors playing an important role in adaption of blockchain in the insurance industry, 12 industry experts were interviewed from both the blockchain technology domain and the insurance sector. In the first round, they were interviewed on some open-ended questions instead of sharing the list of CSFs, which was shortlisted after an extensive literature review by the authors, so that unbiased opinions could be obtained, and the same can also be used to see if the list of CSFs developed by us needs some modification.

As discussed, in this study, we have decided to evaluate the inputs shared by these experts from the blockchain and insurance industry so as to see if the list of CSFs can be made more robust. The list of questions asked in the interview was as follows:

- You have been working with the insurance sector for a long time now. How has been your experience been?
- How do you explain the most important features of this industry? How does the insurance sector work?
- What are the most important characteristics of the insurance sector as compared to other service sectors specially in finance domain?
- Any problems you faced while you were working in this sector? Can we talk about that in detail? Was that/those resolved?
- If not, how do you see the problems can be/could have been solved?
- Have you noticed/heard about/read/come across any recent advancement/ technology which can disrupt the Insurance sector in India?
- Have you heard about blockchain? What do you think about this technology and its application in the insurance sector?

The outcomes of the interviews are as follows:

Co-ordination among Participating Stakeholders: Blockchain/digital ledger technology implementation involves multiple organizations working across business processes. Its implementation would need a change in the internal organizational process to align with the newly designed blockchain process. Financial entities' mindset to protect their own turf makes collaboration difficult.

Data Sharing: Blockchain implementation requires data sharing among participants in real time. Not all organizations would be willing to share complete data transparently to external entities.

Scalability: Blockchain, by its architecture, requires recording of transactions at multiple databases in real time. This recording of authentic data followed by a consensus mechanism becomes a bottleneck for scalability. Traditionally, a credit card transaction can record around 100,000 transactions per second. Any enterprise blockchain solution should have scalability in a similar range.

Bridging Legacy System and New Blockchain Network to Work Together: This bridging would need additional hardware and software implementation, resulting in an increase in cost. Non-interoperability Among Multiple Blockchain Networks: In a collaborative digital age, there is an increasing need for interaction among multiple stakeholders. In some scenarios, regular interaction would be required between stakeholders participating in multiple blockchain networks. Currently, inter-blockchain/DLT interaction has not been standardized, thus hindering its adoption.

5.6.1 Blockchain Use Case in Insurance (Enterprise Applications)

5.6.1.1 General Insurance

1. Vehicle insurance
   a. Claim processing
   b. KYC for customer onboarding
   c. Fraud analytics – industry-wide database
2. Travel insurance
   a. Flight insurance
   b. Road accident insurance
3. Agriculture insurance
   a. Underwriting based on a historical database of production per acreage in the past.
   b. Claim processing based on weather updates.

5.6.1.2 Health Insurance

Centralized blockchain-based database of customers' prescriptions, diagnoses, and ailments history. The database will help in better risk assessment, pricing, and avoiding adverse selection.

5.6.1.3 Life Insurance

1. Employment history and educational institutions database on blockchain network – prevents adverse selection, better pricing, and avoids fraud.
2. Health databases on blockchain help insurance companies in tracking chronic diseases and performing better risk assessment.

5.6.1.4 Security

1. Insurance against any protocol hack and breach.
2. Insurance against smart contract exploits.

Blockchain holds the potential to enhance how insurers document risk, elevating the speed, precision, and transparency of our procedures. The possible applications of blockchain suggest advantages in terms of revenue, cost efficiency, and operational effectiveness. In the near future, underwriters could harness the capabilities of peer-to-peer networks, employing AI and analytics to enhance underwriting practices. A notable recent application of blockchain focuses on identifying losses through sensors, streamlining the claims registration process. However, for this chain to be effective, all involved parties must be connected end to end.

Moreover, in the realm of retail, there are additional opportunities for insurance companies to collaborate with larger ecosystems, simplifying life for customers. This collaboration can not only facilitate the seamless auto-reporting of claims but also validate the legitimacy of the claim and promptly disburse funds to the customer through a predetermined account or method.

5.7 RESULTS AND CONCLUSION

We found that finding factors to make the adoption of technology effective was a long-researched area, but due to the reason that the Blockchain, being a very new technology altogether, various models developed and used by researchers were not helpful for our study. We reviewed a number of research works, but no single model had all the CFSs included in the framework that could suit our study, and hence, we had to combine the TOEH model with economic factors to make it a fully adaptable model for critical evaluation and ranking of the CSFs behind the implementation of Blockchain.

5.8 SCOPE FOR FURTHER RESEARCH

Since blockchain technology is being adopted gradually in many fields, gauging the real-time benefits will take some time. The positive side of every technology is always highlighted first, but a real measure comes with the tools to scale negatives also. This research is conducted to measure the impact of CSFs perceived by the authors and industry experts. We firmly believe that additional factors may be identified in due course of time, especially when the implementation of this technology in the insurance business has been there and both positive and negative implications are measurable.

REFERENCES

Ayuni, N. W. D., Dewi, K. C., & Suwintana, K. (2019). Human Organization Technology Fit (HOT Fit) as evaluation model in E-learning system of bali state polytechnic. *iCASTSS, 354*133–137. https://doi.org/10.2991/icastss-19.2019.29

Balasubramanian, S., Shukla, V., Sethi, J. S., Islam, N., & Saloum, R. (2021). A readiness assessment framework for Blockchain adoption: A healthcare case study. *Technological Forecasting and Social Change, 165*(July 2020), 120536. https://doi.org/10.1016/j.techfore.2020.120536

Barlyn, S. (2017). AIG teams with IBM to use blockchain for 'smart' insurance policy. *Reuters*, 1–9. https://www.reuters.com/article/us-aig-blockchain-insurance-idUSKBN1953CD/

Boynton, A. C., & Zmud, R. W. (1984). Assessment of critical success factors. *Sloan Management Review, 25*(4), 17–27.

Bris, A., & Cordon, C. (2018, October 16). Is blockchain all hype? A financier and supply chain expert discus. *The Conversation.* https://theconversation.com/is-blockchain-all-hype-a-financier-and-supply-chain-expert-discuss-106584

Brophy, R. (2019). Blockchain and insurance: A review for operations and regulation. *Journal of Financial Regulation and Compliance, 28*(2), 215–234. https://doi.org/10.1108/JFRC-09-2018-0127

Chen, W., Xu, Z., Shi, S., Zhao, Y., & Zhao, J. (2019). A survey of blockchain applications in different domains. ArXiv.

Davis, F. D. (1 September 1989) Perceived Usefulness, Perceived Ease of Use, and User Acceptance of Information Technology. *MIS Quarterly, 13*(3), 319–340. https://doi.org/10.2307/249008

Delone, W., & Mclean, E. (2003). The DeLone and McLean model of information systems success: A ten-year update. *Journal of Management Information Systems, 19*(4), 9–30.

DeLone, W. H., & McLean, E. R. (1992) Information Systems Success: The Quest for the Dependent Variable. Information Systems Research 3(1):60–95.

Dutta, P., Choi, T. M., Somani, S., & Butala, R. (2020). Blockchain technology in supply chain operations: Applications, challenges and research opportunities. *Transportation Research Part E: Logistics and Transportation Review, 142*(May), 102067. https://doi.org/10.1016/j.tre.2020.102067

Flynn, D. J., & Arce, E. A. (1997). A CASE tool to support critical success factors analysis in IT planning and requirements determination. *Information and Software Technology, 39* (5), 311–321. https://doi.org/10.1016/S0950-5849(96)01150-0

Gatteschi, V., Lamberti, F., Demartini, C., Pranteda, C., & Santamaria, V. (2018). To blockchain or not to blockchain: That is the question. *IT Professional, 20*(2), 62–74. https://doi.org/10.1109/MITP.2018.021921652

Hans, R., Zuber, H., Rizk, A., & Steinmetz, R. (2017). Blockchain and smart contracts: Disruptive technologies for the insurance market. *AMCIS 2017 - America's Conference on Information Systems: A Tradition of Innovation, 2017-Augus*(August), 1–10.

Kar, A. K., & Navin, L. (2021). Diffusion of blockchain in insurance industry: An analysis through the review of academic and trade literature. *Telematics and Informatics, 58*(May 2020), 101532. https://doi.org/10.1016/j.tele.2020.101532

Kimani, D., Adams, K., Attah-Boakye, R., Ullah, S., Frecknall-Hughes, J., & Kim, J. (2020). Blockchain, business and the fourth industrial revolution: Whence, whither, wherefore and how? *Technological Forecasting and Social Change, 161*(May), 120254. https://doi.org/10.1016/j.techfore.2020.120254

Li, J. H., Ray, P., Seale, H., & MacIntyre, R. (2012). An E-Health readiness assessment framework for public health services - Pandemic perspective. *Proceedings of the Annual*

Hawaii International Conference on System Sciences, 2800–2809. https://doi.org/10.1109/HICSS.2012.95

Li, Z., Zhong, R. Y., Tian, Z. G., Dai, H. N., Barenji, A. V., & Huang, G. Q. (2021). Industrial blockchain: A state-of-the-art Survey. *Robotics and Computer-Integrated Manufacturing*, *70*(October 2019). https://doi.org/10.1016/j.rcim.2021.102124

Liu, J. L., Wu, X. Y., Yu, W. J., Wang, Z. C., Zhao, H. L., & Cang, N. M. (2019). Research and design of travel insurance system based on blockchain. *ICIIBMS 2019 - 4th International Conference on Intelligent Informatics and Biomedical Sciences*, 121–124. https://doi.org/10.1109/ICIIBMS46890.2019.8991444

Mendoza, L. E., Marius, A., Pérez, M., & Grimán, A. C. (2007). Critical success factors for a customer relationship management strategy. *Information and Software Technology*, *49*(8), 913–945. https://doi.org/10.1016/j.infsof.2006.10.003

Nam, K., & Seong, N. (2020). A study on influencing factors for customer satisfaction and the continuing use of social network services in financial industry. *Enterprise Information Systems*, *00*(00), 1–25. https://doi.org/10.1080/17517575.2020.1739342

Orji, I. J., Kusi-Sarpong, S., & Gupta, H. (2020). The critical success factors of using social media for supply chain social sustainability in the freight logistics industry. *International Journal of Production Research*, *58*(5), 1522–1539. https://doi.org/10.1080/00207543.2019.1660829

Pagano, A. J., Romagnoli, F., & Vannucci, E. (2019). Implementation of blockchain technology in insurance contracts against natural hazards: A methodological multi-disciplinary approach. *Environmental and Climate Technologies*, *23*(3), 211–229. https://doi.org/10.2478/rtuect-2019-0091

Peng, X., & Huang, H. (2020). Fuzzy decision making method based on cocoso with critic for financial risk evaluation. *Technological and Economic Development of Economy*, *26*(4), 695–724. https://doi.org/10.3846/tede.2020.11920

Porwal, A, Chaturvedi, A & Gupta, S (March 2022) *Artificial Intelligence with Special Reference to Blockchain Technology: A future of Accounting, Impact of Artificial Intelligence on Organizational Transformation*, John Wiley and Sons Inc and Scrivener Publishing LLC, USA, ISBN 978-1-119-71017-2, 289–303, DOI: 10.1002/9781119710301.ch17

Preissl, B., Rickert, C., & Hansjor, B. (1999). Applicability of success factors in US electronic commerce to the German market. *Electronic Markets*, *9*(1–2), 65–72. https://doi.org/10.1080/101967899359265

Raikwar, M., Mazumdar, S., Ruj, S., Sen Gupta, S., Chattopadhyay, A., & Lam, K. Y. (2018). A blockchain framework for insurance processes. *2018 9th IFIP International Conference on New Technologies, Mobility and Security, NTMS 2018 - Proceedings, 2018-Janua*, 1–4. https://doi.org/10.1109/NTMS.2018.8328731

Rashid, A., Masood, T., Erkoyuncu, J. A., Tjahjono, B., Khan, N., & Shami, M. Ud Din. (2018). Enterprise systems' life cycle in pursuit of resilient smart factory for emerging aircraft industry: a synthesis of Critical Success Factors'(CSFs), theory, knowledge gaps, and implications. *Enterprise Information Systems*, *12*(2), 96–136. https://doi.org/10.1080/17517575.2016.1258087

Rezai-Rad, M., Vaezi, R., & Nattagh, F. (2012). E-health readiness assessment framework in Iran. *Iranian Journal of Public Health*, *41*(10), 43–51.

Roriz, R., & Pereira, J. L. (2019). Avoiding insurance fraud: A blockchain-based solution for the vehicle sector. *Procedia Computer Science*, *164*, 211–218. https://doi.org/10.1016/j.procs.2019.12.174

Sarfaraz, J. (2017). Unified Theory of Acceptance ad Use of Technology (UTAUT) model – Mobile banking. *Journal of Internet Banking and Commerce*, *22*(3), 1–20. http://www.icommercecentral.com

Shardeo, V., Patil, A., & Madaan, J. (2020). Critical success factors for blockchain technology adoption in freight transportation using fuzzy ANP-modified TISM approach. *International Journal of Information Technology and Decision Making*, *19*(6), 1549–1580. https://doi.org/10.1142/S0219622020500376

Sila, I. (2013). Factors affecting the adoption of B2B e-commerce technologies. In *Electronic Commerce Research* (Vol. 13, Issue 2). https://doi.org/10.1007/s10660-013-9110-7

Singer, A. W. (2019). Can blockchain improve insurance? *Risk Management*, *66*(1), 20–25. http://search.proquest.com/openview/cc9af0d1e9a80f81b6ed202ebb6de345/1?pq-orig site=gscholar&cbl=47271

Singh, S. (2020). An integrated model combining ECM and UTAUT to explain users' post-adoption behaviour towards mobile payment systems. *Australasian Journal of Information Systems*, *24*, 1–27. https://doi.org/10.3127/ajis.v24i0.2695

Tarr, J. A. (2018). Distributed ledger technology, blockchain and insurance: Opportunities, risks and challenges. *Insurance Law Journal*, *29*(3), 254–268.

Upadhyay, N. (2020). Demystifying blockchain: A critical analysis of challenges, applications and opportunities. *International Journal of Information Management*, *54*(May), 102120. https://doi.org/10.1016/j.ijinfomgt.2020.102120

Venkatesh, Viswanath and Morris, Michael G. and Davis, Gordon B. and Davis, Fred D., User Acceptance of Information Technology: Toward a Unified View (September 1, 2003). MIS Quarterly, Vol. 27, No. 3, pp. 425–478, 2003, Available at SSRN: https://ssrn.com/abstract=3375136

Walsh, C., O'Reilly, P., Gleasure, R., McAvoy, J., & O'Leary, K. (2020). Understanding manager resistance to blockchain systems. *European Management Journal*. https://doi.org/10.1016/j.emj.2020.10.001

WEB REFERENCES

https://www.iii.org/publications/insurance-handbook/economic-and-financial-data/world-insurance-marketplace

https://www.irdai.gov.in/

6 Integrating Sustainability and Digitalization in the Banking Industry

Bindu K. Nambiar, Kartikeya Bolar, and Arunima K. V.

6.1 INTRODUCTION

In the contemporary epoch, the buzzwords that have been dominating global dialogues and shaping modern life are sustainability and digitalization (Hilty & Aebischer, 2015). These two aspects are no longer stand-alone concepts but are intertwined intricately in our day-to-day lives (Geels, 2018). They have revolutionized not only the way we live and work but also the way businesses operate, creating an enormous impact on the global ecosystem (Bibri & Krogstie, 2017).

The United Nations Environment Programme (UNEP), with its persistent efforts, has been at the forefront of fostering a sustainable planet in the burgeoning digital era (UNEP, 2018). Recognizing the influence of these two trends, the UNEP is striving to build a harmonious balance between the two, ensuring the sustainable use of digital technologies (Bibri & Krogstie, 2017). In conjunction with UNEP's efforts, governments and private sectors worldwide are equally investing time and resources to create a new value system that fosters digital transformation (Schwab, 2017). The objective is to set up a digital ecosystem of planetary data that would cater to a wide array of sectors, including finance, healthcare, education, and more, while upholding the sustainability norms (Bibri, 2018). One of the primary sectors undergoing this seismic shift is the banking industry (Bányai, 2018). For decades, banks have been employing technology to streamline operations, enhance customer experience, and respond to changing market dynamics (Birch & Chiappetta, 2018). However, the digital technologies that have been a boon for the banking industry pose a significant challenge in terms of sustainability (Köhler et al., 2019). These technologies, while revolutionizing the banking industry, consume substantial resources, including energy and rare earth metals, among others (Frolov et al., 2019). Additionally, the carbon emissions from these technologies pose a grave threat to the environment (Belkhir & Elmeligi, 2018).

Looking into the future, the conundrum that banks face is their digital infrastructure's sustainability over the next decade (Köhler et al., 2019). How can banks continue their digital transformation journey while ensuring the sustainability of their operations? It is a critical question that needs to be addressed urgently and

 DOI: 10.1201/9781003638506-6

strategically (Bibri & Krogstie, 2017). Leveraging the next ten years to embark on a path that marries digital transformation with sustainability is the need of the hour (Schaltegger & Wagner, 2017). This journey will not only protect the planet and its inhabitants but also fortify the banking ecosystem, enabling it to contribute significantly to the global sustainability goals (UNEP, 2018).

In this research endeavor, we propose to explore the intricate relationship between sustainability and digitalization in the banking industry (Bányai, 2018). We seek to delve deeper into this interaction and suggest ways to integrate these two elements for mutual benefit (Bibri, 2018). The aim is not to view sustainability and digitalization as two separate ends of a spectrum but to recognize their interdependence and integrate them in a way that strengthens the banking industry while safeguarding the environment (Bibri & Krogstie, 2017).

Our focus will primarily revolve around identifying ways in which digitalization can contribute to sustainability goals. We will scrutinize various digital initiatives and strategies, studying their environmental impact and determining how they can be restructured or reoriented to contribute positively toward sustainability (Hilty et al., 2015). Conversely, we will also investigate how sustainability can guide digitalization initiatives. This involves understanding the principles of sustainability and exploring how they can be embedded in digital transformation strategies, thereby steering the digital journey of banks toward a more sustainable future (Schaltegger & Wagner, 2017).

The integration of sustainability and digitalization in the banking industry is a complex yet critical issue that requires comprehensive research and strategic implementation (Bányai, 2018). Through this research, we aim to provide a roadmap for the banking industry to navigate through its digital transformation journey, keeping sustainability at the helm (Geels, 2018). By doing so, we aspire to create a future where digital banking does not just stand for convenience and efficiency but also signifies a commitment to a greener and more sustainable planet (Schwab, 2017).

6.2 BACKGROUND

The banking industry has been a major beneficiary of the digital revolution, leveraging technology to enhance customer experience and streamline operations (Gomber et al., 2018). However, the marriage between banking and digital technology has raised valid concerns regarding its impact on sustainability. Digital technologies are resource-intensive, consuming significant amounts of electricity, rare earth metals, and other resources, all while generating substantial carbon emissions (Dwivedi et al., 2022). In an era where sustainability is increasingly taking center stage, it is crucial to assess the impact of digitalization on sustainability and identify potential pathways through which digital capabilities can contribute to sustainability goals (Parida et al., 2019).

The first step in this complex endeavor is assessing the impact of digitalization on sustainability. This step involves examining the digital infrastructure and services that significantly impede sustainability. For instance, the carbon emissions associated with digital banking transactions, data centers, and technology waste come under intense scrutiny in this stage. As a complex web of interconnected operations and procedures, the digital banking ecosystem's various aspects contribute to its carbon footprint in varying degrees. High-energy-consuming data centers, wastage from obsolete technology, and the environmental cost of powering countless digital transactions daily significantly contribute to the carbon emissions associated with digital banking.

Bank of America provides a prominent example of a banking institution recognizing and addressing these issues. Committed to achieving net-zero greenhouse gas emissions from its operations and supply chain by 2050, the bank has set ambitious targets to reduce operational emissions by 50% by 2025 and transition to 100% renewable electricity for its operations by the same year. Additionally, Bank of America is endeavoring to reduce paper usage by urging customers to adopt digital banking, thereby minimizing the environmental impact of paper-based marketing materials and transactions. This illustrates how banks can not only identify but also actively address their sustainability challenges.

The second step in this process is identifying areas where digital capabilities can positively impact sustainability. This involves discerning how digitalization can contribute to various sustainability goals. For instance, digitalization can reduce paper usage in banking, promoting responsible investing and ethical financial practices, and even aid in reducing energy consumption. Moreover, digitalization can help realize social inclusion goals and bolster customer engagement with sustainability initiatives.

An example of this is the use of digital platforms to promote responsible investing. Digital platforms can provide customers with real-time information about the environmental, social, and governance (ESG) performance of companies. Armed with this knowledge, customers can make informed investment decisions that align with sustainability goals. Additionally, digital platforms can act as powerful tools for promoting financial literacy and inclusion. By providing customers with easy access to financial education and resources, these platforms can democratize access to financial services, contributing to broader social sustainability goals.

The third and final step of this process involves developing a comprehensive digital sustainability framework. This framework, based on the analysis from the first two steps, will aid banks in achieving sustainability goals through digitalization. The framework will include parameters outlined in our proposal and provide clear guidelines on integrating sustainability into digital strategies. In addition to this, the framework will also help banks measure their progress toward sustainability goals, highlighting areas of success and identifying areas for improvement.

The above framework will serve as a roadmap for banks navigating their digital transformation journey, ensuring that they maintain a firm commitment to sustainability. This structured approach to integrating sustainability and digitalization will help banks transition from traditional resource-intensive models to more efficient, sustainable models that leverage digital capabilities to not only meet their economic objectives but also contribute to broader social and environmental goals.

Digital technology's integration with the banking industry has raised valid concerns about its sustainability impact. However, by systematically assessing this impact, identifying areas of positive potential, and developing a comprehensive digital sustainability framework, banks can leverage their digital capabilities to contribute to sustainability goals. By doing so, banks will not only address the environmental costs associated with digitalization but will also enhance their role in promoting broader social and economic sustainability goals.

The evolution of the digital age has sparked numerous conversations around sustainability, giving rise to key research studies that attempt to gauge and enhance digital sustainability. Two such important studies are the Digital Sustainability Index developed in partnership with The Centre of Digital Enterprise (CODE) at the University of Auckland Business School, and the Korn Ferry Digital Sustainability Index 2017. These indices not only help us understand the current state of digital sustainability in corporations but also pave the way for new strategies and interventions.

6.3 THE DIGITAL SUSTAINABILITY INDEX

The Digital Sustainability Index, conceived in partnership with CODE at the University of Auckland Business School, provides a robust and comprehensive measurement framework based on key parameters that assess how well a corporation is incorporating sustainable digital practices. Each parameter focuses on different aspects of a company's approach to digital sustainability and provides insights into their motivations, capabilities, ecosystem collaborations, and governance structures.

6.3.1 The Digital Sustainability Index Parameters

The first parameter of the index is 'Orientation'. This aspect scrutinizes the motivational drivers that encourage corporations to embrace sustainable behaviors and Digital Sustainability. This can include an analysis of corporate policies, mission statements, and strategic objectives that underline a company's commitment to sustainability. It also involves understanding the corporate culture, employee engagement, and the leadership's role in promoting digital sustainability. Are employees incentivized to adopt sustainable practices? Are there tangible rewards for teams that demonstrate excellent, sustainable performance? By assessing these aspects, the index aims to understand the intrinsic and extrinsic factors that motivate a company to strive for digital sustainability. Identifying gaps in this parameter may reveal inconsistencies between stated commitments and actual practices within an organization.

Banks that prioritize digital sustainability in their orientation demonstrate a commitment to incorporating sustainable practices into their core values, mission statements, and strategic objectives. This implies that they recognize the importance of aligning their operations with environmental and social responsibilities. By emphasizing sustainability in their orientation, banks can enhance their reputation, attract environmentally conscious customers, and gain a competitive advantage in a market that increasingly values sustainable practices.

The second parameter, 'Digital', evaluates the capabilities, performance, and use of digital resources and innovation that deliver business solutions impacting the company's sustainability outcomes. This involves understanding how a company leverages digital technology to improve sustainability outcomes. Are they utilizing energy-efficient data centers? Are they adopting cloud technologies to minimize their carbon footprint? Are they leveraging artificial intelligence and machine learning to optimize resource consumption? How does the company's use of digital resources and innovations contribute to its sustainability goals? Understanding a company's digital resources and their contribution to sustainability goals allows for an assessment of their technological readiness and innovation in achieving sustainable outcomes. Identifying gaps here could highlight areas where companies may need to invest in digital solutions or develop more advanced strategies.

The digital parameter emphasizes the importance of leveraging digital resources and innovation to drive sustainability outcomes. For banks, this means adopting technologies that enable energy efficiency, such as implementing energy-efficient data centers or utilizing cloud computing to reduce their carbon footprint. Banks can also leverage artificial intelligence and machine learning to optimize resource consumption, improve risk management, and enhance operational efficiency. The implication is that by embracing digital technologies, banks can enhance their sustainability performance, reduce costs, and provide innovative solutions that align with the changing expectations of customers and regulators.

The third parameter is 'Partnerships'. This focuses on the ecosystem enablers that help organizations access and leverage the best capabilities, know-how, tools, and platforms to achieve sustainability impact. In a hyper-connected world, no company operates in a vacuum. They collaborate with suppliers, partners, governmental bodies, and non-profit organizations, all of which form their ecosystem. Assessing partnerships and understanding how they enable companies to leverage resources for sustainability provides valuable insights. Identifying gaps in this parameter may indicate limitations in collaboration or missed opportunities to access relevant expertise and resources for sustainable initiatives.

Collaboration and partnerships play a significant role in driving sustainability in the banking industry. Banks can form strategic alliances with suppliers, fintech companies, governmental bodies, and non-profit organizations to access the best capabilities, tools, and platforms for sustainability initiatives. These partnerships can provide banks with valuable insights, resources, and expertise, enabling them to develop innovative sustainable solutions and expand their impact. The implication is that banks that foster strong partnerships and collaborations can amplify their sustainability efforts, share best practices, and contribute to broader industry-wide sustainability goals.

The final parameter is 'Governance'. This evaluates how organizational governance contributes to the success of Digital Sustainability. Governance here refers to the rules, processes, and practices that determine how a company is directed and controlled. It includes how decisions are made, who makes them, and how accountability is ensured. How is sustainability incorporated into these governance structures? Are there designated teams or individuals responsible for sustainability? How is their performance measured and accountability ensured? Assessing governance helps understand how sustainability is embedded in decision-making processes and how performance is measured. Identifying gaps in governance may reveal shortcomings in establishing clear roles, responsibilities, or performance metrics related to digital sustainability.

Governance is crucial for ensuring the effective implementation of digital sustainability strategies in banks. Clear governance structures that incorporate sustainability considerations into decision-making processes, designate responsible teams or individuals, and establish accountability mechanisms are essential. Banks should measure and report on their sustainability performance, align executive compensation with sustainability goals, and establish robust risk management frameworks. The implication is that strong governance ensures that sustainability initiatives are embedded throughout the organization, resulting in improved transparency, risk management, and long-term value creation.

6.3.2 Digital Sustainability Index Framework

The Digital Sustainability Index provides a comprehensive framework for understanding how companies are incorporating sustainable digital practices. By evaluating the orientation, digital capabilities, partnerships, and governance of a company, the index helps gauge their commitment to digital sustainability and the effectiveness of their strategies. This index goes beyond providing a snapshot of a company's current state of digital sustainability and offers several additional benefits and capabilities.

Firstly, the index enables comparative analysis, allowing companies to benchmark their performance against industry standards and peers. By comparing their scores in each parameter with those of other companies in the same sector, organizations can identify areas of strength and areas that require improvement. This comparative analysis helps companies understand their relative position in the market and encourages healthy competition and best practice sharing.

Secondly, the index serves as a diagnostic tool, enabling companies to identify specific areas where they can enhance their digital sustainability efforts. For instance, a low score in the governance parameter suggests a need to strengthen sustainability governance structures and accountability mechanisms. By pinpointing improvement areas, the index guides companies in developing targeted strategies and allocating resources effectively.

Furthermore, the index facilitates progress tracking over time. As a company implements initiatives to improve digital sustainability, it can use the index to monitor advancements and evaluate the effectiveness of its strategies. By conducting periodic assessments using the same parameters, organizations can measure their journey toward digital sustainability goals and make necessary adjustments to ensure continuous improvement.

The index also supports companies in their external reporting efforts by incorporating the parameters into sustainability reports or disclosure frameworks. This enhances transparency and accountability, providing stakeholders with a comprehensive overview of a company's digital sustainability performance. External reporting using the index builds trust among customers, investors, and other stakeholders.

Moreover, the Digital Sustainability Index promotes standardization in assessing digital sustainability practices across industries. By establishing a common framework, it facilitates collaboration and knowledge-sharing among companies, industry associations, and policymakers. Standardization enables companies to learn from each other's experiences, share best practices, and collectively work toward advancing digital sustainability goals on a broader scale.

6.4 THE KORN FERRY DIGITAL SUSTAINABILITY INDEX 2017

The Korn Ferry Digital Sustainability Index 2017 is another critical research instrument designed to evaluate the extent of digital sustainability within organizations. Digital sustainability refers to an organization's capacity to leverage digital technologies to drive performance and maintain relevance in a rapidly evolving digital environment. The index uses five parameters – agility, connectivity, discipline and focus, empowerment and alignment, and openness and transparency – to measure digital sustainability.

6.4.1 The Korn Ferry Digital Sustainability Index 2017 Parameters

Agility is a critical parameter for banks in the context of digital sustainability as it addresses their ability to respond swiftly and effectively to market changes, customer demands, and emerging digital trends. In today's rapidly evolving digital landscape, the banking industry faces constant disruption and transformation. Banks need to be agile to navigate these changes and remain competitive.

For banks, agility means having the capacity to adapt their strategies, products, and services to meet evolving customer expectations and industry trends. It involves embracing change as an opportunity rather than a threat and being proactive in identifying and capitalizing on emerging digital advancements. Agile banks are quick to recognize market shifts and customer needs, enabling them to make swift decisions and implement innovative ideas that align with the demands of the digital era. By emphasizing agility as a parameter, the index recognizes the importance of banks being able to respond promptly to the dynamic nature of the digital landscape. It enables banks to evaluate their capability to navigate this landscape effectively and seize opportunities for growth and innovation. This evaluation helps banks identify areas where they may need to improve their agility, such as streamlining decision-making processes, fostering a culture of innovation, or investing in agile technologies and infrastructure.

Being agile is particularly crucial for banks in delivering enhanced customer experiences through digital channels. As customers increasingly expect seamless and personalized digital banking services, agility enables banks to meet these expectations by adapting their offerings to rapidly evolving customer preferences. This parameter prompts banks to assess their agility in meeting customer demands and leveraging digital technologies to deliver convenient, user-friendly, and personalized experiences.

Furthermore, agility plays a vital role in banks' ability to stay competitive amid changing market dynamics. It allows banks to seize opportunities for growth and innovation, whether through launching new products and services, partnering with fintech firms, or exploring emerging digital trends. By assessing their agility, banks can identify areas where they can improve their responsiveness, embrace change more effectively, and foster a culture of continuous innovation.

The second parameter of the index, 'Connectivity', holds significant importance for banks as it evaluates their ability to create ecosystems and establish networks of people from both within and outside the organization. In the digital age, no bank can operate in isolation and expect to thrive in a competitive landscape. To succeed, banks must cultivate diverse connections with various stakeholders, including suppliers, partners, customers, and even competitors. Creating a connected organization allows banks to leverage these networks and tap into their collective wisdom and resources. By fostering meaningful connections, banks can generate innovative ideas, gather valuable market intelligence, and share knowledge and best practices. This parameter emphasizes the importance of leveraging digital tools and technologies to enhance connectivity, both internally among employees and externally with external stakeholders.

Internally, connectivity enables banks to break down silos and facilitate collaboration and communication among different departments and teams. By utilizing digital platforms and tools, banks can create virtual spaces for employees to connect, share insights, and collaborate on projects. This enhances cross-functional collaboration, accelerates decision-making processes, and promotes a culture of innovation within the organization. Externally, connectivity allows banks to establish and nurture relationships with external stakeholders. This includes forging partnerships with fintech companies, collaborating with technology providers, and engaging with customers through various digital channels. By leveraging digital tools, such as social media platforms, online communities, and collaboration tools, banks can strengthen their external networks and engage in the co-creation of value with their stakeholders.

The parameter of connectivity in the index encourages banks to evaluate how effectively they use digital tools to foster connectivity within their organization and with external partners. It prompts banks to assess their digital infrastructure, communication channels, and collaboration platforms to ensure they are enabling effective connectivity and knowledge sharing. This parameter recognizes that connectivity is not limited to one-way interactions but also involves active engagement and participation. Banks are encouraged to actively seek opportunities to engage with stakeholders, listen to their feedback, and involve them in the co-creation of products, services, and solutions. By fostering connectivity, banks can enhance their understanding of customer needs, stay attuned to market trends, and drive innovation that aligns with the evolving demands of their stakeholders.

The parameter of 'Discipline & Focus' in the index plays a crucial role in evaluating the clarity and determination of businesses in understanding the essence of digital transformation and the necessary steps to successfully navigate the digital landscape. It recognizes that digital transformation is not solely about adopting new technologies but involves a holistic approach that encompasses reimagining business models, transforming organizational cultures, and reinventing customer relationships.

For banks, digital transformation requires a disciplined and focused approach. It begins with having a clear vision of what digital transformation means for the organization and understanding the magnitude of the changes required. This includes recognizing the need to embrace digital technologies, redesign processes, and develop new capabilities to meet the evolving demands of customers in the digital era.

A disciplined and focused organization is one that understands the path it needs to embark on for digital transformation. It identifies the challenges and barriers that may arise along the way and develops strategies to overcome them. This parameter assesses the level of clarity in a bank's digital transformation vision, evaluating how well-defined and understood it is across the organization. Moreover, this parameter highlights the importance of focus in achieving digital objectives. Digital transformation is a journey that requires persistence and commitment. A focused organization maintains a clear alignment between its digital transformation vision and its actions. It establishes measurable goals, identifies key performance indicators, and tracks progress toward achieving digital objectives. This focus ensures that the organization remains on track, making the necessary adjustments and investments to drive successful digital transformation. By assessing the discipline and focus parameter, banks can evaluate their readiness and commitment to digital transformation. It prompts them to critically examine their digital transformation vision, ensuring that it is well-defined and communicated throughout the organization. It also encourages banks to assess their level of focus in implementing the necessary changes and aligning their strategies, resources, and investments with their digital objectives.

The fourth parameter of the index, 'Empowerment & Alignment', holds immense significance as it focuses on the alignment and empowerment of employees within a business. This parameter seeks to evaluate how well employees understand their future roles within the organization, how they can contribute to the digital transformation journey, the skills they need to achieve their goals, and the support available to them.

Employee empowerment and alignment are critical for achieving digital sustainability within banks. Digital transformation often entails significant changes in roles, responsibilities, and workflows. It requires employees to embrace new technologies, acquire new skills, and adapt to evolving customer expectations. For successful digital transformation, employees must be empowered to drive change and align their efforts with the overall digital strategy of the organization. This parameter measures how effectively a bank communicates its digital vision to its employees. It assesses the clarity of the communication, ensuring that employees have a deep understanding of the organization's digital goals, strategies, and expectations. Clear and consistent communication is vital for fostering a shared sense of purpose and enabling employees to align their actions with the digital transformation objectives.

The parameter evaluates how well a bank empowers its employees with the necessary skills and resources to thrive in the digital era. This includes providing training programs, upskilling initiatives, and continuous learning opportunities to enhance digital capabilities. Empowered employees possess the knowledge and confidence to embrace digital tools and technologies, contribute innovative ideas, and drive digital initiatives within their respective roles. Additionally, the parameter assesses how effectively a bank aligns the roles and responsibilities of employees with the overall digital strategy. This involves ensuring that employees understand how their individual contributions contribute to the organization's digital goals and how their work supports the larger digital transformation efforts. Alignment fosters a sense of ownership and accountability among employees, leading to increased engagement and commitment to digital sustainability.

By evaluating the empowerment and alignment parameter, banks can identify gaps in employee understanding, skills, and resources. It helps banks recognize areas where communication and training may need improvement, enabling them to bridge the gap between the digital vision and employee capabilities. Empowering employees and aligning their roles and responsibilities with the digital strategy enhances their effectiveness in driving digital sustainability and ensures a cohesive and coordinated approach to digital transformation throughout the organization.

Finally, the parameter of 'Openness & Transparency' holds great significance as it assesses how open businesses are in understanding that their brands are public property and the importance of being transparent about their ethics, responsibilities, decisions, and practices. In today's era, where corporate responsibility and ethical considerations are gaining heightened importance among stakeholders, companies are expected to demonstrate openness and transparency in their operations. Being open and transparent means recognizing that the public has a vested interest in understanding a company's values, actions, and impact on society and the environment. It involves being forthcoming about the ethical principles that guide the organization, taking ownership of its actions and their consequences, and proactively communicating with stakeholders about its decisions, policies, and practices. This parameter examines the extent to which a company embraces openness and transparency in its operations. It evaluates how effectively a company communicates its values and ethical standards to internal and external stakeholders. This includes being transparent about the company's corporate social responsibility initiatives, sustainability practices, and the measures it takes to address environmental and social impacts.

Furthermore, the parameter assesses the transparency of a company in its dealings, both internally and externally. It examines the extent to which a company openly communicates with employees, customers, suppliers, and other stakeholders. This involves providing clear and accessible information, fostering open dialogue, and responding to inquiries and concerns in a timely and transparent manner.

In the digital space, where information spreads rapidly and reputations can be shaped or damaged quickly, managing reputation with openness and transparency is vital. This parameter evaluates how well a company manages its digital reputation, including its online presence, response to feedback and criticism, and its approach to handling issues or crises in a transparent manner. By evaluating the openness and transparency parameter, companies, including banks, can demonstrate their commitment to responsible and ethical practices. It enables them to build trust and credibility among stakeholders, including customers, investors, employees, and the broader public. Openness and transparency not only enhance a company's reputation but also contribute to long-term sustainability and success by fostering positive relationships, attracting and retaining customers, and creating a culture of integrity and trust.

The Korn Ferry Digital Sustainability Index 2017 provides a comprehensive assessment tool for understanding how companies embrace digital transformation and maintain digital sustainability. By evaluating agility, connectivity, discipline and focus, empowerment and alignment, and openness and transparency, the index offers a comprehensive view of a company's digital health. It helps organizations identify strengths and weaknesses in their digital transformation journey and provides insights to improve their digital strategies.

6.5 COMPARING THE KORN FERRY DIGITAL SUSTAINABILITY INDEX 2017 TO THE DIGITAL SUSTAINABILITY INDEX

While both the Korn Ferry Digital Sustainability Index and the Digital Sustainability Index offer valuable frameworks for evaluating digital sustainability, there are potential gaps that need to be addressed. One such gap is the need for more industry-specific and customized assessments, particularly in the banking sector. Banks operate in a highly regulated environment, requiring frameworks that consider industry-specific challenges, regulations, and compliance requirements. This highlights the need for industry-specific adaptations of existing frameworks to ensure a tailored assessment of digital sustainability in the banking sector.

Another gap is the need for a stronger focus on customer-centricity within the banking industry. While both frameworks touch on customer experience, there is a need for a more explicit emphasis on how banks leverage digital technologies to enhance customer engagement, deliver personalized services, and provide seamless omnichannel experiences. Customer expectations are rapidly evolving, and prioritizing customer-centricity is crucial in digital sustainability strategies.

Additionally, the frameworks may not explicitly address cybersecurity, data privacy, and the interconnectedness of banks with other stakeholders in the digital ecosystem. With banks relying more on digital technologies and data-driven insights, it becomes essential to evaluate how well they protect customer data, mitigate cybersecurity risks, adhere to data privacy regulations, and leverage partnerships to drive digital innovation.

6.6 DIGITAL TRANSFORMATION FOR SUSTAINABILITY FRAMEWORK

To bridge these gaps, a more comprehensive Digital Transformation for Sustainability Framework in the banking industry is proposed. Parameters such as transformation potential, access potential, enabling potential, adaptive potential, and governance are crucial in evaluating digital sustainability. This framework focuses on the strategic use of digital technologies, enhancing access for all stakeholders, building a competent workforce, meeting evolving customer expectations, and ensuring effective governance. By evaluating these parameters, banks can assess their readiness for digital transformation, identify areas for improvement, and develop a roadmap for a truly digital and sustainable banking industry.

Let us discuss these parameters in detail, reflecting on their meaning and impact on digitalization in banking.

6.6.1 Transformation Potential

The 'Transformation Potential' parameter evaluates the bank's strategic utilization of digital technologies to enhance and innovate banking operations. It involves identifying opportunities where digital technologies can be employed to either replace existing operations or integrate with analog resources to create new capabilities. By embracing digital transformation, banks can achieve various benefits. Firstly, they can enhance operational efficiency by automating manual processes, reducing paperwork, and streamlining workflows. This not only saves time but also minimizes errors and improves overall productivity. Secondly, digitalization allows banks to optimize costs by eliminating redundant or underutilized resources. For instance, closing down physical branches with low footfall and shifting toward digital channels can lead to significant cost savings in terms of maintenance and staffing. Moreover, digital transformation enables banks to deliver better services and experiences to their customers. Developing a digital platform for account opening, for example, simplifies and accelerates the onboarding process, resulting in improved customer experience and increased customer acquisition. By leveraging digital technologies, banks can offer self-service options, personalized recommendations, and convenient access to banking services, thereby enhancing customer satisfaction and loyalty. Furthermore, digital transformation can unlock new revenue streams and business opportunities for banks. By exploring innovative solutions and partnerships, banks can create novel offerings and expand their market reach. For instance, collaborating with fintech companies to provide digital payment solutions or launching mobile banking applications can attract new customer segments and drive revenue growth.

6.6.2 Access Potential

The 'Access Potential' parameter focuses on expanding banking opportunities for external parties and maximizing the benefits of digital transformation. It entails enhancing accessibility to banking services through increased connectivity and the provision of seamless digital experiences. To improve access potential, banks need to ensure reliable and efficient connectivity. This can involve enhancing the network infrastructure to provide high-speed internet connectivity at branches, ensuring that customers can access online banking services without disruptions. For example, ensuring a minimum internet speed of 8–16 MBPS at branches can facilitate smooth and fast digital interactions. Additionally, upgrading digital servers for all bank applications and websites is crucial for enhancing service delivery. By investing in robust and scalable server infrastructure, banks can handle increasing digital traffic, offer quick response times, and ensure a seamless customer experience. This includes optimizing website performance, minimizing downtime, and ensuring secure and reliable access to digital services. Furthermore, access potential aims to ensure that digital transformation extends to all stakeholders, including customers, partners, and employees. It involves making digital banking solutions user-friendly, intuitive, and easily accessible across various devices and channels. This may include developing user-friendly mobile banking applications, implementing responsive web design, and offering multichannel support to cater to diverse customer preferences. By maximizing access potential, banks can create a more inclusive banking ecosystem. Customers can conveniently access banking services anytime and anywhere, which improves their overall banking experience. Furthermore, improved access potential facilitates financial inclusion by providing underserved populations with access to digital banking services, empowering them to participate in the formal financial system.

6.6.3 Adaptive Potential

The parameter of 'Adaptive Potential' assesses a bank's ability to adapt to evolving customer expectations and changing market dynamics. It reflects the bank's capacity to adjust and refine its conventional business models to align with emerging trends and customer demands. In today's rapidly evolving digital landscape, customer expectations are constantly changing. Customers now seek more personalized and convenient banking experiences, and they expect seamless interactions across multiple channels. Banks with high adaptive potential can proactively identify these evolving needs and modify their value propositions, products, and services accordingly. One aspect of adaptive potential is the ability to tweak conventional business models. This involves rethinking traditional approaches and exploring innovative ways to deliver value to customers. Banks must analyze their existing business models and identify areas where they can introduce digital solutions, automation, or streamlined processes to enhance efficiency and customer satisfaction. For example, investing in a unified platform that integrates various banking services can provide customers with a seamless and convenient experience across multiple touchpoints. Process design economics is another consideration within adaptive potential. It involves evaluating the cost-effectiveness and efficiency of existing processes and identifying opportunities for optimization. Banks need to leverage digital technologies and data analytics to streamline their operations, reduce manual effort, and improve process efficiency. By doing so, they can better meet customer expectations for faster, more efficient, and error-free banking experiences. Anticipating and responding to changing customer needs and market dynamics is a crucial aspect of adaptive potential. Banks must actively monitor market trends, emerging technologies, and customer preferences to stay ahead of the curve. This requires continuous market research, customer feedback analysis, and staying abreast of industry developments. By anticipating and understanding these changes, banks can proactively adapt their strategies, offerings, and customer experiences to maintain a competitive edge. To enhance adaptive potential, banks may need to invest in technology solutions, such as artificial intelligence, machine learning, and data analytics, to gain deeper insights into customer behavior and preferences. These technologies can help banks identify patterns, forecast trends, and personalize their offerings to cater to individual customer needs.

6.6.4 Governance

The parameter of 'Governance' plays a crucial role in ensuring sustainable digitalization within banks. It focuses on how well the principles of digitalization are integrated into the bank's governance approach and organizational structure. Effective governance in the context of digital transformation requires the establishment of a clear and accountable leadership structure. It involves appointing a core team of responsible leaders who can drive the digitalization efforts, set strategic objectives, and provide guidance and support to the organization. These leaders should have a deep understanding of digital technologies, market trends, and customer expectations. The governance approach should be designed to facilitate digitalization rather than hinder it. It should create an environment that encourages innovation, experimentation, and agile decision-making. This requires aligning the governance framework with the goals and requirements of digital transformation, ensuring that policies, procedures, and decision-making processes support the adoption and implementation of digital initiatives. Effective governance also involves managing risks associated with digital transformation. As banks increasingly rely on digital technologies and data-driven processes, they must be proactive in identifying and mitigating emerging risks such as cybersecurity threats, data breaches, and privacy concerns. This requires implementing robust risk management practices, establishing clear protocols for data protection, and regularly monitoring and evaluating the effectiveness of security measures. Furthermore, governance encompasses compliance with regulations and industry standards. Banks operate within a highly regulated environment, and digitalization efforts must align with relevant laws, regulations, and compliance requirements. Governance frameworks should ensure that digital initiatives adhere to regulatory guidelines, protect customer data, and promote ethical practices. Ethics and integrity are integral to sustainable digitalization. Governance frameworks must emphasize maintaining high ethical standards throughout the digital transformation journey. This includes promoting transparency, accountability, and responsible data management practices. Banks should prioritize the ethical use of customer data, ensure informed consent, and establish robust data privacy measures to protect the interests of their customers.

In essence, the process of digital transformation in the banking industry transcends mere technology adoption. It necessitates strategic thinking, substantial cultural shifts, and a readiness to embrace change. Through the evaluation of the parameters discussed in this Digital Transformation Sustainability Framework, banks can assess their preparedness for digital transformation, identify areas that require improvement, and chart a course toward becoming truly digital. It is important to recognize that digitalization is not an ultimate goal but a means to achieve a more efficient, customer-centric, and sustainable banking industry. By evaluating parameters such as transformation potential, access potential, enabling potential, adaptive potential, and governance, banks can gain insights into their digital capabilities and identify opportunities for improvement.

By embracing this framework and evaluating these parameters, banks can gain a comprehensive understanding of their digital health, identify areas for improvement, and develop a roadmap for their digital transformation journey. This will enable them to achieve greater operational efficiency, enhance customer experiences, stay competitive in the digital era, and ultimately contribute to a more efficient, customer-centric, and sustainable banking industry.

6.7 CONCLUSION

As we gaze into the future of the banking industry, the harmonious fusion of sustainability and digitalization emerges as a crucial determinant of success. This integration is far more than just a corporate trend; it forms the crux of the strategic roadmap guiding financial institutions toward longevity, resilience, and relevance in an ever-evolving global landscape.

The past decade has borne witness to an unprecedented acceleration in digital technology adoption within banking. The incorporation of these digital technologies has revolutionized the industry, imbuing it with a level of convenience and efficiency previously thought unattainable. From enabling seamless digital transactions to implementing robust data analytics, from rolling out intuitive mobile banking applications to utilizing AI for personalized services, digital innovations have redefined the banking experience for millions of customers worldwide.

However, like two sides of a coin, the 'Digital Transformation for Sustainability' journey has also cast a significant shadow on sustainability. Digital technologies, while facilitating innovation, also consume considerable resources, from the electricity powering data centers to the rare earth metals used in hardware. The resultant carbon footprint and e-waste have raised legitimate concerns regarding the long-term viability and the environmental impact of our digital age. Therefore, it's imperative that we assess this impact meticulously and unflinchingly.

In this context, identifying areas where digital capabilities can contribute positively to sustainability goals becomes crucial. Digitalization, when thoughtfully executed, can indeed function as a sustainability enabler. Think of paperless transactions reducing deforestation, or digital platforms promoting ethical financial practices. Even simple measures such as email statements and digital signatures can cumulatively have a considerable impact on a bank's carbon footprint.

To steer through this complex landscape, we propose the development of a 'Digital Transformation for Sustainability Framework'. This comprehensive tool aims to assist banks in weaving sustainability into the very fabric of their digital strategies. This framework can facilitate an ongoing, iterative process wherein banks can track their sustainability journey, measure progress, and identify areas necessitating improvement.

In applying this framework, parameters like 'transformation potential' and 'access potential' emerge as useful metrics to gauge a bank's digital sustainability performance. 'Transformation potential' can shed light on whether digital technology is being used as a substitute or in combination, thereby assessing the effectiveness of the digital solutions in place. On the other hand, 'access potential' can determine the bank's proficiency in broadening banking opportunities for external parties through digital transformation, thereby enhancing its societal impact.

However, it's essential to understand that these parameters are not merely evaluative tools; they should serve as guiding principles in shaping a bank's digital strategy. Banks need to prioritize sustainability, not just in their mission statements, but in their operational blueprint. This involves embracing green IT practices, making ethical decisions regarding resource consumption, and focusing on inclusive digital services that reach even the most marginalized sections of society.

As we move further into the digital era, we must remind ourselves of our collective responsibility. The pursuit of convenience and innovation should not come at the expense of our planet and our future generations. Therefore, the banking industry, at the heart of the world economy, must champion the cause of integrating sustainability into digitalization. By doing so, it can set a powerful precedent for other sectors to follow, contributing to a sustainable, inclusive, and prosperous future for all. The integration of sustainability and digitalization in the banking industry is not just a strategic choice; it is a necessity, an imperative that we cannot afford to ignore.

REFERENCES

Bányai, A. (2018). Analysis of digitalization and sustainability. *Business Ethics and Leadership*, 2(2), 14–25.

Belkhir, L., & Elmeligi, A. (2018). Assessing ICT global emissions footprint: Trends to 2040 & recommendations. *Journal of Cleaner Production*, 177, 448–463. https://doi.org/10.1016/j.jclepro.2017.12.239

Bibri, S. E. (2018). The IoT for smart sustainable cities of the future: An analytical framework for sensor-based big data applications for environmental sustainability. *Sustainable Cities and Society*, 38, 230–253. https://doi.org/10.1016/j.scs.2017.12.034

Bibri, S. E., & Krogstie, J. (2017). Smart sustainable cities of the future: An extensive interdisciplinary literature review. *Sustainable Cities and Society*, 31, 183–212. https://doi.org/10.1016/j.scs.2017.02.016

Birch, D., & Chiappetta, M. (2018). Towards ambient accountability in financial services: Shared ledgers, translucent transactions and the legacy of the great financial crisis. *Journal of Digital Banking*, 3(2), 89–97.

Dwivedi, Y. K., Hughes, L., Kar, A. K., Baabdullah, A. M., Grover, P., Abbas, R., & Wade, M. (2022). Climate change and COP26: Are digital technologies and information management part of the problem or the solution? An editorial reflection and call to action. *International Journal of Information Management*, 63, 102456. https://doi.org/10.1016/j.ijinfomgt.2021.102456

Frolov, V., Bański, A., Berezin, E., Zhegalin, A., & Yatsuk, V. (2019). Using digital technologies to achieve sustainable development goals. In G. Kerimray, B. Daniyarov, & A. Inglezakis (Eds.), *Green energy and technology* (pp. 1–27). Springer.

Geels, F. W. (2018). Disruption and low-carbon system transformation: Progress and new challenges in socio-technical transitions research and the multi-level perspective. *Energy Research & Social Science*, 37, 224–231. https://doi.org/10.1016/j.erss.2017.10.010

Gomber, P., Kauffman, R. J., Parker, C., & Weber, B. W. (2018). On the fintech revolution: Interpreting the forces of innovation, disruption, and transformation in financial services. *Journal of Management Information Systems*, 35(1), 220–265. https://doi.org/10.1080/07421222.2018.1440766

Hilty, L. M., & Aebischer, B. (Eds.). (2015). *ICT innovations for sustainability*. Springer. https://doi.org/10.1007/978-3-319-09228-7

Köhler, G., Walnum, H. J., Aall, C., & Throne-Holst, H. (2019). Can digitalization by default be counterproductive? The case of mobility services. *Journal of Cleaner Production*, 234, 132–144.

Parida, V., Sjödin, D., & Reim, W. (2019). Reviewing literature on digitalization, business model innovation, and sustainable industry: Past achievements and future promises. *Sustainability*, 11(2), 391. https://doi.org/10.3390/su11020391

Schaltegger, S., & Wagner, M. (2017). *Managing the business case for sustainability: The integration of social, environmental and economic performance*. Routledge. https://doi.org/10.4324/9781351280525

Schwab, K. (2017). *The fourth industrial revolution*. Crown Business.

UNEP. (2018). *The weight of cities: Resource requirements of future urbanization*. United Nations Environment Programme.

7 Leveraging Higher Education and Sustainability Through Artificial Intelligence

A Holistic Approach

Shivender Rahul and Veena Roshan Jose

7.1 INTRODUCTION

The 21st century is characterized by unprecedented global challenges that threaten human civilization and planetary well-being. Most important are the existential threats of climate change, environmental destruction, biodiversity loss, and ecological breakdown (Steffen et al., 2015). Climate change, manifested through rising temperatures, extreme weather events, and collapsing ecosystems, can undo development gains and exacerbate poverty, migration, and conflict (IPCC, 2022). Unbridled economic growth based on unsustainable production and consumption has breached multiple planetary boundaries, endangering Earth's operating space for flourishing human societies (Rockström et al., 2009; UNEP, 2019).

Tackling these sustainability crises requires rapid systemic shifts at the technological, economic, political, and social levels (Khandelwal et al., 2021). Higher education institutions have a profound obligation to enable this sustainability transition (Cortese, 2003). Beyond imparting subject matter knowledge, universities must foster within students the critical thinking skills, systemic mindsets, ethical orientation, and change agency needed to create regeneration and transformation across society (Sterling et al., 2022). The traditional role of higher education in fostering economic growth must expand toward nurturing holistic development and well-being within planetary boundaries (Walczak et al., 2021).

The United Nations has outlined sustainability education as an important enabler for achieving the Sustainable Development Goals (SDGs) and building just, resilient post-carbon societies (UNESCO, 2017). Education for sustainable development (ESD) calls for pedagogical innovation that provides relational, experiential learning and connects global issues to local real-world complexities in a transdisciplinary manner (Glassey & Haile, 2022). ESD aims to develop sustainability competencies, including anticipation, normativity, collaboration, systems thinking, self-awareness,

 DOI: 10.1201/9781003638506-7

and integrated problem-solving (Rieckmann, 2018). However, progress on ESD has been slow and often relegated to specific programs rather than embedded within formal and hidden curricula holistically.

Transformative sustainability education requires institutions to reorient not just teaching and learning but also governance, operations, campus culture, and societal relationships (Stephens et al., 2008). Leading universities are integrating sustainability across infrastructure, resource management, procurement, human resources, student engagement, and community partnerships (Lozano et al., 2015). Yet, the scale of implementation often remains incremental rather than the radical reimagination needed for thriving future ecosystems (Sharp, 2002).

Meanwhile, the rise of advanced digital technologies is radically transforming society, economics, and education. In particular, recent breakthroughs in artificial intelligence (AI) herald immense opportunities as well as risks. The field of AI encompasses technologies enabling machines to perform tasks requiring human cognition, adapt to uncertainties, learn from data patterns, undertake knowledge work, and interact naturally using language (Kaplan & Haenlein, 2019). AI-based tools are enhancing efficiency, productivity, and innovation across diverse industries, including agriculture, finance, healthcare, and more (Siddiqa et al., 2022). Higher education is equally being reshaped by AI applications for administration, teaching, and learning (West, 2019). Integrating AI also creates new pathways for advancing sustainability within higher education in impactful ways. This chapter examines the nexus between burgeoning AI technologies and the pursuit of urgent sustainability imperatives in higher education institutions. It analyzes how the capabilities of AI systems can optimize ecological sustainability, digital equity, institutional effectiveness, and social justice on campuses. The chapter investigates the implementation challenges and proposes an integration framework. For the purposes of this chapter, the term "Sustainability" is defined using the United Nations holistic framework encompassing environmental, social, and economic dimensions pursued through good governance and peace (United Nations, 1987).

The analysis is grounded in scholarly discourse on AI, ESD, and technology impact assessments. It synthesizes insights from use cases, institutional reports, and global surveys to provide a comprehensive, evidence-based perspective. This timely interrogation of AI's double-edged potential contributes vital insights for leveraging AI responsibly and effectively to accelerate higher education's sustainability transition amid converging digitalization and climate disruptions.

The chapter is structured into five sections. The first section reviews relevant literature on sustainability in higher education and AI applications. The second section outlines the mixed methods research approach. The third section presents key findings on the potential sustainability benefits and risks of AI. The fourth section proposes an AI integration framework for higher education institutions. Finally, the fifth section discusses limitations and draws conclusions on the role of AI in transforming scholarship, student development, and institutional sustainability.

7.2 REVIEW OF LITERATURE

The authors have extensively reviewed the available literature on the subject matter and have synthesized scholarly work across two domains – sustainable higher education and applications of AI – to foreground this study's analytical foundations regarding their convergence.

7.2.1 Sustainability in Higher Education

The role of higher education institutions in advancing sustainable development has received extensive research attention. Education is pivotal for nurturing sustainability consciousness regarding complex ecological and social justice challenges (Barth et al., 2007). Sustainability education fosters cognition of intricate causes of unsustainability like colonialism, capitalism, patriarchy, and racial inequities underpinning climate change and biodiversity collapse (Kopnina, 2020). Beyond cognitive skills, it develops social-emotional competencies like empathy, mindfulness, ethics, and critical self-reflection needed for cooperative resolutions (Shephard, 2015).

Initially, sustainability efforts in universities focused on campus greening through initiatives like emissions reduction, renewable energy, green buildings, water conservation, and waste management (Ralph & Stubbs, 2014). However, scholars argue that isolated green practices fail to foster deeper cultural shifts toward sustainability values in students, faculty, and administration (Jones et al., 2010). Embedding sustainability requires addressing the hidden curriculum of disciplinary silos, competitive individualism, and disconnected theory application prevalent in higher education (Glassey & Haile, 2022). Thus, contemporary scholarship examines the holistic integration of sustainability within formal, informal, and operational aspects of higher education (Lozano et al., 2015). Formal curriculum reviews assess the incorporation of sustainability concepts in courses across programs, including professional disciplines like engineering, business, and law (Christensen et al., 2011). Case studies reveal how pedagogical innovations like experiential learning, problem-based learning, and community projects provide contextual, relatable sustainability education (Sipos et al., 2008). Transdisciplinary programs integrating social and technical disciplines are highlighted for instilling systems thinking abilities (Balsiger et al., 2017).

Research also stresses the need for policies, organizational culture, and stakeholder engagement that align institutional practices with sustainability values espoused in the formal curriculum (Shiel & Williams, 2016). Assessing sustainability learning outcomes for graduates and conducting campus sustainability audits can foster continuous improvements (Yarime & Tanaka, 2012). Scholars conclude that holistic integration of operations, governance, stakeholder perspectives, and formal/informal learning is essential for impactful sustainability education (Cotton et al., 2009). This requires cultural transformation and reorienting the purpose of higher education itself toward ecological flourishing and social justice (Sterling et al., 2022).

7.2.2 AI Applications in Higher Education

The rapid evolution of AI and educational technologies is transforming teaching and learning. Holmes et al. (2019) outline five important capabilities of AI-driven tools: predicting student outcomes using analytics, adapting to individual learners, providing personalized and detailed feedback, customizing learning experiences, and automating administrative processes. Examples include intelligent tutoring systems, adaptive courseware, student success algorithms, and conversational agents (Roll & Wylie, 2016).

Scholarship finds that properly designed and deployed AI can enhance learning, lower costs, and improve access and outcomes, especially for at-risk students (Escueta et al., 2017). Woolf et al. (2013) demonstrate that an AI teaching assistant improved student problem-solving in physics courses. AI-enabled personalized learning is found to be as or more effective than traditional methods (Yu et al., 2017). Beyond academics, AI chatbots provide admissions guidance, course selection advice, and mental health support to students AI also assists educators by automating assessment and feedback, thereby expanding human teaching capacity for higher-value tasks (Zawacki-Richter et al., 2019). However, researchers caution that careless implementation of AI in education risks severe detrimental outcomes. Knox (2020) warns that extensive data harvesting and surveillance jeopardize student privacy. Williamson (2018) notes that edtech innovations often lack rigorous evaluation of their pedagogical and social impacts. Scholars highlight the need for transparency, ethics, and human oversight in AI system design (Knox et al., 2020). Luckin et al. (2016) argue that AI should complement and augment rather than replace human teachers' social role. In general, nuanced examinations revealing AI's pros and cons are vital for informed adoption.

Thus, the existing literature highlights AI and automation's complex tensions between efficiency and precarity, innovation, and inequality. It reveals that the technology itself is not deterministic; thoughtful regulations and policies tailored for socioeconomic contexts can shape more inclusive outcomes. This underscores the need for evidence-based governance and coalitions between technology leaders, policymakers, and civil society to create good AI societies. The risks ahead require collective responsibility. Ultimately, AI should serve shared values and aspirations, not just commercial motives. Its historic potential can only be actualized through foresight, ethics, and compassion toward our common humanity.

7.3 RESEARCH METHODOLOGY

This study employs an explanatory sequential mixed methods approach to assess and validate the role of AI in promoting sustainability across diverse higher education institutions worldwide. Combining qualitative exploratory assessments and quantitative validation provides comprehensive insights into AI's sustainability benefits and adoption barriers that neither method alone could produce (Halcomb & Hickman, 2015). This pragmatic approach leverages the strengths of multiple data sources for a holistic understanding (Feilzer, 2010). Secondly, an analysis of AI integration initiatives, sustainability reports, and published studies from over 20 higher education institutions worldwide was undertaken. The sample included public and private universities, research-intensive campuses, liberal arts colleges, and community colleges in both developed and emerging economies. Content analysis identified common patterns, use cases, and program outcomes regarding AI's sustainability benefits and challenges.

Finally, important findings were derived by triangulating the insights from the qualitative and quantitative strands to develop a comprehensive assessment of the status, opportunities, and concerns involved in leveraging AI for advancing sustainability in higher education institutions.

7.4 RESEARCH FINDINGS

The sequential explanatory mixed methods analysis reveals that AI integration within higher education institutions worldwide has shown promising benefits but also poses formidable risks requiring mitigation for successful adoption. While the applied study generated significant empirical insights, further research through longitudinal assessments can deepen understanding of AI's evolving impact on higher education sustainability over time.

One potential area of inquiry is a multicampus study tracking sustainability outcomes from AI adoption using key performance indicators. As institutions implement and refine AI applications in learning, operations, and administration, regularly collecting data on metrics like energy consumption, retention rates, fossil fuel dependence, and community engagement could reveal long-term patterns. Comparing AI sustainability integration efforts and results between research universities, liberal arts colleges, community colleges, and institutions in both developed and emerging economies may uncover contextual success factors, barriers, and best practices.

Surveying student perceptions of AI technologies through participatory research would provide another valuable viewpoint. Students are the primary stakeholders who experience AI's benefits and risks firsthand. Participatory assessment tools like user experience walks, focus groups, and co-creation workshops could illuminate student needs and concerns. Capturing diverse student voices across socioeconomic, cultural, and accessibility dimensions is critical for holistic evaluation and inclusive AI design. Cross-institution collaborations can enable larger-scale participatory AI impact studies. Additionally, in-depth case studies assessing the curricular integration of AI for sustainability education could offer pedagogical insights. This may examine implementing AI simulations, learning analytics, adaptive courseware, and conversational AI across disciplines from engineering to social sciences. Multi-year analysis can reveal how AI influences the development of systems thinking and other sustainability competencies compared to traditional teaching methods. Facilitating educator communities to share effective practices across institutions can accelerate pedagogical innovation.

Interdisciplinary research partnerships between universities, technology firms, and community organizations can pilot localized AI applications that address specific sustainability challenges. For instance, developing machine learning tools to optimize renewable energy systems tailored for microclimates and demographics of a university town. Or co-creating AI social services for underserved populations by pairing company expertise with community insights. Such collaborative design thinking can yield context-specific AI solutions while also providing students with hands-on experiential learning. Exploring symbiotic partnerships between higher education institutions and industries developing ethical AI systems could enable constructive technology governance.

Universities have the mandate to steward AI for social good, while companies possess the capacity to rapidly translate research into solutions at scale. But balancing academic freedom, public interest, and technological innovation requires new cooperation models. Joint policy labs, adaptive regulatory sandboxes, and open intellectual property ecosystems may foster responsible AI ecosystems.

Given automation's broad societal impacts, interdisciplinary inquiry assessing AI's effects on livelihoods, inclusion, and well-being is imperative. Assessing gender, racial, and ability dimensions across who designs, regulates, benefits from, and could be disadvantaged by AI systems is vital to inform just policies and interventions. Research should move beyond a narrow techno-economic focus toward centering human dignity and the right to meaningful work.

In principle, ongoing critical yet constructive scholarship can shape wise pathways for artificially intelligent technologies to empower universities' missions of fostering sustainability, equity, and human development rather than inadvertently undermining these goals. However, realizing AI's potential requires asking difficult questions, making conscious choices, and continually connecting innovation with ethical responsibility as society enters uncharted waters of exponential technologies.

Hence, this timely mixed-methods study illuminating the promises and perils of AI adoption for advancing sustainability in higher education institutions marks only the inception of an extensive, long-term research agenda as the AI era unfolds. The rise of ubiquitous smart systems calls for vigilance, courage, and wisdom within academia and policy ecosystems to cultivate benevolent technologies for the common good. Through collaborative envisioning and transparent public discourse, universities could yet guide digital transformations toward more just, compassionate, and ecologically thriving futures. With technological change quickening, the window for conscious choice is fleeting but profound. Further research must proceed judiciously yet ambitiously to light the way ahead.

7.5 MAJOR SUSTAINABILITY BENEFITS

The prominent benefits of AI adoption for accelerating higher education sustainability are identified as follows.

7.5.1 Energy and Resource Optimization

The most widely observed sustainability impact of AI involved significantly optimizing energy consumption, water usage, and waste generation through smart systems, sensors, and automation. AI-enabled heating, cooling, and lighting management alone achieved 10–25% reductions in campus energy usage by dynamically responding to occupancy and environmental conditions. Example technologies included machine learning–based Building Management Systems (BMSs), lighting motion sensors, and AI-powered energy storage and microgrid integration. Water conservation was also enhanced through leakage monitoring and intelligent sprinklers.

7.5.2 Enhanced Data-Driven Decision-Making

University administrators highly valued AI's ability to extract insights from vast, real-time data streams on student learning, facilities operations, and urban mobility for more effective planning and policymaking aligned with sustainability goals. For instance, learning analytics helps identify at-risk students needing timely assistance while also improving course outcomes and retention. Participants reported that data-driven infrastructure maintenance enabled extended asset lifespan and optimum capacity utilization. Intelligent dashboards were also created for tracking sustainability key performance indicators.

7.5.2.1 Personalized Student Support

Leveraging predictive modelling and adaptive learning algorithms, AI counselling and tutoring systems provided customized guidance, improving student success, especially for disadvantaged groups. By anticipating individual needs and intervention points, AI-enabled holistic development and reduced dropouts, thereby optimizing student-to-outcome ratios – an important sustainability metric for higher education. AI augmentation also expanded educators' reach and augmented human advising capacity.

7.5.2.2 Simulation-based Sustainability Education

Immersive simulations incorporating real-time data and virtual environments enabled experiential learning about sustainability challenges from climate impacts to social inequities at both the systemic and human levels. Such participatory learning built students' change agency and future thinking capabilities by bridging theory–reality gaps. Augmented field trips also provided onsite visualizations of infrastructure systems for engineering students.

7.5.3 Automating Processes

Routine administrative and operational processes across finance, admissions, learning design, maintenance, monitoring, and reporting were automated by AI to improve efficiency, consistency, and accessibility. By taking over repetitive tasks, AI systems enabled human time to be redirected toward value-adding analytics, services, and teaching for deeper sustainability education. AI was also used for automated documentation of sustainability initiatives.

Alongside advantages, the study also identified the risks that could undermine AI's sustainability potential without mindful management.

7.6 THE MAJOR RISKS IN THE ADOPTION OF AI

Low-income, first-generation, and disabled students with limited technology access or digital literacy skills could face exclusion from AI-enhanced learning and campus experiences, deepening the digital divide. Proactive efforts are needed to provide universal connectivity, skills, and learning tools. The risk of digital marginalization warrants deeper investigation to formulate equitable AI policies and interventions. Mixed methods studies assessing how students across socioeconomic statuses, racial groups, genders, and abilities differentially experience barriers to accessing and utilizing AI systems would highlight needs. Comparing challenges like device availability, internet connectivity, digital fluency, disability accommodation, and cultural aspects could inform the universal design of inclusive AI.

Institutions must also proactively evaluate whether AI adoption improves or worsens completion rates and achievement gaps between privileged and underserved students. Learning analytics and adaptive learning systems could personalize support for disadvantaged learners. However, biased algorithms that overlook student diversity risk harming inclusion. Participatory assessments co-creating AI with marginalized communities can surface unseen biases and barriers. Policy labs engaging technologists, social scientists, and grassroots advocates can prototype reforms addressing algorithmic discrimination and digital accessibility.

Researching effective pedagogies and change management strategies for developing staff and student digital capabilities as AI permeates campuses is equally vital.

Digital fluency programs contextualized to community needs using peer coaching and project-based learning can build computational skills for diverse learners. Curriculum reviews should integrate digital literacy and ethical reasoning across subjects to prepare students to be empowered yet critical citizens in an increasingly automated, predictive society.

As automation transforms employment, analyzing impacts on non-college-educated youth and funding life-long learning programs to enable dignified livelihoods will be critical. AI is forecast to automate repetitive entry-level jobs that lower-income students depend on to afford higher education. Partnerships between universities, technical institutes, and industries can create apprenticeship pipelines for in-demand roles like AI technicians, which require vocational qualifications. Labor market reforms like youth allowances, mid-career transition support, and automation tax funding for community colleges can mitigate economic exclusion.

Researching how intelligent tutoring systems and conversational agents could provide personalized, scalable education to remote regions lacking qualified teachers is worthwhile. While cautioning against simplistic techno-solutionism, AI does hold the potential to expand access and localize learning where teacher shortages persist, given thoughtful design and cultural adaptation. Comparative trials assessing learning outcomes using AI versus traditional classroom teaching could guide context-appropriate, blended strategies. However, any use of AI in education should empower local teachers with digital tools, not replace them.

As campus environments grow increasingly instrumented with sensors, cameras, and learning software, safeguarding student privacy is imperative. Impact audits, participatory design processes, and transparency systems can identify risks early and maintain trust. Research should assess sentiment around 24/7 surveillance, data mining, and algorithmic manipulation to formulate rights-based oversight policies that uphold student agency and dignity. Institution-wide digital ethics curricula to build awareness of technology's influence on freedom, identity, and democracy is vital. Ultimately, realizing AI's benefits while mitigating its risks to equity and inclusion will require scholarship and policies that consciously center marginalized communities rather than merely optimizing technological efficiency. If designed collaboratively, AI could make quality education radically more accessible. However, if imposed blindly, it may encode biases that amplify injustice. By shining light on the shadows, research has a profound opportunity to guide AI systems that enable all students to flourish with compassion and wisdom.

In principle, realizing AI's potential for transformative good demands continual mindfulness, courage, and participatory co-creation between institutions and communities, on campus and beyond. In these times of uncertainty, scholarship must boldly envision technology's highest possibilities, while speaking difficult truths. The future merits nothing less than our full humanity.

7.7 ETHICAL SHORTCOMINGS

AI systems reflect the biases and blind spots of creators. Lack of oversight risks unethical outcomes, including manipulation, surveillance, automation bias, and loss of human dignity. Cultivating a strong ethical culture is critical.

7.7.1 Human Role Reduction

Excessive substitution of educators and advisors with AI risks diminishing the invaluable nurturing of student–teacher relationships and mentoring for holistic development. Wisdom traditions must shape deployment.

7.7.2 Lack of Regulation

The absence of oversight standards and risk assessment protocols for education technology procurement and implementation poses dangers of misuse, cybersecurity lapses, and rights infringement.

7.7.3 Non-Prioritization of Sustainability

With limited resources, sustainability use cases may not receive priority amid pressures for digital transformation and maximizing return on AI investments.

The qualitative risks illuminated tensions that must be addressed through values-based policies and community participation when adopting AI for sustainability, unlike technical aspects.

7.8 AI INTEGRATION FRAMEWORK

Synthesizing the findings, this study proposes INVEST – a comprehensive integration framework for higher education institutions to actualize the promise of AI for driving sustainability while mitigating risks. It comprises:

7.8.1 Inclusive Leadership

Given the cultural dimensions involved, distributed leadership spanning information technology, academics, facilities, finance, and student life must collectively steward the integration vision, adhering to institutional values and ethics. Democratizing technology governance is vital.

7.8.2 Needs Assessment

Each institutional division should evaluate priority pain points where AI could enhance sustainability and learning outcomes based on community input. Assessing technical feasibility and resource trade-offs is key before pilots.

7.8.3 Values Alignment

Proposed AI uses should undergo impact assessment for alignment with espoused values like accessibility, care, equity, transparency, and ecological justice using participatory processes. Course-correcting biases are essential.

7.8.4 Strategic Planning

An interdisciplinary working group of institutional leaders should formulate an AI integration roadmap guided by the needs and values alignment. It should detail phased use case rollout, considering resources and risks.

7.8.5 Technology Evaluation

AI tools and software being procured should be rigorously evaluated for data security, accessibility, integration with existing infrastructure, and mitigating bias, before vendor and technology selection. The phased rollout should assess the efficacy of AI use cases, monitor for unintended harms, and gather continuous community feedback to address concerns through adaptive policies as adoption advances. Institutions must report regularly on measurable sustainability and well-being improvements due to AI implementations, along with lessons learned through transparent public engagement.

This multidimensional INVEST framework can guide higher education leaders seeking to harness AI's potential holistically to further interdependent objectives of ecological sustainability, social equity, and community well-being. It calls for bringing shared wisdom to navigating the promises and perils of the AI era toward more just, compassionate, and flourishing futures.

7.9 DISCUSSION

This pragmatic mixed-methods study generated significant insights regarding the nexus of AI technologies and sustainability imperatives confronting higher education institutions at the onset of the 2020s. Current applications of AI encompassing automation, analytics, personalized advising, infrastructure optimization, and immersive simulation exhibit the considerable potential to advance campus sustainability across its environmental, social, economic, and cultural dimensions. However, the findings also reveal that AI is no silver bullet. Realizing transformational benefits involves grappling with risks of overreach, inequity, and dehumanization inherent in complex socio-technical systems through ethical frameworks and campus co-creation.

7.9.1 Important Study Contributions

Firstly, the study addresses a critical gap at the intersection of emerging AI systems and established sustainability education efforts, which has received limited scholarly attention despite immense implications. While previous studies have examined AI and sustainability separately, this chapter provides an original synthesis of their linkages in higher education.

Secondly, the proposed INVEST framework offers much-needed guidance for integrating AI holistically across academic, operational, and cultural domains in alignment with institutional missions. It brings together technical dimensions like needs analysis and infrastructure integration with human dimensions of participatory governance, values alignment, and ethical oversight for responsible innovation. The framework enables translating sustainability commitments into AI implementation strategies.

Thirdly, by highlighting pitfalls like digital marginalization, loss of wisdom, and over-reliance on technology, the study provides a timely caution as automation in the education sector accelerates. Findings advise that staying true to higher education's highest purposes of human development requires centering human dignity and ecological flourishing when adopting AI. Heeding this warning is critical as increasingly capable technologies hold both constructive and destructive potentials.

The research opens up several fruitful avenues for future inquiry. Comparative studies assessing sustainability outcomes from AI adoption across diverse universities could reveal success factors. Evaluating student perceptions of AI through participatory research would provide their vital perspective. Longitudinal studies tracking AI's sustainability impact over time using key indicators like carbon footprint, equity, and community well-being would yield instructive patterns. As the field matures, developing targeted AI sustainability applications based on institutions' needs and demographics could accelerate adoption. Partnerships with technology firms and foundations can facilitate solutions at scale.

7.10 CONCLUSION

In summary, this chapter reveals that AI holds considerable yet contingent potential to assist higher education's vital transition toward sustainability across scholarship, operations, and campus culture. With thoughtful oversight and ethics, AI tools can enhance learning, efficiency, and ecological modernization. However, pursuing AI's benefits cannot absolve leaders from continually upholding higher education's highest purposes of fostering holistic human development, wisdom cultivation, social justice, and ecological stewardship. Ultimately, the compass for navigating the promises and perils of exponential technologies must come from shared values, ethics, and our common hopes for more just, compassionate, and sustainable futures.

It is to be noted that AI constitutes neither an inherent boon nor a bane for higher education sustainability. Rather, its impacts depend on how institutions interpret, implement, and govern socio-technical changes based on their academic missions and humanity's common destiny within fragile, interdependent ecological systems. With astute foresight and moral courage, university leaders and communities can collaboratively leverage these unprecedented technologies to educate future generations capable of creating societies that are equitable, resilient, and harmonious with the natural world. However, they must listen to unheard voices, question unseen biases, and balance optimization with a wisdom of care.

In the final analysis, the merits of AI in advancing sustainability and equity will be contingent on centering shared hopes over fears, democratic ideals over control, and reconnecting technology with virtue, empathy, and conscience. Higher education institutions must think boldly yet carefully to illuminate the path ahead. Their stewardship can help unveil AI's best potential while averting its gravest perils, cultivating both sustainability and wisdom.

REFERENCES

Balsiger, J., Förster, R., Mader, C., Nagel, U., Sironi, H., Wilhelm, S., & Zimmermann, A. B. (2017). Transformative Learning and Education for Sustainable Development. *GAIA: Ecological Perspectives for Science and Society*, *26*(4), 357–359. https://doi.org/10.14512/gaia.26.4.15

Barth, M., Godemann, J., Rieckmann, M., & Stoltenberg, U. (2007). Developing Key Competencies for Sustainable Development in Higher Education. *International Journal of Sustainability in Higher Education*, 8(4), 416–430. https://doi.org/10.1108/14676370710823582

Christensen, C. M., & Eyring, H. J. (2011), *The Innovative University: Changing the DNA of Higher Education from the Inside Out*, John Wiley & Sons.

Cortese, A. D. (2003), The Critical Role of Higher Education in Creating a Sustainable Future, *Planning for Higher Education*, 31(3), 15–22.

Cotton, D. R. E., Bailey, I., Warren, M., & Bissell, S. (2009). Revolutions and Second-best Solutions: Education for Sustainable Development in Higher Education. *Studies in Higher Education, 34*(7), 719–733. https://doi.org/10.1080/03075070802641552

Escueta, M., Quan, V., Nickow, A. J., & Oreopoulos, P. (2017). Education Technology: An Evidence-based Review (NBER Working Paper No. 23744). National Bureau of Economic Research. https://doi.org/10.3386/w23744

Feilzer, M. Y. (2010). Doing Mixed Methods Research Pragmatically: Implications for the Rediscovery of Pragmatism as a Research Paradigm. *Journal of Mixed Methods Research, 4*(1), 6–16. https://doi.org/10.1177/1558689809349691

Glassey, J., & Haile, S. (2022), Transforming Sustainability Education: The Potential of Critical Pedagogy, *Journal of Sustainability Education*, 28, 1–23.

Halcomb, E., & Hickman, L. (2015). Mixed Methods Research. *Nursing Standard, 29*(32), 41–47. https://doi.org/10.7748/ns.29.32.41.e8858

Holmes, W., Bialik, M., & Fadel, C. (2019). *Artificial Intelligence in Education: Promises and Implications for Teaching and Learning*. Center for Curriculum Redesign.

IPCC (2022), Summary for Policymakers. In *Climate Change 2022: Impacts, adaptation and vulnerability* (pp. 3–33), Cambridge University Press.

Jones, P., Selby, D., et al., (Eds.)., (2010), *Sustainability Education: Perspectives and Practice Across Higher Education*, Routledge.

Kaplan, A., & Haenlein, M. (2019), Siri, Siri, in my Hand: Who's the Fairest in the Land? *On the Interpretations, Illustrations, and Implications of Artificial Intelligence, Business Horizons*, 62(1), 15–25.

Khandelwal, A. R., Chaturvedi, A., & Kanwar, M. V. (2021), Effectiveness Of Virtual Classroom: A Comparative Study of Theoretical and Practical Papers. *Webology*, 1735–188X, 18(4).

Knox, J. (2020), Artificial Intelligence and Education in China, *Learning, Media and Technology*, 45(3), 298–311.

Kopnina, H., (2020), Education for the Future? Critical Evaluation of Education for Sustainable Development Goals, *The Journal of Environmental Education*, 51(4), 280–291.

Lozano, R., Ceulemans, K., et al., (2015), A Review of Commitment and Implementation of Sustainable Development in Higher Education: Results from a Worldwide Survey, *Journal of Cleaner Production*, 108, 1–18.

Luckin, R., Holmes, W., Griffiths, M., et al., (2016), *Intelligence Unleashed, An Argument for AI in Education*, Pearson.

Nedelkoska, L., & Quintini, G. (2018), *Automation, skills use and training* (OECD Social, Employment and Migration Working Papers No. 202, 124 pp.). OECD Publishing. https://doi.org/10.1787/2e2f4eea-en.

Ralph, M., & Stubbs, W. (2014). Integrating Environmental Sustainability into Universities. *Higher Education*, 67(1), 71–90.

Rieckmann, M. (2018), *Learning to Transform the World: Key Competencies in ESD in Issues and Trends in Education for Sustainable Development* (pp. 39–59), UNESCO Publishing.

Rockström, et al. (2009). Planetary Boundaries: Exploring the Safe Operating Space for Humanity. *Ecology and Society*, 14(2), 32. https://doi.org/10.5751/ES-03180-140232 (open access).

Roll, I., & Wylie, R. (2016), Evolution and Revolution in Artificial Intelligence in Education, *International Journal of Artificial Intelligence in Education*, 26(2), 582–599.

Sharp, L. (2002). Green Campuses: The Road from Little Victories to Systemic Transformation. *International Journal of Sustainability in Higher Education, 3*(2), 128–145. https://doi.org/10.1108/14676370210422357

Shephard, K. (2015). *Higher Education for Sustainable Development*. Springer. https://doi.org/10.1057/9781137548412

Shephard, K. (2015a), Higher Education for Sustainability: Seeking Affective Learning Outcomes, *International Journal of Sustainability in Higher Education*, 9(1), 87–98.

Shephard, K. (2015b). *Higher Education for Sustainable Development*. Springer. https://doi.org/10.1057/9781137548412

Shiel, C., & Williams, A. (2015). Working Together, Driven Apart: Reflecting on a Joint Endeavour to Address Sustainable Development Within a University. In W. Leal Filho (Ed.), *Integrative approaches to sustainable development at university level* (pp. 425–447). Springer International Publishing. https://doi.org/10.1007/978-3-319-10690-8_30

Siddiqa, A., Hashem, I. A. T., et al., (2022), The Role of Artificial Intelligence in Achieving Sustainable Development Goals: A Systematic Review, *Mathematics*, 10(4), 596.

Sipos, Y., Battisti, B., & Grimm, K. (2008). Achieving Transformative Sustainability Learning: Engaging Head, Hands and Heart. *International Journal of Sustainability in Higher Education*, *9*(1), 68–86. https://doi.org/10.1108/14676370810842193

Steffen, W., Richardson, K., et al., (2015), Sustainability, Planetary Boundaries: Guiding Human Development on a Changing Plane, *Science*, 347(6223), 1259855.

Stephens, J. C., Hernandez, M. E., et al., (2008), Higher Education as a Change Agent for Sustainability in Different Cultures and Contexts, *International Journal of Sustainability in Higher Education*, 9(3), 317–338.

Sterling, S., Maxey, L., et al., (2022), *The Sustainable University: Progress and Prospects*, Routledge.

UNEP (2019), *Global Environment Outlook – GEO-6: Summary for Policymakers*, Nairobi.

UNESCO (2017), *Education for Sustainable Development Goals: Learning Objectives*, UNESCO.

United Nations. (1987). *Report of the World Commission on Environment and Development: Our common future*. Oxford University Press. https://sustainabledevelopment.un.org/content/documents/5987our-common-future.pdf

Walczak, K., Griffiths, T., et al., (2021), Positionality in the Light of Sustainability–Reflections on the Role of Higher Education Institutions in Sustainability Transformations, *Sustainability*, 13(9), 5063.

West, D. M. (2019), *The Future of Work: Robots*, AI, And Automation, Brookings Institution Press.

Williamson, B. (2018), Brain Data: Meditation, Mindfulness and the Capitalist Appropriation of our Neural Lives, *The Handbook of Neoliberalism* (pp. 322–332), Routledge.

Woolf, B. P., Lane, H. C., et al., (2013), AI Grand Challenges for Education, *AI Magazine*, 34(4), 66–66.

Yarime, M., & Tanaka, Y. (2012), The Issues and Methodologies in Sustainability Assessment Tools for Higher Education Institutions: A Review of Recent Trends and Future Challenges. *Journal of Education for Sustainable Development*, 6(1), 63–77.

Yu, H., Miao, C., Leung, C., White, T. J., & Lau, R. W. H. (2017). Towards AI-powered Personalization in MOOC Learning. *npj Science of Learning*, 2, 15. https://doi.org/10.1038/s41539-017-0016-3

Zawacki-Richter, O., Marín, et al., (2019), Systematic Review of Research on Artificial Intelligence Applications in Higher Education–Where are the Educators? *International Journal of Educational Technology in Higher Education*, 16(1), 1–27.

8 Contribution of AI in Education and Its Future Goals for Sustainable Economy

Vikas Godara, Xi Zhang, Yiqun Kou, Ye Yang, Hang Zheng, and Youmin Hu

8.1 OVERVIEW

Alan Turing's concept of "thinking machine" has now become prominent in many sectors, but not limited to the educational industry, start-ups, entertainment, industries, etc. Being focused on education, AI-based knowledge is communicated through these "thinking machines" (e.g., chatbots, Grammarly, ChatGPT) that are embedded with various elements like sensors, actuators, controllers, and internal programmable systems to meet cognitive behaviour. To achieve that, these machines need to be programmed with efficient rational agents based on the approach of rational thinking and actions. Educational robots with better rational agents can deliver appropriate study materials considering ethical uncertainty.

AI in education can overcome the gaps caused by factors such as national borders, distance, language barriers, insufficient study materials, and instructor–student interactions via virtual classes supported by chatting and other audio/video sharing platforms. This will facilitate better understanding between the instructors and students to share the best knowledge about the field.

Machine Learning (ML) is an essential part of developing requested statistical models for performing intelligent tasks as humans to have better cognitive behaviour developed by AI-based machines for delivering educational information. The education model is based on these behaviour data collected from the learning process of individuals, which includes students, instructors, parents and other stakeholders of educational institutes. Data mining is another approach based on which an intelligent tutoring system (ITS) can be framed, considering human cognitive behaviour. Through learning regression methods, data mining along with AI can enhance the overall process of virtual teaching and deliver knowledge to the needy. Apart from imparting knowledge of various fields through these virtual environments, AI can enhance administrative as well as management tasks in the workplace or institutes.

DOI: 10.1201/9781003638506-8

In colleges or schools, the use of AI tools can bring efficient results in the daily workloads of different departments, neglecting hectic paperwork. AI tools implemented for account sections, student or staff attendance records, instructor–student feedback, student score cards and performances. Students can have better interaction platforms through AI technologies, which include virtual reality (VR), chatbots, audio/video conferencing, file storage and robotic instructors in classes and laboratories. All the tasks in the examination section can be adapted to the aforementioned AI tools and technologies to have better processes, which may include student performance details, test results, activity participation and achievement records, attendance records of staff and students, virtual complaint books, faculty records, previous student records, virtual feedback files and other necessary files. AI can help in generating student performance by checking individual contributions in various fields. AI tools turn out to be helpful in plagiarism checking of assignments, publications or other materials submitted by students to their instructors.

Students can access these online platforms to attend missed lectures or to get answers to their queries whenever and wherever required. All these study materials and student information can be obtained when the need arises (Khandelwal et al., 2021). Many AI tools like chatbots are proving effective in delivering education; for example, Computer-Assisted Language Learning (CALL) is helpful to students for learning new languages by providing various instructions directing them to make correct sentences. On the other hand, AIWBES is becoming a powerful tool in AI-based education, where the role of a teacher is played by machines to deliver lectures similar to those of teachers. This is achieved by studying various teachers of the respective field through AI/ML. It is a model-based approach where an individual student's target, preferences and level of understanding of the subject are taken into consideration. Through this model, a teacher is able to answer various queries from students. Students can improve learning through chatbots or other AI platforms designed with suitable models and structures, thus enhancing the performance of tutors. AI-based materials are customized, considering the capabilities of individuals.

8.1.1 Education–Industry Bridge Through AIEd

Companies dealing with the education sector have launched various platforms to deliver educational requirements through AI platforms. Many start-ups have also come into existence to have their share in these AI-based education systems; on the other hand, industrial start-ups have started utilizing AI-based knowledge sharing among the employees and other stakeholders. With numerous business models, companies focus on developing AI-based products for the needs of educational institutes and get funding from government and other private organizations. Some of the established companies have gained profit in their annual revenue, which was based on providing virtual platforms into educational sector. The era of the pandemic was a great opportunity for the entrepreneur who imagined education to be framed, considering AI to develop virtual platforms for both instructors and students to share information. The process of product development is more focused on AI-based products in the market after the exponentially increased demand in the past few years of the pandemic. AI-based electromechanical products are now an essential part of digital electronics markets. They are equally in demand when compared to smartphones, laptops and other electronic devices.

8.1.2 Government Policies and AI for Sustainability

Government policies should be launched on the basis of the arising need for AI-based education systems. Infrastructures in various countries may be reformed with virtual platforms and other AI tools, which are now a part of educational institutes. In developing countries, governments are still figuring out how to frame new policies and implement them in public sectors due to ethical uncertainty and other factors. Private institutes are ahead in welcoming AI-based education. Public sector institutes should provide required platforms, including AI tools, laboratories and other infrastructures through experts in the field to cope up with the demands of the upcoming generation. Essential training should be provided with sufficient funds at the public institutes. Trained staff should be familiar with ongoing and upcoming technologies, in order to have academic institutions with the required research opportunities and the necessary AI tools. To bridge the gap between industry and academic platforms, the educational model should be designed to meet the needs of both sectors. Research collaboration can build a further scope of development through the advancement of AI in education. Policies should be framed considering equity and sustainability for AI-based education.

8.1.3 Training of Instructors

Instructors or teachers should be familiar with the use of AI platforms to deliver efficient education to the learners. AI should not replace their roles but be the platform for sharing knowledge using various tools and be a helping hand to tutors. Appropriate training platforms should be arranged by different educational centres to encourage skill development sessions for instructors. These programmes further need funds from the government, industrial organisations or other funding agencies. Government investment for enhancing career growth for the trainers by implementing required educational bills can be a stepping stone in the upcoming era of education, where virtual and real-world interaction are highly desirable. Trained instructors can deliver essential information to students across the geopolitical domain with expertise in virtual platforms such as VR, Augmented Reality (AR), Digital people, etc., to overcome the gap of time, space and other constraints.

8.1.4 Challenges of AI in Education

Educational data is still a challenge to collect around the world, specifically in developing and underdeveloped nations. According to UNESCO, countries need to provide more educational data to have better AI structures in the future at each level of education. The value of AI-based education is yet to be accepted by many countries through policymaking, considering ethical uncertainty, funding and other factors in the area. Ethical uncertainty is still a challenge for various countries in the world to accept new AI-based educational standards. Information is now replaced by data ethics, which is required by AI in education. There's need for the formation of enhanced models considering ethics. On the other hand, students' performance at college or school may be judged on the basis of ML algorithms, which may not be able to explain the reason for the low scores acquired. Discrimination is another aspect of ML which needs to be enhanced further through deep learning to be fair enough for everyone without hurting any individual, group or society. Another issue is to prevent delivering wrong information or knowledge to the students, which creates another struggle of updating laws considering the cybersecurity and other data protection laws of a particular state or country. Personal information needs to be protected to avoid data leaking into the wrong hands of attackers on virtual platforms. In order to have more ethically sound AI, focus on advanced research is required, considering the more rational behaviour of "thinking machines", to have transparent, fair enough and acceptable outcomes.

AI should deliver efficiency regarding the cost of jobs in the educational and industrial sectors. This is another issue that hinders AI development at the required pace. AI can help by writing an email or it can assist in writing an email. For example, "help me write" and "help me explain" are examples of prompts, the usage of which all depends on the user. In this example, an elaborate response can give better results. Cognitive AI in the above example helps in accessing the information based on what it has gathered in the system. This approach can be moved towards generative AI to obtain better rational outcomes. ChatGPT, based on generative AI, can give solutions to many wrongdoings in society, for example, "how to make a weapon?", which creates a risk of harmful content. For example, the use of a knife can be explained to cut vegetables or otherwise, which totally depends on the use of synthetic data.

8.1.5 Case Study of Educational Trend in China Through AI Applications

AI development in China is rising rapidly, with applications including but not limited to "Koolearn", "AI class director" for teaching purposes, "Tomorrow Advancing Life (TAL)", an emerging AI education provider. Many online registrations have been recorded at a fast pace in recent years since the pandemic. This overall effort of government organizations and private enterprises to bring together these AI-based educational platforms with efficient public policies in the educational sector has enhanced the overall national economy as a whole. In this section, a case study is presented taking AI-based education held by institutes and their overall impact on tutors, students and other stakeholders. Many factors are to be discussed that influence the learning patterns of upcoming students and the learning and adaptation of new virtual environments by instructors and other staff within the institutes. Also, VR training, digital people, virtual engineering education, and other such concepts at both school and university levels are also introduced. Figure 8.1 shows how the role of an instructor can become efficient by adapting AI tools while working in educational platforms to share workload smartly.

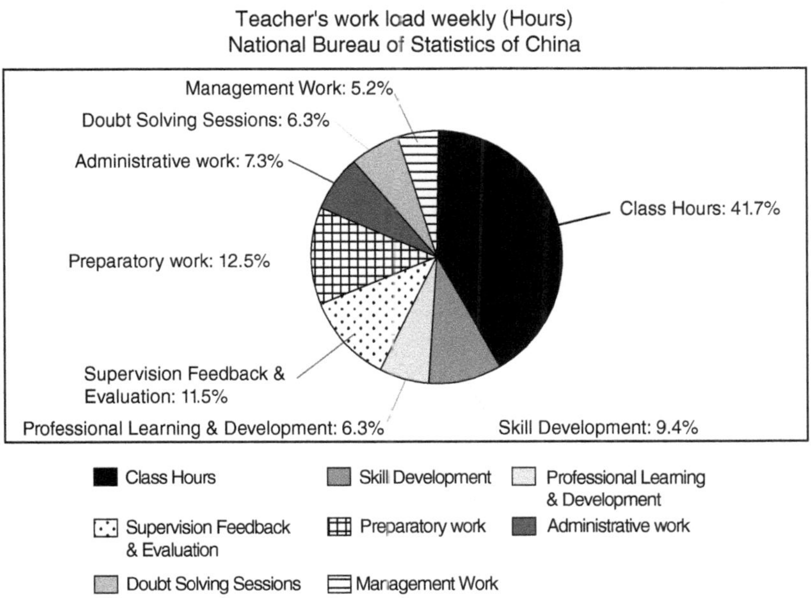

FIGURE 8.1 Pie chart showing teachers' workload in Chinese vocational colleges. It consists of 5.2% management work, 6.3% doubt solving sessions, 7.3% administrative work, 12.5% preparatory work, 11.5 % supervision, feedback and evaluation, 6.3% professional learning and development, 9.4% skill development and 41.7% class hours of total load assigned to an instructor in a week.

8.2 DIGITAL TWIN

A digital twin is a high-fidelity model of physical objects in virtual space. Through the manipulation of a virtual model of learning, students can indirectly grasp the operating principles and the internal structure of a real physical equipment. The concept of digital twin can be traced back to the "information mirroring model" proposed by Michael Grieves of the University of Michigan. Later, NASA defined the concept of "digital twin" in detail and explored key technologies for its application on spacecraft. In order to monitor the operation status of a spacecraft in outer space in real time, NASA proposes to use digital twin technology to build a digital twin of the spacecraft on Earth, which is able to be consistent with the behaviour and status of a spacecraft in outer space. Thus, the behavioural status of a spacecraft in space can be indirectly accessed by monitoring the behavioural status of the digital twin spacecraft on Earth. With the development of sensor technology, computer technology, and AI algorithms, the digital twin technology has been applied to the construction, manufacturing, aerospace, and marine industries. In the field of engineering applications, the digital twin technology still awaits further breakthroughs in model fidelity, prediction reliability, and other key challenges, as engineering applications require high model accuracy and dimensionality, and actual engineering objects often involve multi-physical fields, multi-component coupling, and other issues. However, less-advanced digital twin models are sufficient for educational purposes. This is because education is a means of popularizing basic knowledge, mainly allowing students to understand the composition of physical objects, the operation mechanism, and the essence of which only requires replicating the physical behaviour in virtual space. Therefore, using digital twin technology can effectively improve the quality and efficiency of education, thus improving the industry's economy.

As shown in Figure 8.2, Chinese scholars currently classify the maturity of digital twins into six strata. The first four layers of technology have been relatively mature, while the last two layers are still being explored. In the L0 layer, virtual objects merely simulate physical objects offline, while in the L1 layer, virtual objects can synchronize the behavioural state of physical objects, that is, the virtual model can be driven by real-time physical data, and after the L2 layer, the bidirectional relationship between the virtual and real objects starts to be constructed, and the virtual space not only maps the behaviours of the physical objects but also controls the physical objects, and in the L3 layer, the virtual space through the high-fidelity virtual models preview future events, so that the state of the physical object after the execution of the task can be predicted. In the L4 layer, through the preview of the virtual model and the optimization strategy formulation, the operational performance of the

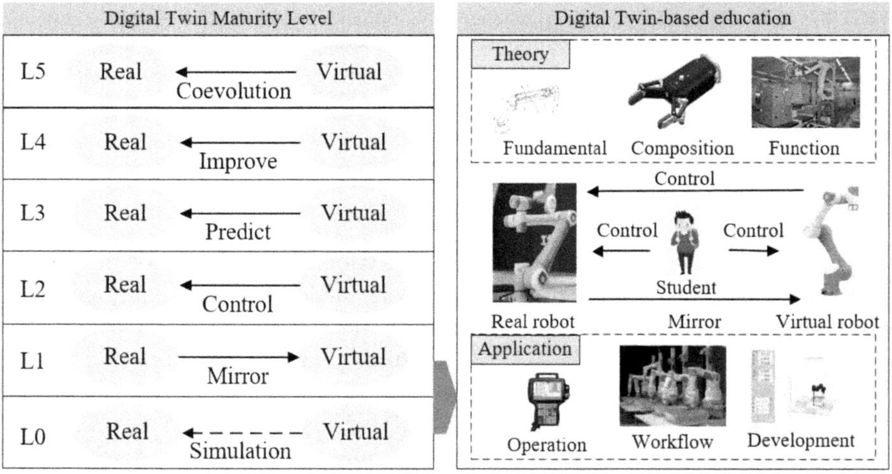

FIGURE 8.2 Conceptual Diagram of the Digital Twin Maturity Model and Digital Twin-Based Educational Approach. The left-side of the picture shows digital maturity levels from L0 to L5 with interaction between the virtual and real environment at each level. Where L0 shows simulation between virtual and real world, bidirectional relation (mirror) at L1, control at L2, predictive behaviour at L3, improvement through optimization strategy at L4 and finally L5 shows symbiotic evolution between virtual and physical domains. The right-hand picture shows digital-based education with the first row reflecting a theoretical approach showing fundamental, composition and function. The middle layer shows the relation between the student and the robot arm with control through virtual and real domains. The bottom layer of the right-hand side picture shows the application with operation, workflow and development images of the robot arm at the workplace.

physical equipment can be improved, and at the final L5 layer, the virtual object and the physical object can evolve symbiotically. Therefore, it can be seen that the application of the digital twin technology to the education industry currently requires only L1–L2 layers of the digital twin model to achieve the goal well.

The use of digital twin technology can assist students in learning theory and applying practice. Traditional practical education is often where the teacher first leads students to learn the theoretical part with pictures, videos and other ways and then uses teaching aids to make students familiar with the operation. This approach has certain shortcomings: (1) relying solely on images, videos and other instructional materials limits students' understanding, reduces learning efficiency and makes the teaching method's effectiveness heavily dependent on the instructor's lecturing skills; (2) physical teaching aids often require a large teaching space and there is a certain degree of wear and tear, and at the same time, it is difficult for the students to recognise the internal composition of the equipment that is difficult to dismantle, and it is even more difficult to learn the functioning of the mechanism; (3) the teacher–student ratio is crucial. The number of teachers is positively correlated with the number of students, and if the number of teachers is limited, the teaching cycle will be prolonged; (4) for factory training, teaching activities often affect normal production and delivery of goods. By using the digital twin education to establish a virtual scene and virtual model, students can learn the composition and operation process of the equipment through the operation of the virtual scene and virtual model, which can reduce the demand for teacher resources, teaching aids, teaching space resources, but also improve the efficiency and effectiveness of teaching and reduce the risk of teaching and learning safety.

The main features of the digital twin-based approach to teaching and training are as follows.

8.2.1 Multi-Level Cognition

By constructing virtual models, digital twins break the limits of physical space and constraints. For example, trainees can learn the composition of multiple levels by disassembling the virtual model. Take the workshop as an example. The trainee can cognize the composition of the workshop (machine tools, industrial robotic arms, conveyor belts, AGV carts, etc.) by roaming in the virtual space, and then cognize the composition of subsystems (control systems, structures, actuation systems, drives, etc.) by disassembling the workshop units such as the industrial robotic arms; by disassembling them to the component level, the trainee can have a full cognition of the composition of the whole system. In addition to the hierarchical relationship, the system can set up supplementary information for the virtual object and display it in the form of Kanban boards, cue boxes, highlights, etc., so as to enhance the cognitive efficiency of the trainees. For example, when disassembling a system, the relevant information about the component can be instantly accessed through simple Kanban information, or the associated information can be automatically accessed through the built-in searcher. Thus, the basic knowledge of the object, its composition, operation process and function can be fully learned.

8.2.2 Virtual–Real Integration Operation Practice

In the educational scenario, learners can observe the operation of physical and virtual objects at the same time, which means that learners can deepen their knowledge of the operation of physical objects through the content presented in the virtual scenario. For example, when students learn how to operate a robotic arm, a digital twin robotic arm system will be constructed next to the physical robotic arm, which integrates a virtual model, a virtual scene and auxiliary teaching information, and the students can monitor the physical equipment through the virtual system while indirectly controlling the physical equipment through the control module in the system. Since the virtual system has a basic geometric model, kinematic model, etc., it is also possible to learn the robot arm's workflow and functional development process in the guide.

8.3 DIGITAL PEOPLE

Digital people is an interactive virtual model based on artificial intelligence, big data and other technologies to simulate human behaviour, thinking and emotion. In the field of education, digital people can simulate real teachers and provide students with a personalized and intelligent educational experience. Since its exploration in the 1960s, the definition and connotation of digital man have continued to evolve. In 1964, Boeing began manufacturing the Digital people for use in simulated flight experiments, through which researchers could simulate and evaluate the performance of various aerospace components and systems. At this time, Digital people is defined by researchers as a computer model that relies on specific rules to represent human characteristics through simulation. In 1994, the National Library of Medicine completed an innovative research effort called the Visible Human Project, which aimed to create a complete, detailed digital mannequin. The project digitally documents every part of the human body, from the epidermis to the gut. The digital man in this study is a highly detailed, structured computer model that accurately reflects the various characteristics and details of the human body. Through this model, researchers can simulate, analyse and understand the structure and function of the human body, which in turn provides valuable resources for medical research and education. With the rapid development of artificial intelligence, big data, Internet of Things and other technologies in the early 21st century, digital people technology has made a qualitative leap. Nowadays, digital people can not only imitate human beings in appearance but also have advanced functions such as emotional expression, interactive communication, and autonomous learning. Luka, for example, released Replika, a deep learning-based chat companion, in 2017. It can not only provide users with daily conversational communication but also learn to adapt their preferences through interaction with users, thereby providing more in-depth and meaningful communication and improving users' mental health and self-awareness. Nowadays, digital people have been widely used in business, social networking, games, film and television, medicine and other fields with their advantages of high fidelity, strong interactive ability, AI-based decision-making and high-speed learning mechanism. The wide application of digital people in these fields has brought valuable enlightenment to the field of education and guided its potential value under the digital teaching system.

The traditional education model has obvious disadvantages in some aspects: (1) insufficient learning immersion. Most of the time, traditional education relies on the teacher's dictation, pictures and videos and courses with strong application are difficult to simulate both real environments and interactions with work peers in real settings. (2) There are limitations in educational assessment methods. Traditional educational assessment methods are mostly based on written exams and assignments, which make it difficult to fully grasp students' learning progress and status. (3) Neglect of individual differences. Traditional education methods often adopt a "one-size-fits-all" approach to designing curricula and teaching programmes, ignoring the individual differences and learning needs of each student.

Utilising digital people technology can enhance the learning experience and effectiveness of students. A digital people-based approach to education has the following distinguishing features:

8.3.1 Virtual Guidance in Real Situations

Digital people, combined with digital rendering and large language modelling technologies, can serve as interactive partners and guides in virtual learning scenarios. Students in such an environment can not only obtain a more realistic simulation of the learning state but also receive timely feedback and guidance provided by the digital person to ensure targeted and efficient learning. For example, when students conduct chemistry experiments in a virtualized laboratory, the digital person can provide real-time operational advice, point out possible errors and provide relevant chemical knowledge to deepen students' understanding.

8.3.2 Deep Data-Driven Learning Process Tracking

Equipped with data collection technology, Digital people can collect all-around learning information from students, including learning behaviour, interaction records and outcome assessment. This in-depth data-driven approach not only accurately grasps students' learning progress but also provides educators with a more precise feedback mechanism. Compared with traditional learning progress tracking, this approach can provide a more detailed understanding of students' learning status and help educators provide targeted guidance.

8.3.3 Intelligent Planning for Personalized Learning

Combined with data mining technology, digital people can develop a personalized learning plan for each student. Unlike traditional "one-size-fits-all" teaching methods, this approach ensures that each student has access to their own learning resources and strategies to maximize learning results. For example, based on students' answer data, the digital person can analyse which knowledge points students have difficulty mastering, and recommend more in-depth and detailed learning materials or teaching videos based on these knowledge points to help them overcome their difficulties (Figure 8.3).

Overall, Digital people is bringing an unprecedented level of personalized and intelligent experiences to the education space, ensuring that students achieve optimized learning outcomes in a more authentic, in-depth learning environment.

FIGURE 8.3 Comparison between traditional educational approach and digital human-based educational approach. The picture is divided into six sections. The top-left corner image shows an instructor conducting a lecture with a text mentioning "traditional education relies heavily on teachers and lack and immersive learning experience". The top-right corner image shows user experience with VR technique with a text mentioning "As learning guides, digital humans provide students with realistic simulations of learning experiences". The middle part of the image shows two sections, where the left one has an image of four students sitting and reading on a table with a text mentioning "the methods of educational assessment are limited". The right section of the middle image shows a student touching a virtual screen through VR technology with a text mentioning " Collect learning information from all aspects and keep track of students' learning progress". The bottom left corner image shows a table with the weekly class schedule with a text mentioning "Ignored individual differences and learning needs of each student". The bottom right corner image shows an instructor sitting at a table and reading a text mentioning "Develop personalized learning plans for each student".

8.4 EXTENDED REALITY (XR)

Extended Reality (XR) is a general term for immersive virtual and real-world interaction technologies such as virtual reality (VR), AR, and mixed reality (MR).

Virtual reality (VR): the scene provided by VR to the user is completely virtual, and it brings an immersive experience to the user by creating a completely virtual world through a computer.

Augmented reality (AR): different from VR, AR is characterised by virtual objects being superimposed on the real-world picture to complete the effect of enhancing the real-world information.

Mixed Reality (MR): MR can provide interactive experiences that combine the real and virtual worlds. It can either superimpose virtual information or objects in the real world (e.g., AR) or provide an immersive and completely virtual environment (e.g., VR). In a mixed-reality world, virtual and real things can coexist and interact.

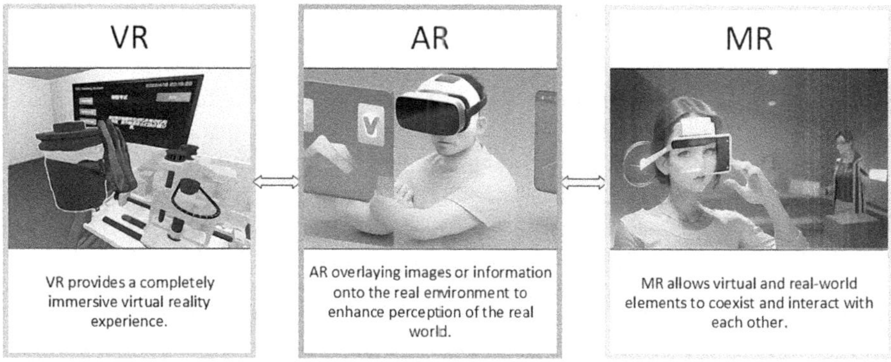

FIGURE 8.4 Demonstrations of Applications for VR, AR, and MR. The left image shows VR technology displayed via a monitor with a text mentioning "VR provides completely immersive virtual reality experience". The middle image shows AR technology where a human is wearing a VR headset in front of a screen with a text mentioning "AR overlaying images or information onto real environment to enhance perception of the real world". The right image shows MR technology where a girl is wearing a headset with a text mentioning "MR allows virtual and real-world elements to coexist and interact with each other.

Users can choose and adjust according to their own needs. This means that users can see and interact with virtual objects through the trait device, and these virtual objects can also interact with the real environment (Figure 8.4).

Since the 1960s, the concept and practice of XR have gradually come into public view. In 1968, Ivan Sutherland and Bob Sproull launched the "Sword of Damocles", the first VR headset; although its performance was limited, it had a digital reality simulation function, so that people could initially explore 3D virtual space. In 1986, Dr. Tom Furness created the "Super Cockpit" program, which projected avionics data, 3D maps, IR and radar images through a touch-sensitive panel, combined with a helmet, flight suit and gloves, to ensure that the user went deep into three-dimensional virtual space. In the 1990s, Tom Caudell coined the term "augmented reality" (AR) to describe the overlay of digital information on real equipment in aircraft production. In the 21st century, along with technological advances, the advent of the Oculus Rift in 2010 marked the arrival of a new era of VR. Oculus Rift presents a more realistic virtual world for users, subverting people's existing cognition of VR. In 2016, Microsoft launched HoloLens, a combination of VR and AR technology, allowing users to interact with virtual 3D objects in the actual environment, heralding a new XR era. As technology changes, devices such as the Valve Index take the XR experience to the next level. The application of XR in entertainment, medical, industrial and other fields has also attracted the attention of scholars in the field of education. They believe that the application of XR can bring profound changes to traditional teaching methods and provide a new perspective on the field of education, thus demonstrating the potential value of immersive learning environments for learners under the digital teaching system.

Here are some examples of how XR is used in education.

8.4.1 Historical and Cultural Reappearance

Through the use of VR technology, students have the opportunity to explore the complexities of history in an immersive way. They can truly "return" to the scene of historical events and intuitively feel the life and culture of that era. For example, students can visit the ancient Hanging Gardens of Babylon, see the magnificent hanging plants and feel the glory of ancient civilization; Or walk into the streets of China's Tang Dynasty, observe the daily life of Chang'an citizens, hear ancient Chinese dialogues and taste the culture and art of that era. This immersive way of learning has transformed History from tedious book knowledge into a vivid historical journey.

8.4.2 Art and Design

With the blessing of XR technology, art and design education will be more vivid and three-dimensional than ever before. Students are free to paint, sculpture or design in three dimensions without being restricted by physical space. In this new world, they can create vivid three-dimensional paintings or carve out textured sculptures. In addition, students can experiment with different design concepts to create artworks that are both real and abstract in the virtual world. This way of learning can not only stimulate students' creativity but also provide them with a broader and deeper artistic experience (Figure 8.5).

FIGURE 8.5 Image shows an instructor discussing in the class using VR technology, where everyone is wearing VR headsets at the School of Mechanical Science and Engineering of Huazhong University of Science and Technology, Wuhan, China.

8.4.3 Special Education

XR technology will bring revolutionary changes to the field of special education. For students with special needs, such as autistic or hearing-impaired students, traditional teaching methods often fail to meet their learning needs. However, with XR technology, educators can provide these students with a more personalized and immersive learning experience. For students with autism, for example, virtual reality environments can create a safe, non-stressful learning space, which can help them concentrate better. For hearing-impaired students, XR technology can help them better understand and master knowledge through visual and tactile feedback. All of this proves that XR technology opens a new door for special education, providing equitable and efficient learning opportunities for all students.

8.5 CASE OF EDUCATIONAL TREND IN CHINA THROUGH AI APPLICATIONS

8.5.1 Data Analysis for a Particular Domain of Education/Specific AI Application in Education (Definitions, Explanations, Details, and Page Numbers)

In the field of Vocational Education, efforts are being made to build virtual simulation training bases to overcome the high investment, high attrition, high risk and difficulty in implementing, observing, reproducing and other practical training problems of traditional teaching methods. In 2010, the Ministry of Education issued the "Outline of the National Medium and Long-Term Educational Reform and Development Plan (2010–2020)", the first time for the development of China's virtual teaching plan. In 2013–2020, the Ministry of Education and other departments issued a series of policies to vigorously promote the construction of virtual simulation teaching centres, virtual simulation experimental teaching projects and virtual simulation training teaching bases, which promoted the development of China's virtual teaching industry. In China, according to "Vocational Education Improvement and Excellence Action Plan (2020–2023)", the future will select about 100 demonstrative virtual simulation training bases, and application scenarios, including such as VR+ mechanics, corporate training, etc. The representative scene includes a 3D interactive teaching system and a flight simulation training system, etc. In addition, the State Grid, Audi, Midea and other enterprises continue to introduce VR technology for employee training, and now Manheng Technology, One Transmission and other enterprises have also realized the commercialization of related business. In foreign countries, the United States Inclusion launched an aircraft maintenance VR training system to help the development of VR + mechanics; in 2021, the Bank of America announced that there are nearly 4,300 financial centres in North America with VR technology for soft skills training, involving about 50,000 employees; Walmart deployed VR equipment on a large scale in 2018, the use of the Oculus Rift and STRIVR content for millions of employees to carry out VR training, including new products and STRIVR content, VR training, including new products and services, and customer response training. According to statistics, as of 2020, China's digital talent gap is already close to 11 million; in order to adapt to the needs of the industry, many vocational colleges and universities have accelerated the pace of digital transformation, such as the Sichuan Southwest Aviation Vocational College, and created a VR/AR training room to train the disassembly of drones and carry out the cognitive teaching of the landing gear, aircraft engines and so on.

Huatai Securities expects China's virtual reality + education and training market space to be 2.2 billion Yuan in 2021 and is expected to reach 24.2 billion Yuan in 2026, with a CAGR of 62% in 2022–2026. VR Gyro expects China's VR/AR + education and training market space to be 2.2 billion Yuan in 2021 and is expected to reach 6.1 billion Yuan in 2023. Huatai Securities expects the global virtual reality + education and training market space to be of $12.8 billion in 2021, $101.9 billion in 2026, with a CAGR of 51% in 2022–2026, and they estimate the overseas virtual reality + education and training market space to be of $12.5 billion in 2021 and $98.4 billion in 2026, with a CAGR of 51% in 2022–2026. The global market share of China's VR/AR + education and training reached $6.1 billion in 2023, with a CAGR of 62% in 2022–2026. The global market share of virtual reality + education and training is 3.4%. ReportLinker believes that the global VR+ education market space is $6.4 billion in 2021, and is expected to reach $32.9 billion in 2026, with a CAGR of 39% from 2022 to 2026. The Business Research Company reports that the global AR+ education market is $6.4 billion in 2021, and is expected to reach $32.9 billion in 2026, with a CAGR of 39% from 2022 to 2026. It will be 6.4 billion in 2021, and expected to reach $69 billion in 2026, with a CAGR of 49% in 2022–2026 (Table 8.1).

TABLE 8.1
Global Virtual Reality Market Measurement

	Year						
Economy	**2021**	**2022E**	**2023E**	**2024E**	**2025E**	**2026E**	**22-26CAGR**
China's virtual reality + education and training space (billions of dollars)	22	37	61	99	157	242	62%
YOY		68%	65%	62%	58%	54%	
Market share in China	2.5%	2.6%	2.8%	3.0%	3.2%	3.4%	
global VR education and training market space (USD billion)	64	92	131	182	249	329	39%
YOY		45%	42%	39%	36%	33%	
Global AR education and training market space (USD billion)	64	109	181	291	454	690	49%
		70%	65%	61%	56%	52%	

(Continued)

Table 8.1 (Continued)

	Year						
Economy	**2021**	**2022E**	**2023E**	**2024E**	**2025E**	**2026E**	**22-26CAGR**
Global virtual reality + education and training space (USD billion)	128	202	312	473	702	1019	51%
YOY		58%	55%	52%	48%	45%	
Overseas virtual reality + education and training market space (USD billion)	125	196	303	459	679	984	51%
YOY		57%	54%	51%	48%	45%	

8.6 CONCLUSIONS AND FUTURE TRENDS

The objective of this chapter is to address the role of AI in the educational sector and its influence on future trends in upcoming educational standards. The impact of AI on the economic growth of the world by enhancing infrastructure equipped with AI tools and bridging the gap between industry and academics by adopting AI tools developed in the market for future needs. Trained instructors are in demand in educational whereabouts with upgraded skills in the domain of artificial intelligence, which can be achieved by advanced learning to cope with handling instructions and other necessary actions to be taken to share information. Government policies need to be revised in various educational organizations, considering ethical, legal, cybersecurity, and data mining to impart knowledge at various levels, which can minimize the gap between industries and educational institutes, thus impacting the future economic growth of the world as a whole. Further studies are required in the field of deep learning, data mining, and to have more generative AI, considering ethical uncertainty and other privacy and legal concerns around the world by experts in the respective domain. Technologies with AI have a vital role in bringing sustainable economic growth around the world in the coming era.

ACKNOWLEDGEMENT

The work here is partly supported by the National Key Research and Development Program of China (No. 2023YFD2100905). The authors are heartily thankful to their supervisor, Professor Hu Youmin, for encouragement, guidance, and support.

BIBLIOGRAPHY

Allen, G. C. (2019). Understanding China's AI strategy clues to Chinese strategic thinking on artificial intelligence and national security. *Centre for New American Security*. Retrieved from https://www.cnas.org/publications/reports/understanding-chinas-ai-strategy

Artificial Intelligence Index Report (2022). The economy and education. *Artificial Intelligence, Index Stanford Institute for Human Centered Artificial Intelligence, Stanford University*. Retrieved from https://aiindex.stanford.edu/wp-content/uploads/2022/03/2022-AI-Index-Report_Chapter4.pdf

Avella, J. T., Kebritchi, M., Nunn, S., & Kanai, T. (2016). Learning analytics methods, benefits, and challenges in higher education: A systematic literature review. *Online Learn*, 20(2), 1–17.

Cairns, M. M. L. (2017). Computers in education: The impact on schools and classrooms. *Life Schools Classrooms* (pp. 603–617). Singapore: Springer.

Caudell, T. P., & Mizell, D. (1990). Augmented reality: An application of heads-up display technology to manual manufacturing processes. *Proceedings of twenty-fifth Hawaii International Conference on System Sciences* (Vol. 2).

Chassignol, M., Khoroshavin, A., Klimova, A., & Bilyatdinova, A. (2018). Artificial intelligence trends in education: A narrative overview. *Procedia Computer Science*, 136, 16–24.

Chaudhry, M., & Kazim, E. (2021). Artificial Intelligence in Education (AIEd): A high-level academic and industry note 2021. *AI and Ethics*, 2(2),1–9.

Chen, L., Chen, P., & Lin, Z. (2020). Artificial intelligence in education: A review. *IEEE Access* 8, 75264–75278.

Cimino, C., Negri, E., & Fumagalli, L. (2019). Review of digital twin applications in manufacturing. *Computers in Industry*, 113, 103130.

Estevez, J., Garate, G., & Graña, M. (2019). Gentle introduction to artificial intelligence for high-school students using scratch. *IEEE Access*, 7, 179027–179036.

Flamm, K. (1988). *Creating the Computer: Government, Industry, and High Technology.* Washington, D.C. : Brookings Institution Press.

Furness, T.A. (1986). The super cockpit and its human factors challenges. *Proceedings of the Human Factors and Ergonomics Society Annual Meeting* (Vol. 30(1), pp. 48–52). Los Angeles, CA: Sage Publications.

Guo, C., Xu, Z., Fang, C., & Qin, B. (2022). China survey report on the online learning status of high schools during the COVID-19 pandemic. *ECNU Review of Education* (pp. 1–13): Sage Publications

Jones, A., & Castellano, G. (2018). Adaptive robotic tutors that support selfregulated learning: A longer-term investigation with primary school children. *International Journal of Social Robototics*, 10, 357–370.

Khandelwal, A. R., Chaturvedi, A., & Kanwar, M. V. (2021). Effectiveness of virtual classroom: A comparative study of theoretical and practical papers. *Webology*, 18(4), 1735–188X.

Kim, Y., Soyata, T., & Behnagh, R. F. (2018). Towards emotionally aware AI smart classroom: Current issues and directions for engineering and education. *IEEE Access*, 6, 5308–5331.

Knox, J. (2020). Artifificial intelligence and education in China. *Learning, Media and Technology*, 45(3),1–14.

Lin, P., Wooders, A., Wang, J. T., & Yuan, W. M. (2018). Artificial intelligence, the missing piece of online education? *IEEE Engineering Management Review*, 46(3), 25–28.

Mubayi, P., Cheng, E., Terry, H. P., Tilton, A., & Hou, T. (2017). China's rise in artificial intelligence: The new China. *Goldman Sachs Equity Research Report*. Retrieved from https://knowen-production.s3.amazonaws.com/uploads/attachment/file/3075/China%2BRise%2Bin%2BAI.pdf

Oppenheimer, R. (1966). William Fetter, E.A.T., and 1960s Computer graphics collaborations in Seattle. Retrieved from https://www.academia.edu/7801224

Pokrivcakova, S. (2019). Preparing teachers for the application of AI-powered technologies in foreign language education. *Journal of Language and Cultural Education*, 7(3), 135–153.

Rivas, A. (2019). Artificial intelligence in education: Challenges and opportunities for sustainable development. *United Nations Educational, Scientific and Cultural Organization, 7, place de Fontenoy, 75352 Paris 07 SP, France.*

Russel, S., & Norvig, P. (2010). *Artificial Intelligence: A Modern Approach* (3rd ed.). New Jersey: Prentice Hall Press.

Sharma, R. C., Kawachi, P., & Bozkurt, A. (2019). The landscape of artificial intelligence in open, online and distance education: Promises and concerns. *Asian Journal of Distance Education*, 14(2), 1–2.

Spitzer, M. V., Ackerman, J. M., Scherzinger, L. A., & Whitlock, D. (1996). The visible human male: A technical report. *Journal of the American Medical Informatics Association*, 3(2), 118–130.

Squirrel AI Learning (2019). Retrieved July 11, 2019, https://www.prnewswire.com/news-releases/squirrel-ai-learningattended-ieee-icdm-2018-to-show-the-charm-of-ai-education-300776050.html

Sutherland, I.E. (1968). A head-mounted three dimensional display. *1968 Fall Joint Computer Conference* (pp. 757–764). New York: ACM Press.

Tao, F., Qi, Q., Wang, L., & Nee, A.Y.C. (2019). Digital twins and cyber–physical systems toward smart manufacturing and Industry 4.0: Correlation and comparison. *Engineering*, 5(4), 653–661.

Tao, F., & Qi, Q. (2018). New IT driven service-oriented smart manufacturing: Framework and characteristics. *IEEE Transactions on Systems, Man, and Cybernetics Systems*, 49(1), 81–91.

Tao, F., Sui, F., Liu, A., & Qi, Q. (2018). Digital twin-driven product design framework. *International Journal of Production Research*, 57(1), 1–19.

Wartman, S. A., & Combs, C. D. (2018). Medical education must move from the information age to the age of artificial intelligence. *Academic Medicine*, 93(8), 1107–1109.

Zhang, K., & Aslan, A.B. (2021). AI technologies for education: Recent research & future directions. *Computers and Education: Artificial Intelligence*, 2 (2), 100025. https://www.prnewswire.com/news-releases/squirrel-ai-learning-attended-ieee-icdm-2018-to-show-the-charm-of-ai-education-300776050.html

9 Transforming Human Resource Management Research

A Bibliometric Analysis on the Impact of Artificial Intelligence

Ruchika Joshi and Mrinalini Kacker

9.1 INTRODUCTION

Industry 4.0 has led to the origin of intelligent technologies in Human Resource Management (HRM). One such technology is Artificial Intelligence (AI). The advancement in technology has permitted concepts like AI to impact various sections of society (Bolander, 2019). The AI software market is anticipated to expand up to $118.6 billion by 2025 (Seal, 2019). AI is the biggest revolution worldwide. Jia et al. (2018) concluded that AI is also known as machine intelligence, which has the ability to simulate human behaviour. John McCarthy, in 1956, originated the term AI. Several departments of various organizations have initiated the process of introducing AI-based tools, but HRM is one department which is still struggling to incorporate them fully (Vrontis et al. 2022). Human resources is anticipated to be one of the most significant assets of an organization and directing it well is a crucial task for the organization's success. With the momentous rise of AI, it is anticipated that the task of managing human resources can be handled efficiently. But it is only the huge organizations that have been able to implement AI in their HR practices, whereas for the rest its still a developing innovation (Palos et al., 2022).

With the development of AI, organizations have reconstituted profoundly. AI is likely to influence HRM significantly. Few studies have pointed out the functions of HRM, where the application of AI and its tools have been initiated range from resume shortlisting from consultancy websites and video-conferencing interview (Nawaz, 2020), employee performance prediction (Nazri et al., 2019), Fuzzy Artificial Neural Networks (ANN) techniques for data mining, solving several issues in HR (Masum et al., 2018), machine learning for effective decision-making and

DOI: 10.1201/9781003638506-9

forecasting turnover (Hamilton et al., 2022). All the functions of HR are prominently cohesive with each other, and one function influences the others evidently; advancement in technology is likely to influence all the functions. The expected function of AI in human resources practices is to streamline complex decision-making while improving the quality of decision-making (Qamar et al., 2021). AI has several functions to play in organizations as suggested in a study: to improve the efficacy and efficiency of human resources practices by making the management techniques flexible and precise (Nankervis et al., 2021). Chowdhury et al. (2023) presented a new picture that companies must look beyond the technical aspects and try to enhance the non-technical aspects, like employee skills, competencies, organization innovation, etc., to gain the fruits of adopting AI in organization HR practices.

Evolution in technology has frequently impacted the HR processes and their management. Technology has changed the nature of work and the ways of achieving it; more employees are working from home and in virtual teams from across the globe (Nankervis et al., 2021). Speculations are that Industrial Revolution 4.0 is likely to bring a major revolutionary development in the traditional ways of HRM due to advancements in technology. Russel and Norvig (2014) revealed that AI and its tools have the ability to acclimatize easily to new surroundings and assist management by providing real-time HR advice, eliminating the requirement of an HR professional. With technology also comes the hurdles of securing one's present and future job opportunities. Though it is arguable whether the effect of technology on individual jobs in HR will be a boon or a curse, Herweijer et al. (2018) concluded that HR employees must enhance the benefits of technology development, as well as the industrial revolution and try to either curtail or efficiently handle the negative effects on the workplace and employees. Though AI and its tools are comparatively upcoming concepts in the workplace, the opportunities it poses in the domain of HRM are vast, as discussed by Strohmeier and Piazza (2015). As per an Ernst and Young (2018) report on the use of AI in HR, a vast extent of daily opportunities are related to the initiation of machine learning (one of the AI tools) in HRM. Modernization brings opportunities and hurdles. In reference to AI and HRM, AI tools assist in reducing repetitive tasks, allowing HR personnel to focus on work that is more critical and entails special talents and capabilities, as concluded by Pillai and Sivathanu (2020). Tambe et al. (2019) identified four major challenges associated with using technological tools for HR-related tasks and suggested possible solutions to the issues via appropriate usage of AI tools and techniques. Strohmeier et al. (2015) explored the possibility of AI technology in HRM and revealed that though the potential of AI in HRM is an unexplored charter, not every technique of AI is suitable and can be applied to all the operations of HRM. Qamar et al. (2021) studied the application of AI in human resources–related tasks and concluded that the usage of AI can lead to unethical activities, as with AI tools, the possibility of biasedness and unfairness can be on the rise. Further, a review study to analyse the various ways through which AI is revolutionizing HRM was done by Tewari and Pant (2020) and revealed that though AI tools and techniques assist in making the managers task efficient, it has some challenges associated with it as well; like the possibility of AI taking over human jobs, financial burden on organization for introducing AI tools and techniques in their work space. A conceptual framework model related to AI application in HRM was provided by Jia et al. (2018) by analysing two organizations and concluding that AI is used in HRM in recruitment, such as face recognition of the prospective candidate and in training and development technology assists in a convenient and efficient way to learn. Rykun (2019) reported that approximately 61% organizations are applying AI to enhance HRM practices. Past studies (Albert, 2019; Black and Van, 2020; Geetha and Bhanu, 2018; Van Esch et al., 2019) have reported that, in reference to the opportunities for companies, AI assists in effective recruitment and selection. Kong et al. (2021) concluded that the use of AI creates a feeling of fear and anxiety among employees regarding their careers, as they can be easily replaced by machines. The lack of well-qualified individuals with the required ability to handle the technology is a challenge in AI applications in HRM, as reported by Abdeldayem and Aldulaimi (2020). Moreover, enhanced exposure of the organization can lead to the risk of a security breach, as concluded by Malik et al. (2022).

The available research and literature on the diverse applications of AI in HRM are sparse (Pan et al., 2022). With growing applications of AI in HRM, it has attracted the interest of researchers and industrialists. There is constant advancement being made in the domain of HRM due to advances in technology. Keeping all these aspects in mind, this study aims to identify the contribution of key authors, journals, countries, and sub-areas around HRM, where AI has the potential to be implemented and is emerging in HRM. In this context, a bibliometric analysis was conducted. On the basis of the aforementioned purpose of this study, we came across the following research questions:

RQ1. Who are the contributing authors, journals, and territories in this area of research?
RQ2. What are the most recent developments in the HRM application of AI?
RQ3. What research titles are being addressed by various academicians in reference to AI and HRM?
RQ4. What are the sub-areas of HRM around which AI can be applied?

9.2 RESEARCH METHODOLOGY

For a bibliometric study, it is necessary to establish a research protocol for answering the research questions. The protocol should explain the details that encompass the collection of data and its synthesis. Further, it should explain the search strategy and criteria for the inclusion or exclusion of articles.

9.2.1 Data Collection

For the purpose of this study, data have been extracted from one of the most popular databases – Scopus. The initial search produced a total of 1,143 research articles. However, after applying the exclusion criteria, only 325 research articles were selected.

9.2.2 Search Strategy

Relevant keywords were realized to conduct this study. The search was limited to keywords, abstracts, and titles of articles in the Scopus database. The primary keywords used to extract data were:

"Human Resource Management" AND "Artificial Intelligence"

9.2.3 Inclusion and Exclusion Criteria

Articles that are included in this study are articles published in the Scopus database from 2003 to the present in the English language. Articles that were part of a conference proceeding, workshops, editorials, and meetings are not included. Any article that does not meet the language or time criteria is excluded from the study.

9.2.4 Extraction and Synthesis of Data

The principal aim of this phase was to acquire pertinent literature for the review. Information pertaining to the subject matter, location, title, source, authors, publication and citation was extracted from the Scopus database. A performance analysis was performed utilizing Google Sheets, while a keyword analysis was conducted using Vos Viewer (Table 9.1).

TABLE 9.1
The Systematic Procedure Followed to Arrive at the Final Result of 325 Articles for Analysis

Filtering Criterion	Accept	Reject
Search Criterion		
Search Engine: Scopus	1143	
Search date: 20th January, 2023		40
Search item: Artificial Intelligence and Human Resource Management	1103	
Year: 2003–2023		758
Document type: Articles and Reviews	345	
Language Screening: English documents only		20
Total accepted papers	325	

9.3 BIBLIOMETRIC ANALYSIS

9.3.1 Research Growth

There has been a substantial growth in the publication over the last five years, with 2022 registering 112 publications. 2023 has already registered 15 publications in the first month in the Scopus database (Figure 9.1).

9.3.2 Top Source Journal

HRM has registered the most publications by any journal. It has a total of nine research articles published, followed by *Automation in Construction* with five publications (Figure 9.2).

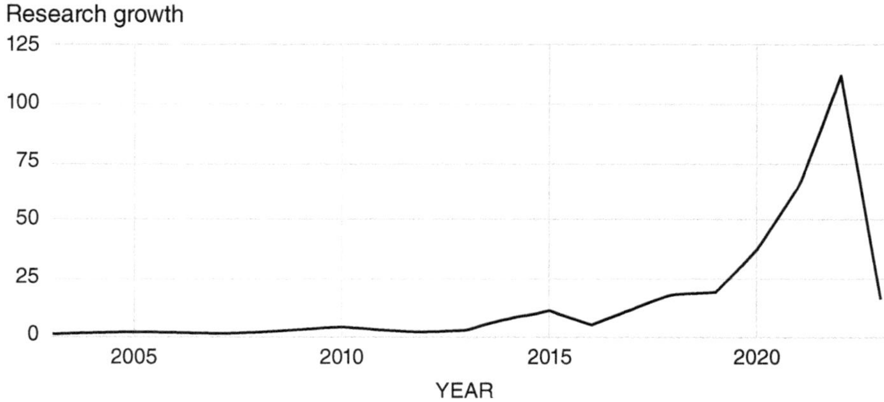

FIGURE 9.1 Documents by year.

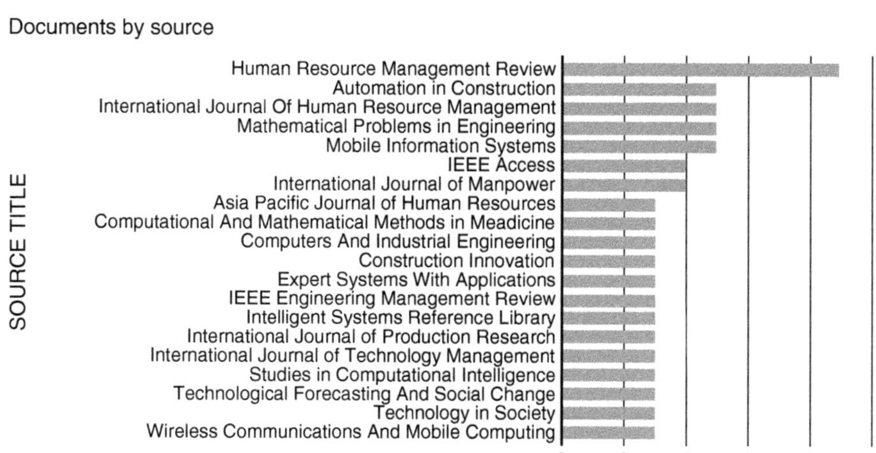

FIGURE 9.2 Documents by source.

9.3.3 Top Contributing Authors

The maximum contribution in this field has been made by two authors. Budhwar, P. and Malik, A have contributed significantly to this new area of research with five publications each (Figure 9.3).

9.3.4 Discipline/Subject-Wise Analysis

Though it may seem obvious that most of the documents are categorized in computer science and management, the impact of AI on HRM seems multidisciplinary, as evidenced in Figure 9.4.

9.3.5 Country-Wise Analysis

Although many countries have contributed to the literature on HRM and AI, China leads with a total of 70 publications, followed by the United States with 57 publications (Figure 9.5).

9.3.6 Citation Analysis

The document h-index for the Scopus database is 29. The citation growth in recent years is shown in Figure 9.6.

9.3.7 Keyword Analysis

Network analysis has been an important tool to visualize the bibliometric data. For this study, the VOSviewer has been used for this purpose. Data taken from the

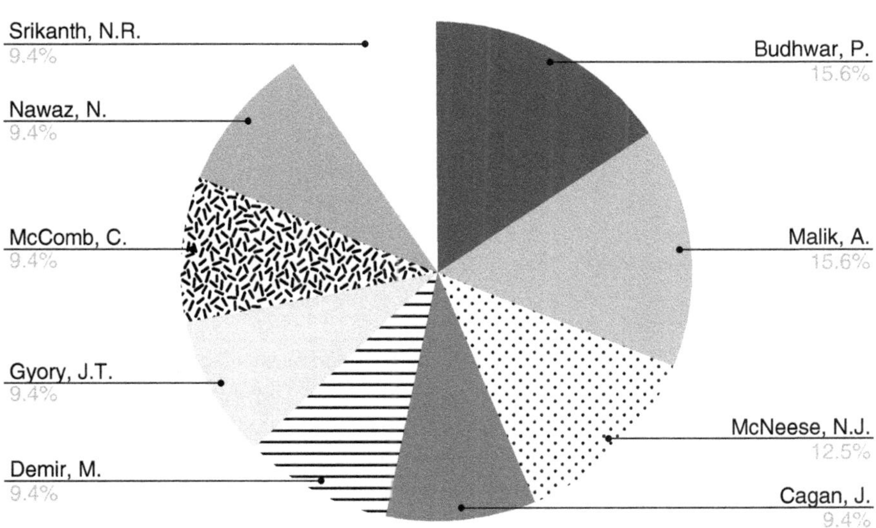

FIGURE 9.3 Documents by authors.

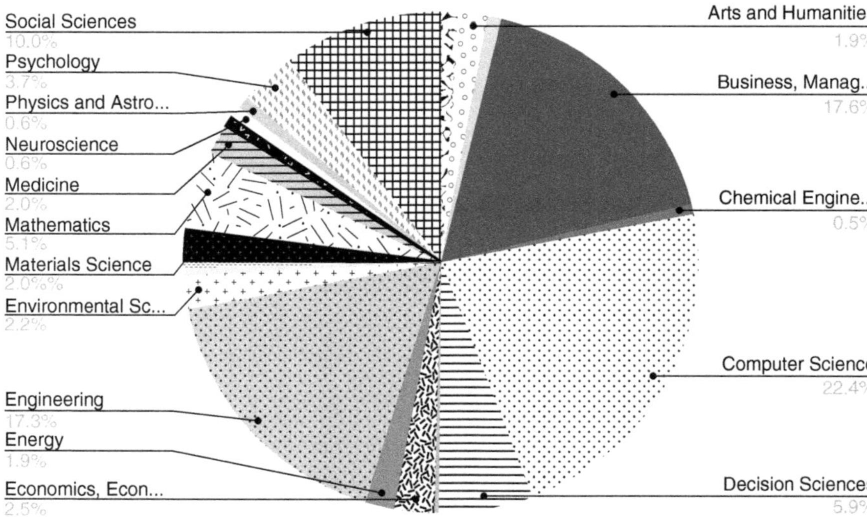

FIGURE 9.4 Document by discipline.

FIGURE 9.5 Document by discipline.

Scopus database has been used, and a co-occurrence analysis has been done using keywords. The results are visualized in Figure 9.7. The keyword co-occurrence analysis network visualization reflects a keyword such that the bigger the size of the node, the more frequently that keyword has occurred. The keywords that are linked together usually occur together and the colours represent common themes. The most common words based on this analysis are the keywords identified for the analysis.

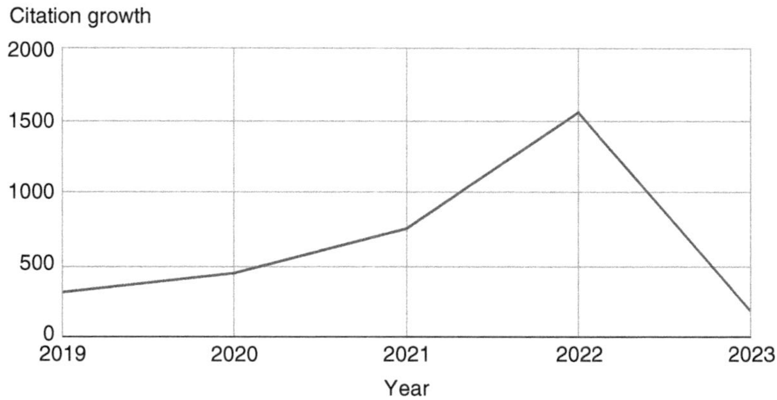

FIGURE 9.6 Citation growth in recent years.

FIGURE 9.7 Visualization of keyword networks.

9.4 DISCUSSION

AI has the potential to shape the future of HRM. Researchers have started contributing to this field. It has registered a substantial growth over the past few years (see Table 9.2). Academicians from varied domains have contributed to the study of AI, its related tools and HRM.

RQ1: The analysis revealed that China (70) tops the list in terms of research contribution in the subject area, followed by the United States (57) and India (33). Authors with prominent contributions in the field are Malik, A., Budhwar, P., McNeese, N.J. (See 3.3). The *Human Resource Management Review* is the journal with pioneering articles on the research area (See 3.2).

RQ2: AI has been making a substantial contribution in the area of HRM. As per our analysis, a prominent growth was observed in the research section between the years 2015 and 2020 (See Table 9.2). From gauging

TABLE 9.2
The Research Titles Being Addressed by Various Academicians in Reference to AI and HRM

Author	Theme/Context	Title
Del Giudice et al. (2023)	Comprehend the ways AI affects HRM. The study tried to answer the question whether AI will replace humans in workplace.	"Toward the human – Centered approach. A revised model of individual acceptance of AI."
Prikshat et al. (2023)	Developed an integrated and coherent model for HRM-AI assimilation	"AI-augmented HRM: Antecedents, assimilation and multilevel consequences."
Pereira et al. (2023)	An SLR study of 60 papers to explore the impact of AI on workplace outcomes.	"A systematic literature review on the impact of artificial intelligence on workplace outcomes: A multi-process perspective."
Karki and Hadikusumo, (2023)	A predictive model using ML to hire competent project manager for project handling.	"Machine learning for the identification of competent project managers for construction projects in Nepal."
Aleem et al. (2023)	An SLR on 198 papers using AI and ML to identify themes in HR for remote working and future research prospects.	"Remote work and the COVID-19 pandemic: An artificial intelligence-based topic modelling and a future agenda."
Ore and Sposato (2022)	Qualitative analysis of the opportunities AI presents to recruiters in the context of recruitment and selection	"Opportunities and risks of artificial intelligence in recruitment and selection."
Rozman et al. (2022)	Multidimensional model to assess the effect of AI in various aspects of HR.	"Integrating artificial intelligence into a talent management model to increase the work engagement and performance of enterprises."

an employee's disposition in the work environment to aiding in the recruitment process to determining an employee's training and development requirements in light of their previous achievements, AI is slowly and gradually spreading its horizons, as concluded by Maity (2019). Aleem et al. [30] suggested the emerging trends of AI in HRM for future academicians. As per our analysis, AI and its tools have the possibility to emerge substantially in disciplines like Business and Management, Computer Science, Engineering, and Social Sciences primarily.

RQ3: Research in the field of AI and HRM is gaining the interest of researchers all around the globe. The studies are limited to not just the role/impact of AI on HRM, but it extends to all the sub-areas of HRM, with technology being a part of HR practices. In order to list out the research areas being focused on by various researchers in the domain, a thematic analysis was done, and recent studies are presented in Table 9.2.

RQ4: The results derived reflected that AI as a tool is used in various sub-areas of HRM, such as: machine learning, project management, decision-making, recruitment, training, and development. Though AI is becoming a significant tool in HRM, studies have primarily focused on applying AI in the HR domain one at a time, specifically to extract the maximum benefit reported by Qamar et al. (2021). Aleem et al. (2023) demonstrated eight sub-areas where AI-related studies can be applied. Pan and Froese (2022) revealed the various functions of HRM by studying the impact of AI on HRM, AI and jobs, AI and gender equality.

9.5 CONCLUSION

We used a bibliometric analysis of 345 documents discovered via Scopus to throw light on the application/utility of AI in HRM. The upcoming evolution in the field of HRM is because of technology adaptation. A noticeable upward trend from the year 2015 was registered in terms of research publication till we discovered major significant and contributing journals and authors. HRM has the maximum number of publications in its journal, followed by *Automation in Construction*. Also, Budhwar, P. and Malik, A. are the major contributing authors, followed by McNeese, N.J. In terms of country contribution, China tops the list, followed by the United States. The prominent keywords used by the researchers to delve deeper into the study are recruitment, HRM, decision-support system, human resources, project management, apart from AI and HRM. In terms of the future scope of studies related to the application of AI in HRM, previous studies point towards the direction that AI can be used in the workplace to identify the training and development needs of employees, filtering out potential candidates, and bringing efficiency in the recruitment process and in learning about the behavioural/emotional switch in employees.

The study assists in deriving valuable insights into the use of technology (AI) in HRM by carefully studying the existing literature. The study divulges that research related to AI in HRM has a vast scope, yielding a sign for organizations and managers to adopt AI and its tools in their HR operations to enhance their work efficiency and productivity. Researchers are trying to advance further significantly to inspect deeper into the effect of AI in HR practices, indicating that there is a wide scope of study in this area.

9.6 FUTURE SCOPE

Due to the sparse and sporadic bibliometric studies primarily on AI and HRM, our study can be an addition to the already scattered studies. This study can be helpful for future researchers who wish to explore the subject in-depth; the study provides an insight into the emerging trends in AI and HRM, which can assist prospective researchers in deciding the direction of their study.

Also, the practical implication of the study is that it guides the management of an organization to comprehend the upcoming potential of AI in HRM. The study acts as a guide for the management of companies and industries to comprehend the potential AI holds in HR and related areas.

REFERENCES

Abdeldayem, M. M., & Aldulaimi, S. H. (2020). Trends and opportunities of artificial intelligence in human resource management: Aspirations for public sector in Bahrain. *International Journal of Scientific and Technology Research*, *9*(1), 3867–3871. 10.12691/jbms-12-2-3

Albert, E. T. (2019). AI in talent acquisition: A review of AI-applications used in recruitment and selection. *Strategic HR Review*, *18*(5), 215–221. https://doi.org/10.1108/shr-04-2019-0024

Aleem, M., Sufyan, M., Ameer, I., & Mustak, M. (2023). Remote work and the COVID-19 pandemic: An artificial intelligence-based topic modeling and a future agenda. *Journal of Business Research*, *154*, 113303. https://doi.org/10.1016/j.jbusres.2022.113303

Black, J. S., & van Esch, P. (2020). AI-enabled recruiting: What is it and how should a manager use it? *Business Horizons*, *63*(2), 215–226. https://doi.org/10.1016/j.bushor.2019.12.001

Bolander, T. (2019). What do we loose when machines take the decisions? *Journal of Management and Governance*, *23(4)*, 849–867. https://doi.org/10.1007/s10997-019-09493-x

Chowdhury, S., Dey, P., Joel-Edgar, S., Bhattacharya, S., Rodriguez-Espindola, O., Abadie, A., & Truong, L. (2023). Unlocking the value of artificial intelligence in human resource management through AI capability framework. *Human Resource Management Review*, *33*(1), 100899. https://doi.org/10.1016/j.hrmr.2022.100899

Del Giudice, M., Scuotto, V., Orlando, B., & Mustilli, M. (2023). Toward the human–centered approach. A revised model of individual acceptance of AI. *Human Resource Management Review*, *33*(1), 100856. https://doi.org/10.1016/j.hrmr.2021.100856

Ernst & Young (2018). The new age: artificial intelligence for human resource opportunities and functions. https://hrlens.org/public/storage/categories/resorces/ey-the-new-age-artificial-intelligence-for-human-resource-opportunities-and-functions-1.pdf

Geetha, R., & Bhanu, S. R. D. (2018). Recruitment through artificial intelligence: A conceptual study. *International Journal of Mechanical Engineering and Technology*, *9*(7), 63–70.

Hamilton, R. H., & Davison, H. K. (2022). Legal and ethical challenges for HR in machine learning. *Employee Responsibilities and Rights Journal*, *34*(1), 19–39. https://doi.org/10.1007/s10672-021-09377-z

Herweijer, C., Combes, B., Johnson, L., McCargow, R., Bhardwaj, S., Jackson, B., & Ramchandani, P. (2018). Enabling a sustainable Fourth Industrial Revolution: How G20 countries can create the conditions for emerging technologies to benefit people and the planet (No. 2018-32). *Economics Discussion Papers*. https://hdl.handle.net/10419/177283

Jia, Q., Guo, Y., Li, R., Li, Y., & Chen, Y. (2018). A conceptual artificial intelligence application framework in human resource management. *Association for Information Systems AIS Electronic Library (AISeL)*. https://aisel.aisnet.org/iceb2018/91

Karki, S., & Hadikusumo, B. (2023). Machine learning for the identification of competent project managers for construction projects in Nepal. *Construction Innovation*, *23*(1), 1–18. https://doi.org/10.1108/CI-08-2020-0139

Kong, H., Yuan, Y., Baruch, Y., Bu, N., Jiang, X., & Wang, K. (2021). Influences of artificial intelligence (AI) awareness on career competency and job burnout. *International Journal of Contemporary Hospitality Management*, *33*(2), 717–734. https://doi.org/10.1108/IJCHM-07-2020-0789

Maity, S. (2019). Identifying opportunities for artificial intelligence in the evolution of training and development practices. *Journal of Management Development*, *38*(8), 651–663. https://doi.org/10.1108/JMD-03-2019-0069

Malik, N., Tripathi, S. N., Kar, A. K., & Gupta, S. (2022). Impact of artificial intelligence on employees working in industry 4.0 led organizations. *International Journal of Manpower*, *43*(2), 334–354. https://doi.org/10.1108/IJM-03-2021-0173

Masum, A.-K., Beh, L.-S., Azad, A.-K. and Hoque, K. (2018). Intelligent human resource information system (i-HRIS): A holistic decision support framework for HR/ excellence. *The International Arab Journal of Information Technology*, *15*(1), 121–130.

Nankervis, A., Connell, J., Cameron, R., Montague, A., & Prikshat, V. (2021). 'Are we there yet?' Australian HR professionals and the Fourth Industrial Revolution. *Asia Pacific Journal of Human Resources*, *59(1)*, 3–19. https://doi.org/10.1111/1744-7941.12245

Nawaz, N. (2020). Artificial intelligence applications for face recognition in recruitment process. *Journal of Management Information and Decision Sciences*, *23*(S1), 499–509.

Nazri, M.Z.A., Ghani, R.A., Abdullah, S., Ayu, M. and Nor Samsiah, R. (2019), "Predicting academician publication performance using decision tree", *International Journal of Recent Technology and Engineering*, *8*(2),180–185.

Ore, O., & Sposato, M. (2022). Opportunities and risks of artificial intelligence in recruitment and selection. *International Journal of Organizational Analysis*, *30*(6), 1771–1782. https://doi.org/10.1108/IJOA-07-2020-2291

Palos-Sánchez, P. R., Baena-Luna, P., Badicu, A., & Infante-Moro, J. C. (2022). Artificial intelligence and human resources management: A bibliometric analysis. *Applied Artificial Intelligence*, *36*(1). https://doi.org/10.1080/08839514.2022.2145631

Pan, Y., & Froese, F. J. (2022). An interdisciplinary review of AI and HRM: Challenges and future directions. *Human Resource Management Review*, 100924. https://doi.org/10.1016/j.hrmr.2022.100924

Pan, Y., Froese, F., Liu, N., Hu, Y., & Ye, M. (2022). The adoption of artificial intelligence in employee recruitment: The influence of contextual factors. *The International Journal of Human Resource Management*, *33*(6), 1125–1147. https://doi.org/10.1080/09585192.2021.1879206

Pereira, V., Hadjielias, E., Christofi, M., & Vrontis, D. (2023). A systematic literature review on the impact of artificial intelligence on workplace outcomes: A multi-process perspective. *Human Resource Management Review*, *33*(1), 100857. https://doi.org/10.1016/j.hrmr.2021.100857

Pillai, R., & Sivathanu, B. (2020). Adoption of artificial intelligence (AI) for talent acquisition in IT/ITeS organizations. *Benchmarking: An International Journal*, *27*(9), 2599–2629. https://doi.org/10.1108/BIJ-04-2020-0186

Prikshat, V., Malik, A., & Budhwar, P. (2023). AI-augmented HRM: Antecedents, assimilation and multilevel consequences. *Human Resource Management Review*, *33*(1), 100860. https://doi.org/10.1016/j.hrmr.2021.100860

Qamar, Y., Agrawal, R. K., Samad, T. A., & Chiappetta Jabbour, C. J. (2021). When technology meets people: The interplay of artificial intelligence and human resource management. *Journal of Enterprise Information Management*, *34*(5), 1339–1370. https://doi.org/10.1108/JEIM-11-2020-0436

Rožman, M., Oreški, D., & Tominc, P. (2022). Integrating artificial intelligence into a talent management model to increase the work engagement and performance of enterprises. *Frontiers in Psychology*, *13*, 1014434. https://doi.org/10.3389/fpsyg.2022.1014434

Russell, S. J., & Norvig, P. (2014). *Artificial Intelligence: A Modern Approach*. Pearson.

Rykun, E. (2019). Artificial Intelligence in HR Management–What Can We Expect? *The Boss Magazine*. https://thebossmagazine.com/ai-hr-management/

Seal, A. (2019), "The most interesting AI statistics of 2019", available at: https://www.vxchnge.com/blog/ai-statistics

Strohmeier, S., & Piazza, F. (2015). Artificial intelligence techniques in human resource management—A conceptual exploration. *Intelligent Systems Reference Library*, *87*, 149–172. https://doi.org/10.1007/978-3-319-17906-3_7

Tambe, P., Cappelli, P., & Yakubovich, V. (2019). Artificial intelligence in human resources management: Challenges and a path forward. *California Management Review, 61(4)*, 15–42. https://doi.org/10.1177/0008125619867910

Tewari, I., & Pant, M. (2020). Artificial intelligence reshaping human resource management: A review. In *2020 IEEE international conference on advent trends in multidisciplinary research and innovation (ICATMRI)* (pp. 1–4). IEEE. https://doi.org/10.1109/ICATMRI51801.2020.9398420

Van Esch, P., Black, J. S., & Ferolie, J. (2019). Marketing AI recruitment: The next phase in job application and selection. *Computers in Human Behavior, 90*, 215–222. https://doi.org/10.1016/j.chb.2018.09.009

Vrontis, D., Christofi, M., Pereira, V., Tarba, S., Makrides, A., & Trichina, E. (2022). Artificial intelligence, robotics, advanced technologies and human resource management: A systematic review. *The International Journal of Human Resource Management, 33*(6), 1237–1266. https://doi.org/10.1080/09585192.2020.1871398

10 AI's Potential to Reduce Emissions in the Retail Industry

Manjushree G. S. and Nayana N.

10.1 INTRODUCTION

10.1.1 Overview of the Growth of E-Commerce and Its Impact on the Environment

The evolution of e-commerce (van Gelder, 2023) over the past few decades has been nothing short of remarkable. What began as a novel concept has now become a global economic powerhouse, with profound implications for the environment:

1. *Explosive Growth*: E-commerce's rapid expansion is fueled by internet access, mobile tech, and changing shopping habits, reshaping global retail.
2. *Convenience*: E-commerce offers unmatched ease, spurring a surge in online shopping.
3. *Environmental Impact*: Growth leads to increased carbon emissions and packaging waste.
4. *Last-Mile Delivery Challenges*: Individual deliveries contribute to traffic congestion, fuel use, and emissions, particularly in cities.
5. *Packaging Waste*: Excessive packaging leads to non-recyclable waste accumulation.
6. *Energy Consumption*: Data centers and warehouses consume substantial energy.
7. *Sustainability Initiatives*: Many e-commerce firms implement green practices, like eco-friendly packaging and optimized delivery (Brewer, 2019).
8. *Role of AI*: Artificial intelligence (AI) optimizes routes, reduces energy use, and supports sustainable sourcing, mitigating e-commerce's environmental impact.

As e-commerce continues to grow, understanding and addressing its environmental impact is crucial. The integration of AI-driven solutions and sustainability initiatives can help balance the convenience of online shopping with responsible environmental practices.

DOI: 10.1201/9781003638506-10

10.1.2 The Potential for AI to Help Combat Carbon Emissions in E-commerce

AI holds significant promise in mitigating the carbon emissions associated with the e-commerce industry. Its capabilities extend across various facets of the e-commerce supply chain, offering environmentally friendly solutions:

1. *Last-Mile Delivery Optimization*: AI route algorithms reduce carbon emissions by finding efficient delivery routes based on real-time data, traffic, and customer locations.
2. *Demand Forecasting*: AI predicts consumer demand accurately, helping e-commerce firms optimize inventory, thus reducing expedited deliveries and excess stock, lowering carbon emissions.
3. *Energy Efficiency*: AI improves energy usage in e-commerce warehouses by monitoring and controlling systems like lighting and HVAC, reducing wasteful consumption and the facilities' carbon footprint.
4. *Sustainable Sourcing*: AI identifies eco-friendly sourcing options, promoting products with lower environmental impacts like those made with renewable energy or sustainable materials.
5. *Consumer Behavior Analysis*: AI analyzes browsing and purchase history to predict future preferences, helping e-commerce companies stock products accurately and reduce waste.
6. *Carbon Accounting*: AI software calculates and monitors a company's real-time carbon emissions, empowering e-commerce businesses to make informed decisions about reducing environmental impact.
7. *Sustainable Packaging*: AI optimizes packaging design to minimize materials and waste, suggesting eco-friendly options to reduce the environmental impact of e-commerce shipments.
8. *Supply Chain Optimization*: AI enhances supply chain management by predicting disruptions and optimizing logistics, leading to more efficient and environmentally friendly transportation, thus reducing emissions (Andrews, 2023).
9. *Dynamic Pricing*: AI adjusts product prices based on demand, encouraging off-peak purchases and optimizing transportation resource utilization.

The potential of AI in combating carbon emissions in e-commerce is vast. By optimizing routes, reducing energy consumption, and promoting sustainable practices, AI can play a crucial role in ensuring that e-commerce evolves into a more environmentally responsible industry. It contributes to a greener future while maintaining the convenience and efficiency that consumers expect from online shopping (Kalkandha, 2024).

10.1.3 Purpose of the Chapter

The purpose of this chapter is to explore the dynamic interplay between the rapidly growing e-commerce industry and its environmental impact, with a specific focus on carbon emissions and waste. We aim to:

1. *Raise Awareness*: Highlight the environmental challenges of e-commerce's rapid growth, including carbon emissions and waste (Huang et al., 2014).
2. *AI's Role*: Emphasize AI's potential to address these challenges, revolutionizing e-commerce for sustainability.
3. *Insights*: Discuss AI's use in reducing carbon emissions, from delivery optimization to warehouse energy efficiency.
4. *Sustainability Initiatives*: Explore e-commerce's green practices like eco-packaging and sustainable sourcing.
5. *Recommendations*: Offer advice for e-commerce to adopt AI and sustainable practices for environmental impact reduction.
6. *Responsible E-commerce*: Advocate for eco-conscious e-commerce practices alongside convenience.

By addressing these key points, our chapter aims to contribute to the discourse on sustainable e-commerce and inspire positive change within the industry, ultimately leading to an eco-friendlier and more efficient digital marketplace.

10.2 THE CARBON FOOTPRINT OF E-COMMERCE AND ITS ENVIRONMENTAL IMPACT AND SUSTAINABLE SOLUTIONS

10.2.1 Definition of Carbon Footprint

The Carbon Footprint of E-commerce refers to the environmental impact caused by online shopping and the associated activities, such as packaging, transportation, and data centers

10.2.2 Some Positive Aspects of E-commerce Related to Carbon Footprints

1. *Reduced Travel*: Online shopping often eliminates the need for individual trips to physical stores, reducing fuel consumption and emissions associated with transportation.
2. *Efficient Fulfilment*: Some e-commerce companies use advanced logistics and inventory management systems, which can optimize delivery routes and reduce unnecessary packaging, minimizing carbon emissions.
3. *Local Sourcing*: E-commerce platforms can enable consumers to source products from local or regional businesses, potentially reducing the carbon footprint of long-distance shipping.
4. *Green Initiatives*: Some e-commerce companies are actively working to reduce their carbon footprint by investing in renewable energy, using eco-friendly packaging, and adopting sustainable practices (Kalkandha, 2023).
5. *Data-Driven Improvements*: E-commerce platforms can use data analytics to optimize inventory management and reduce waste, leading to lower carbon emissions.
6. *Remote Work*: The growth of e-commerce has also led to more remote work opportunities, reducing commuting-related emissions.

10.2.2.1 Several Ways of Contribution of E-commerce to Carbon Emissions

1. *Shipping and Delivery*: The transportation of products from fulfillment centers to customers' doorsteps often involves trucks, airplanes, and delivery vehicles powered by fossil fuels, which release carbon dioxide (CO_2) and other greenhouse gases into the atmosphere (United States Environmental Protection Agency (US EPA), 2023).
2. *Packaging*: The packaging materials used for shipping, such as cardboard boxes and plastic, contribute to carbon emissions due to their production, which typically involves energy-intensive processes and emissions. Additionally, over-packaging can lead to more waste.
3. *Warehousing and Fulfilment Centers*: E-commerce companies operate large warehouses and fulfilment centers that require heating, cooling, and lighting. These facilities consume significant amounts of energy, contributing to emissions.
4. *Returns*: The process of returning products purchased online can result in additional carbon emissions, as items may be shipped back to distribution centers, leading to extra transportation and packaging.
5. *Data Centers*: E-commerce platforms rely on data centers to run their websites and manage online transactions. These data centers consume substantial amounts of energy, often derived from fossil fuels.
6. *Consumer Behavior*: The convenience of online shopping can lead to increased consumption and a "click and ship" culture, where consumers order individual items with separate deliveries, which can be less efficient than traditional shopping in physical stores.
7. *Waste Generation*: E-commerce generates packaging waste and electronic waste from devices like smartphones and computers used for online shopping.

Efforts are being made to reduce the carbon footprint of e-commerce, such as optimizing delivery routes, improving packaging sustainability, and using renewable energy in fulfillment centers. However, the overall impact on carbon emissions largely depends on the efficiency of these measures and consumer choices.

10.2.3 The Positive Environmental Impact of E-Commerce

1. *Reduced Retail Space*: E-commerce reduces the need for extensive physical retail spaces, which can lower energy consumption related to heating, cooling, and lighting in brick-and-mortar stores.
2. *Optimized Inventory*: E-commerce businesses often employ data-driven inventory management, which can reduce overstocking and waste, leading to a more sustainable supply chain (Andrews, 2023).
3. *Encouraging Local Markets*: E-commerce can support local and small-scale producers and reduce the environmental impact of long-distance shipping, promoting regional sustainability.
4. *Paperless Transactions*: Online transactions typically involve electronic payments and digital receipts, reducing the need for paper and lowering the impact on forests.

10.3 THE CHALLENGE OF CARBON EMISSIONS AND WASTE IN E-COMMERCE

10.3.1 Overview of the Current State of Carbon Emissions and Waste in E-commerce

1. *Shipping and Transportation*: E-commerce heavily relies on shipping and transportation, which can contribute to carbon emissions. Companies have been working to reduce emissions through strategies like optimizing delivery routes and using electric or hybrid vehicles.
2. *Warehousing and Energy Use*: E-commerce requires vast warehouses, which consume energy for lighting, heating, and cooling. Companies have been adopting energy-efficient technologies and renewable energy sources to reduce their carbon footprint.
3. *Packaging*: Packaging materials, particularly single-use plastics, can be a significant source of waste and emissions. Many e-commerce companies are transitioning to more sustainable packaging options.

Note: Top 100 cities globally only.
*Average commute for representative city.

10.3.2 Waste in E-Commerce

1. *Packaging Waste*: Excessive packaging and single-use plastics have been a concern. Efforts are being made to reduce waste through minimal packaging and the use of recyclable or biodegradable materials.
2. *Product Returns*: E-commerce often faces higher return rates compared to traditional retail, leading to increased product waste. Companies are improving sizing guides and return policies to reduce returns.
3. *Electronic Waste*: E-waste is a growing issue as consumers frequently upgrade electronic devices. Proper recycling and disposal programs are essential.
4. *Food Waste*: For e-commerce groceries, managing food waste is a challenge due to perishable items. Companies are working on better inventory management and donations of surplus food.

Since the 2010s, there has been a growing awareness of the environmental impact of e-commerce, and many companies have made commitments to reduce their carbon emissions and waste. They are adopting sustainable practices, such as carbon-neutral shipping and more responsible sourcing.

10.4 THE ROLE OF ARTIFICIAL INTELLIGENCE IN REDUCING CARBON EMISSIONS AND WASTE

10.4.1 Overview of AI in E-Commerce

AI in E-commerce is transforming the way businesses operate and enhancing customer experiences. Here's an overview:

1. *Personalization*: AI algorithms analyze customer data to provide tailored product recommendations, personalized content, and customized shopping experiences. This leads to increased customer engagement and higher conversion rates.
2. *Chatbots and Virtual Assistants*: AI-powered chatbots and virtual assistants offer 24/7 customer support, helping users with inquiries, product searches, and even order processing. They enhance customer service efficiency.
3. *Visual Search*: Visual AI enables customers to search for products using images or photos. It simplifies the shopping process, making it more intuitive.
4. *Inventory and Demand Forecasting*: AI helps retailers optimize inventory management by forecasting demand more accurately, reducing overstocking and understocking issues.
5. *Pricing Optimization*: AI algorithms can adjust prices in real-time based on various factors, such as demand, competitor prices, and customer behavior, maximizing revenue.
6. *Fraud Detection and Security*: AI is crucial in identifying and preventing fraudulent activities, protecting both customers and businesses from cyber threats.(Rehman, 2019)
7. *Recommendation Systems*: AI-driven recommendation engines analyze user behavior and preferences to suggest products, increasing cross-selling and upselling opportunities.
8. *Supply Chain Management*: AI enhances logistics and supply chain operations by optimizing routes, predicting delivery times, and reducing shipping costs.
9. *Voice Commerce*: Voice-activated AI, such as virtual assistants, enables voice shopping, making it more convenient for users.
10. *Customer Insights*: AI tools process large volumes of data to provide insights into customer behavior, helping businesses make informed decisions and tailor marketing strategies.
11. *Augmented Reality (AR) and Virtual Reality (VR)*: AI-driven AR and VR technologies enable customers to virtually try products before purchasing, enhancing the online shopping experience.
12. *Dynamic Pricing*: AI can set prices based on real-time data, competitor pricing, and customer behavior, allowing businesses to remain competitive.
13. *Behavioral Analysis*: AI analyses customer behavior on e-commerce websites, helping companies understand how customers interact with their platform and make improvements.
14. *Content Generation*: AI can generate product descriptions, reviews, and other content, reducing the need for manual content creation.
15. *Recommendation Systems for Customer Service*: AI-powered chatbots and virtual assistants are increasingly used for customer support, answering questions and resolving issues more efficiently.

The integration of AI in e-commerce is ongoing, offering businesses the potential to streamline operations, improve customer engagement, and remain competitive in a rapidly evolving industry.

10.4.1.1 Optimizing Delivery Routes with AI

AI algorithms can significantly improve the efficiency of delivery routes in e-commerce by using advanced data analytics and optimization techniques. Here's a detailed explanation of how this works:

1. *Data Collection*: AI algorithms start by collecting vast amounts of data, including historical delivery routes, traffic patterns, weather conditions, and real-time information such as road closures or accidents.
2. *Route Planning*: The algorithms analyze this data to identify the most efficient routes for delivery, taking into account factors like distance, traffic congestion, and delivery time windows.
3. *Machine Learning*: AI systems often employ machine learning to continuously improve route optimization. They learn from past data and adapt to changing conditions, making them more effective over time.
4. *Real-time Updates*: AI can provide real-time updates to drivers, helping them adjust their routes in response to unexpected events or changes in traffic conditions.
5. *Geospatial Technologies*: AI utilizes geospatial technologies such as GPS and geographic information systems (GIS) to pinpoint precise delivery locations and optimize routes accordingly.

10.4.1.2 Benefits of Optimizing Delivery Routes

Optimizing delivery routes using AI offers several benefits:

1. *Cost Savings*: Efficient routes reduce fuel consumption, maintenance costs, and the wear and tear on delivery vehicles, resulting in significant cost savings.
2. *Faster Deliveries*: AI-optimized routes can lead to quicker deliveries, improving customer satisfaction and potentially enabling same-day or next-day delivery options.
3. *Reduced Emissions*: By minimizing travel distance and reducing idling time, optimized routes can lead to a decrease in carbon emissions, promoting environmental sustainability.
4. *Improved Accuracy*: AI helps prevent delivery errors by ensuring that packages are delivered to the right location on time.
5. *Enhanced Customer Experience*: Faster, more reliable deliveries contribute to a better overall customer experience, which can lead to higher customer retention rates and loyalty.

10.4.2 Examples of E-commerce Companies Implementing AI for Delivery Route Optimization (Hardesty, 2022)

1. *Amazon*: Amazon is a pioneer in using AI for delivery route optimization. Their system, known as Amazon Flex, uses AI to determine the most efficient routes for delivery drivers, allowing them to make deliveries more quickly and efficiently.

2. *Flipkart*: One of India's leading e-commerce platforms, Flipkart, uses AI algorithms to optimize delivery routes, reduce delivery times, and improve customer experiences. This technology helps them efficiently serve a vast customer base across the country.
3. *Snapdeal*: Snapdeal, another prominent Indian e-commerce company, leverages AI to enhance delivery logistics. They focus on improving delivery accuracy and efficiency, particularly in the context of the diverse and complex Indian market.
4. *Grofers*: As an online grocery delivery platform, Grofers employs AI for route optimization to ensure timely deliveries of fresh produce and groceries to customers' doorsteps. This optimization is critical for meeting the unique demands of grocery e-commerce.
5. *Zomato*: While primarily known for food delivery, Zomato also uses AI-driven route optimization to ensure that food orders reach customers as quickly as possible. This is essential for the food delivery sector in India, which demands fast and reliable service.
6. *BigBasket*: BigBasket, a leading online grocery store, utilizes AI to optimize delivery routes, minimize delays, and ensure efficient delivery of fresh groceries to households across India.
7. *UPS*: The United Parcel Service (UPS) in the United States employs On-Road Integrated Optimization and Navigation (ORION), an AI system that optimizes delivery routes, reducing fuel consumption and improving delivery times.

These examples illustrate how AI-driven route optimization has become a standard practice in the e-commerce and logistics industry, helping companies deliver packages faster, more accurately, and with reduced operational costs.

10.4.2.1 Predictive Analytics for Reducing Unnecessary Shipments

Predictive analytics involves using historical and real-time data, statistical algorithms, and machine learning to forecast future events or behaviors. In e-commerce, predictive analytics can help reduce unnecessary shipments in the following ways:

1. *Demand Forecasting*: Predictive analytics analyzes historical sales data, market trends, and other relevant information to forecast future demand accurately. This enables e-commerce companies to stock products more efficiently, reducing the need for excess inventory and the associated shipping of unnecessary items.
2. *Return Predictions*: By analyzing past return data and customer behavior, predictive analytics can identify potential return cases before they happen. This allows companies to proactively address issues, reduce returns, and prevent unnecessary shipments related to returns.
3. *Customer Behavior Analysis*: Predictive analytics assesses customer behavior, such as browsing patterns, cart abandonment, and purchase history, to predict what customers are likely to buy. This enables companies to optimize their inventory and avoid unnecessary shipments of unwanted products.

4. *Real-time Inventory Management*: Predictive analytics can continually monitor inventory levels and reorder products automatically when they are running low. This reduces the risk of running out of stock, which can result in missed sales and unnecessary expedited shipping.

10.4.2.2 Benefits of Predictive Analytics

Predictive analytics offers several advantages in reducing unnecessary shipments in e-commerce:

1. *Cost Reduction*: By accurately forecasting demand and optimizing inventory, e-commerce companies can reduce the costs associated with holding excess inventory and expedited shipping.
2. *Enhanced Customer Experience*: Predictive analytics ensures products are available when customers want them, improving the overall shopping experience and reducing instances of out-of-stock items.
3. *Sustainability*: Fewer unnecessary shipments mean a reduced carbon footprint, contributing to environmental sustainability.
4. *Operational Efficiency*: Predictive analytics streamlines supply chain operations by helping companies plan and manage inventory more efficiently.

10.4.2.3 Examples of E-Commerce Companies Using Predictive Analytics for Reducing Unnecessary Shipments

1. *Amazon*: Amazon uses predictive analytics to optimize its inventory and demand forecasting, reducing unnecessary shipments of overstocked items.
2. *Alibaba*: Alibaba leverages predictive analytics for inventory management and demand forecasting, which helps it avoid unnecessary shipments and reduce costs.
3. *Walmart*: Walmart uses predictive analytics to forecast demand and optimize inventory levels, preventing unnecessary shipments and ensuring products are available when customers need them.
4. *Zalando*: Zalando, a European e-commerce fashion platform, employs predictive analytics to predict fashion trends and customer preferences, reducing unnecessary shipments of unpopular clothing items.

10.4.3 How AI Can Help with Green Energy Usage

AI plays a crucial role in optimizing green energy usage in various ways:

1. *Energy Forecasting*: AI models can analyze historical data, weather patterns, and energy consumption to forecast energy demand accurately. This helps utilities and grid operators optimize the use of renewable energy sources like wind and solar.
2. *Grid Management*: AI systems monitor the energy grid in real-time, balancing supply and demand. They can predict and respond to fluctuations in renewable energy production, ensuring a stable power supply.

3. *Energy Efficiency*: AI-powered systems can analyze energy consumption patterns in buildings and industrial processes. They identify areas for improvement and suggest energy-saving measures, reducing overall energy usage.
4. *Demand Response*: AI can manage demand response programs, where electricity consumers reduce their usage during peak periods. This reduces stress on the grid and allows for a higher proportion of green energy in the energy mix.
5. *Renewable Energy Integration*: AI assists in integrating renewable energy sources into the grid efficiently. It can predict when solar or wind energy production will be at its peak and adjust grid operations accordingly.

10.4.3.1 Examples of Companies Using AI for Green Energy Usage

1. *Google*: Google utilizes AI for data center energy management. Their DeepMind AI system optimizes cooling systems, leading to a 30% reduction in energy consumption.
2. *Siemens*: Siemens employs AI to improve the efficiency of wind turbines. AI algorithms analyze weather data to optimize turbine performance and increase energy output.
3. *Tesla*: Tesla uses AI in its Powerwall and Powerpack energy storage systems. These systems store excess energy from solar panels and optimize when to release it, reducing dependence on non-renewable energy sources.
4. *Enel Green Power*: This global renewable energy company employs AI to forecast wind and solar energy production. These forecasts help them integrate renewable energy into the grid more effectively.
5. *Schneider Electric*: Schneider Electric offers AI-powered solutions for building management and energy efficiency. Their EcoStruxure platform uses AI to reduce energy waste and improve sustainability.

These companies demonstrate how AI is a valuable tool for green energy usage, helping to enhance energy efficiency and reduce reliance on fossil fuels, ultimately contributing to a more sustainable future.

10.5 CASE STUDIES

10.5.1 Amazon's AI-Powered Delivery Route Optimization

Routing problems, like the traveling salesman problem, present computational challenges. Drivers often deviate from computed routes due to their insights into road conditions, traffic, and parking.

10.5.1.1 Amazon's Approach

- Amazon's Last Mile team optimizes delivery routes, emphasizing safety, experience, sustainability, and efficiency.

10.5.1.2 Collaboration with MIT

- Amazon collaborates with MIT's Center for Transportation & Logistics to organize the Amazon Last Mile Routing Research Challenge (Hardesty, 2022).
- The challenge invites academic teams to train machine learning models to predict routes preferred by experienced drivers.

10.5.1.3 Data Utilization

- Participants access Amazon's historical data, including delivery locations, package details, travel times, and distances.
- They also receive over 4,000 driver-determined routes to enrich their models in detecting deviations from conventional routes.

10.5.1.4 Competition Rewards

- Monetary rewards are offered, with $100,000 for first place, $50,000 for second, and $25,000 for third.
- Exceptional teams may secure research positions at Amazon's Last Mile division or present their work at MIT.

10.5.1.5 Encouraging Innovation

"We are encouraging participants to develop innovative approaches leveraging artificial intelligence, machine learning, deep learning, computer vision, and other non-conventional methods," says Julian Pachon, Director and Chief Scientist at Amazon Last Mile. The goal is to surpass traditional optimization-driven operations research methods in terms of solution quality and computational efficiency.

10.5.2 Predictive Analytics for Inventory Management at Zara

In the dynamic realm of supply chain management, staying competitive requires more than mere responsiveness to demand; it demands the capacity to anticipate it. Predictive analytics is a transformative technology revolutionizing inventory management, enhancing operational efficiency, and ushering in a paradigm shift. Here you can explore the profound impact of predictive analytics on inventory optimization and its revolutionary implications across industries (Steady Pace – Market Research, 2023).

1. *The Dilemma of Inventory Management*: Traditional supply chains grapple with balancing insufficient inventory, leading to missed sales and excess inventory, causing extra expenses, requiring perceptive discernment and anticipatory acumen.
2. *The Role of Predictive Analytics as a Strategic Disruptor*: Powered by machine learning algorithms, predictive analytics offers a data-centric approach to inventory management, proactively anticipating demand instead of reacting to it.

3. *The Accuracy of Forecasting*: Predictive analytics utilizes historical data and market trends for precise predictions, surpassing previous accuracy levels, optimizing inventory levels to avoid shortages and overstock situations.
4. *Dynamic Inventory Optimization*: Predictive analytics enables real-time adjustments to inventory management, responding swiftly to changing demand patterns, seasonal shifts, and external influences.
5. *Cost Reduction and Capital Efficiency*: Accurate demand anticipation optimizes stocking levels, reducing excess inventory, freeing up resources, and improving cash flow and profitability.
6. *Supplier Collaboration Enhancement*: Predictive analytics facilitates collaboration with suppliers by providing demand forecasts, streamlining the supply chain, and minimizing procurement time.
7. *Risk Mitigation*: Predictive analytics identifies potential supply chain risks, aiding strategic risk mitigation efforts in response to events like weather events or geopolitical circumstances.

10.5.2.1 Several Businesses have Adopted Predictive Analytics for Inventory Optimization with Impressive Results

- Amazon uses predictive analytics to forecast customer demand and enhance inventory management, contributing to expedient and dependable delivery services.
- Walmart employs predictive analytics to ensure product availability according to customer demands and locations.
- Zara anticipates fashion trends and controls inventory management effectively, responding to evolving customer preferences (Breeur Solutions, n.d.).

10.5.2.2 The Forthcoming Era: Perpetual Enhancement and Innovation

The continuous evolution of technology will amplify predictive analytics' role in supply chain management. Emerging technologies like AI and the Internet of Things (IoT) will enhance predictive analytics, resulting in more intelligent and adaptable supply chains.

In Conclusion, by embracing predictive analytics for inventory optimization, businesses can unlock unprecedented efficiency, reduce costs, enhance customer satisfaction, and stay ahead in an ever-evolving marketplace. The future belongs to those who can predict it.

10.5.3 Alibaba's Initiatives for Green Energy Optimization

Alibaba has been actively involved in green energy optimization and sustainability initiatives.

10.5.3.1 Alibaba's Green Energy Optimization Efforts

- *Data Centers and Cloud Computing*: Alibaba Cloud, the cloud computing arm of Alibaba Group, has focused on optimizing energy use in data centers. They have invested in energy-efficient infrastructure and have committed to using more renewable energy to power their data centers, reducing their carbon footprint.

- *Sustainable Operations*: Alibaba Group has adopted sustainable practices in its daily operations, including energy-efficient buildings and green logistics. They've worked on reducing energy consumption and emissions throughout their supply chain.
- *Renewable Energy Procurement*: Alibaba has made efforts to procure renewable energy to power its operations. They have invested in wind and solar projects, aiming to use more clean energy sources in their operations.
- *Carbon Neutrality Commitment*: Alibaba announced its commitment to becoming carbon neutral by 2030. They have set targets to reduce carbon emissions and offset the remaining emissions by investing in carbon offset projects.
- *Environmental Responsibility*: Alibaba has been transparent about its environmental initiatives and has disclosed information about its sustainability goals and progress in annual sustainability reports.

10.6 ADVANTAGES AND DISADVANTAGES OF AI-EQUIPPED TOOLS AND APPLICATIONS

10.6.1 Advantages

1. *Increased Efficiency*: AI can automate repetitive tasks, allowing for quicker and more accurate completion of tasks. This can significantly boost productivity.
2. *Data Processing and Analysis*: AI algorithms can process and analyze large volumes of data much faster and more accurately than humans. This enables businesses to make data-driven decisions.
3. *Personalization and Customization*: AI can analyze user behavior and preferences to tailor experiences. This is evident in recommendation systems (like those used by Netflix or Amazon) or personalized marketing strategies.
4. *Predictive Analytics*: AI can analyze patterns in data to make predictions about future trends or events. This is invaluable for businesses in planning and strategy development.
5. *Enhanced Healthcare*: AI can assist in diagnosing diseases, interpreting medical images, and even suggesting treatment plans. This leads to more accurate and timely healthcare.
6. *Improved Safety and Security*: AI can monitor and analyze video feeds, identify anomalies, and alert authorities in real-time. This is used in applications like surveillance systems and fraud detection.
7. *Automation in Manufacturing*: AI-powered robots and machines can perform tasks that are dangerous, repetitive, or require high precision. This leads to increased safety and efficiency in manufacturing processes.
8. *Language Translation*: AI-driven translation tools are making it easier for people to communicate across language barriers. This is particularly useful in global business and travel.

9. *24/7 Availability*: AI-driven systems can operate around the clock without needing breaks, providing continuous service or support.
10. *Environmental Impact*: AI can optimize energy consumption, reduce waste, and contribute to sustainable practices in industries like transportation, agriculture, and energy.

10.6.2 Disadvantages

1. *High Initial Costs*: Implementing AI systems can be expensive, especially for smaller businesses or organizations with limited budgets.
2. *Lack of Human Touch*: In some cases, AI-driven interactions may lack the personal touch and empathy that human interactions provide. This can be a drawback in customer service or caregiving scenarios.
3. *Limited Understanding of Context*: AI systems can struggle to understand context or nuance, leading to misinterpretations or incorrect responses, especially in complex or ambiguous situations.
4. *Job Displacement*: As AI automates certain tasks, there is concern about potential job displacement. Some roles may be replaced by machines, leading to workforce restructuring.
5. *Ethical and Bias Concerns*: AI systems can inherit biases present in the data they're trained on, leading to unfair or discriminatory outcomes. This is a significant concern in areas like hiring, criminal justice, and lending.
6. *Overreliance on Technology*: Depending too heavily on AI can lead to a lack of human expertise in critical areas, potentially causing problems if the technology fails or makes incorrect decisions.
7. *Difficulty in Interpretability*: Some AI models, particularly deep learning models, can be highly complex and difficult to understand or explain, making it challenging to trace decisions back to their source.
8. *Regulatory and Legal Challenges*: There may be legal and regulatory hurdles to implementing AI systems, especially in industries with strict compliance requirements (e.g., healthcare, finance).
9. *Inability to Handle Unforeseen Situations*: AI systems are typically trained on historical data, which means they may struggle with novel or unprecedented situations.
10. *Potential for Technological Bias*: If the development and implementation of AI are not carefully monitored, it can lead to technological biases that reinforce existing social, economic, or cultural disparities.

10.6.3 Evaluation of the Effectiveness of AI Tools and Applications in Reducing Carbon Emissions and Waste

The effectiveness of AI tools and applications in reducing carbon emissions and waste depends on various factors, including the specific technologies used, the industries involved, and the implementation strategies. Here is an evaluation of the effectiveness based on key areas of application.

- *Monitoring Emissions*: Companies can utilize AI-powered data engineering to automatically track emissions across their carbon footprint. They collect data from operations, including corporate travel and IT equipment, and throughout the value chain, such as materials suppliers and transporters. AI can leverage new data sources like satellites and intelligently estimate missing data to improve accuracy.
- *Predicting Emissions*: Predictive AI forecasts future emissions, considering current reduction efforts, new methodologies, and future demand. This enables more accurate setting, adjustment, and achievement of reduction targets.
- *Reducing Emissions*: Prescriptive AI and optimization provide detailed insights into the value chain, improving efficiency in production, transportation, and other areas. This leads to reduced carbon emissions and cost savings.

To conclude, AI can optimize energy consumption in various sectors, including manufacturing, transportation, and buildings. Smart grids, predictive maintenance, and demand–response systems can lead to significant energy savings.

10.6.3.1 Ethical Concerns Around AI in E-Commerce

1. *Privacy and Data Security*: AI systems often require access to vast amounts of user data to operate effectively. There's a concern about how this data is collected, stored, and protected.
2. *Transparency and Explainability*: AI algorithms can be highly complex, making it challenging to understand how they make decisions. Lack of transparency can lead to distrust and concerns about hidden biases.
3. *Bias and Fairness*: AI models can inherit biases present in training data, leading to discriminatory outcomes. This can impact pricing, recommendations, and product availability.
4. *Consumer Manipulation*: AI-powered systems can personalize experiences, potentially leading to targeted marketing tactics that manipulate consumer behavior or create filter bubbles.
5. *Informed Consent*: Users may not fully understand the extent to which AI is used in e-commerce platforms. Providing clear information and obtaining informed consent can be a challenge.
6. *Job Displacement*: Automation through AI in e-commerce may lead to job loss in traditional retail, potentially exacerbating socioeconomic disparities.
7. *Algorithmic Accountability*: Who is accountable if an AI system makes a harmful or biased decision? Establishing responsibility and liability can be challenging.
8. *Counterfeit and Fraud Detection*: While AI is used to detect counterfeit products, there may be ethical considerations in how sellers and products are flagged, potentially leading to unfair treatment.
9. *Environmental Impact*: AI-powered logistics optimization and predictive inventory management can lead to increased demand for energy and resources, potentially contributing to environmental issues.
10. *Customer Profiling and Segmentation*: AI can create highly detailed customer profiles, potentially raising issues of privacy and data ownership.

10.7 THE FUTURE OF AI IN E-COMMERCE FOR REDUCING CARBON EMISSIONS AND WASTE

10.7.1 Overview of the Prospects of AI in the E-Commerce Industry

AI has significant and promising prospects in the e-commerce industry, offering numerous opportunities to enhance customer experiences, streamline operations, and drive revenue growth. Here's an overview of the prospects of AI in e-commerce:

1. *Personalization and Customer Experience*: AI can analyze user data to provide personalized product recommendations, content, and offers, creating a more engaging and satisfying shopping experience.
2. *Search and Navigation*: AI-powered search engines and chatbots can help customers find products more easily and answer their questions, reducing friction in the shopping process.
3. *Inventory Management*: AI can optimize inventory levels, reducing out-of-stock and overstock situations, which can lead to cost savings and increased sales.
4. *Pricing Optimization*: AI can dynamically adjust prices based on factors like demand, competition, and historical data, helping e-commerce businesses maximize revenue.
5. *Supply Chain and Logistics*: AI can optimize routing, forecasting, and demand planning, reducing shipping costs, minimizing delivery times, and improving overall supply chain efficiency.
6. *Customer Service*: AI-driven chatbots and virtual assistants can provide 24/7 customer support, answer common queries, and free up human agents to handle more complex issues.
7. *Fraud Detection and Security*: AI can identify and prevent fraudulent activities, such as payment fraud or account takeovers, enhancing the security of e-commerce platforms.
8. *Visual Search and Image Recognition*: AI can enable users to search for products using images, enhancing the shopping experience and simplifying product discovery.
9. *Voice Commerce*: AI-powered voice assistants can facilitate voice-based shopping, making it more convenient for customers and potentially opening up new revenue streams.
10. *Returns and Refunds*: AI can streamline the returns process, helping e-commerce businesses reduce the costs and complexity associated with returns while ensuring customer satisfaction.
11. *Customer Feedback Analysis*: AI can analyze customer reviews and feedback to provide insights into product improvements and areas of concern.
12. *Competitive Intelligence*: AI can monitor competitors' pricing, product offerings, and customer sentiments, enabling e-commerce businesses to make more informed decisions.
13. *Localization and Translation*: AI can facilitate international expansion by providing accurate translation and localization of product listings and content.

14. *Sustainability and Ethical Sourcing*: AI can help e-commerce businesses optimize supply chains for sustainability and ethical sourcing, meeting growing consumer demands for responsible practices.
15. *AI-Driven Marketing*: AI can enhance marketing efforts by targeting the right audience, creating personalized campaigns, and optimizing advertising spend.

10.7.1.1 Potential for Reducing Carbon Emissions and Waste

Reducing carbon emissions and waste is crucial for mitigating the impacts of climate change and promoting sustainable practices. There are numerous strategies and technologies with significant potential for achieving these goals:

1. *Renewable Energy Sources*: Transitioning from fossil fuels to renewable energy sources like solar, wind, and hydropower can significantly reduce carbon emissions from energy production.
2. *Energy Efficiency Measures*: Implementing energy-saving technologies and practices in industries, buildings, and transportation can lead to substantial reductions in energy consumption and emissions.
3. *Transportation Electrification*: Shifting to electric vehicles (EVs) for transportation, especially in combination with a transition to renewable energy, can drastically reduce emissions from the transportation sector.
4. *Public Transportation and Active Transportation*: Investing in and promoting public transit systems, as well as supporting walking and cycling infrastructure, can lead to reduced reliance on individual cars and lower emissions.
5. *Carbon Pricing and Taxes*: Implementing policies that put a price on carbon emissions can incentivize businesses and individuals to reduce their emissions.
6. *Waste Reduction and Recycling*: Promoting waste reduction practices and improving recycling infrastructure can divert a significant amount of waste from landfills, reducing methane emissions and conserving resources.
7. *Circular Economy Practices*: Implementing circular economy models that focus on reuse, repair, and recycling can minimize waste generation and resource consumption.
8. *Sustainable Agriculture and Forestry*: Adopting sustainable agricultural and forestry practices, such as agroforestry and no-till farming, can sequester carbon and reduce emissions associated with land use.
9. *Carbon Capture and Storage (CCS)*: Deploying CCS technologies in industries like power generation, cement production, and heavy manufacturing can capture and store carbon emissions.
10. *Afforestation and Reforestation*: Planting trees and restoring forests can sequester significant amounts of carbon dioxide from the atmosphere.
11. *Smart Grids and Energy Storage*: Implementing smart grid technologies and energy storage solutions can improve the efficiency and reliability of energy systems, reducing emissions.

12. *Green Building Standards*: Constructing buildings that meet high energy efficiency standards can significantly reduce emissions associated with heating, cooling, and electricity use.
13. *Digitalization and IoT in Industry*: Utilizing digital technologies and the IoT in industries can optimize processes, reduce energy consumption, and minimize waste production.
14. *Behavioral Changes and Awareness*: Promoting sustainable behaviors and educating the public about the impacts of their choices can lead to reduced consumption and waste.
15. *Innovation and Technology Advancements*: Investing in the research and development of new technologies and solutions for clean energy, waste reduction, and emission reduction can drive significant progress.

10.7.1.2 Challenges and Future Research Areas

1. *Technological Innovation and Adoption*: Developing and implementing advanced technologies to reduce emissions and waste across various industries can be resource-intensive and face resistance from established practices.
2. *Cost and Financial Considerations*: Many sustainable technologies and practices may have high upfront costs, making it challenging for some businesses and industries to invest in them.
3. *Policy and Regulation*: Achieving effective and enforceable policies and regulations at the local, national, and international levels can be complex, and implementation may face political and economic hurdles.
4. *Behavioral Change and Consumer Awareness*: Encouraging individuals and communities to adopt sustainable behaviors and make environmentally conscious choices can be difficult due to entrenched habits and limited awareness.
5. *Data Availability and Quality*: Access to accurate and comprehensive data is crucial for informed decision-making, but obtaining reliable data on emissions, waste generation, and resource use can be challenging.

10.7.2 Future Research Areas

1. *Carbon Capture and Storage (CCS)*: Advancing CCS technologies and exploring innovative approaches for capturing and storing carbon emissions from various sources.
2. *Circular Economy and Material Efficiency*: Investigating strategies to maximize resource use and minimize waste generation through practices like recycling, reusing, and remanufacturing.
3. *Behavioral Economics and Decision Science*: Understanding consumer behavior and decision-making processes to design interventions that promote sustainable choices.
4. *Blockchain and Sustainable Supply Chains*: Exploring the potential of blockchain technology to enhance transparency, traceability, and accountability in global supply chains.

5. *Climate Adaptation and Resilience*: Developing strategies and technologies to adapt to climate change impacts and enhance resilience in vulnerable communities and ecosystems.
6. *Bioenergy and Carbon Neutral Technologies*: Investigating bioenergy technologies and carbon-neutral approaches to energy production and utilization.

10.8 REGULATORY AND INDUSTRY INITIATIVES TO COMBAT CARBON EMISSIONS AND WASTE IN E-COMMERCE

10.8.1 Global Initiatives to Combat Carbon Emissions and Waste

1. *Paris Agreement*: The Paris Agreement is a landmark global initiative that aims to limit global warming to well below 2 degrees Celsius above pre-industrial levels. It encourages countries to reduce carbon emissions, which indirectly affects industries, including e-commerce.
2. *United Nations Sustainable Development Goals*: Several of the United Nations Sustainable Development Goals (SDGs) address environmental sustainability, and these goals influence global initiatives to reduce carbon emissions and waste.
3. *Carbon Pricing*: Some countries and regions have implemented carbon pricing mechanisms, such as carbon taxes or cap-and-trade systems, to incentivize businesses, including e-commerce, to reduce their carbon emissions.
4. *Circular Economy Initiatives*: Global efforts promote a circular economy, which encourages the recycling and reuse of products and materials, ultimately reducing waste generation.
5. *Montreal Protocol on Substances that Deplete the Ozone Layer*: An international treaty designed to protect the ozone layer by phasing out the production and consumption of ozone-depleting substances. While primarily focused on ozone protection, it also contributes to reducing greenhouse gas emissions (Simonchuk & Romasheva, 2021).

10.8.2 Examples of Industry Initiatives

Industry initiatives for combatting carbon emissions and waste in the e-commerce sector are gaining traction as companies recognize the importance of sustainability in their operations. Here are some examples of such initiatives:

1. *The Climate Group's EV100*:
   - *Description*: EV100, initiated by The Climate Group, urges companies to transition their fleets to EVs, particularly relevant for e-commerce delivery fleets.
   - *Impact*: Committing to EV transition significantly reduces carbon emissions linked with last-mile delivery.

2. Amazon's Climate Pledge:
    - *Description*: Amazon's Climate Pledge pledges to achieve net-zero carbon emissions by 2040, a decade earlier than the Paris Agreement targets. It involves transitioning to renewable energy and investing in sustainable technologies.
    - *Impact*: As a major e-commerce entity, Amazon's Climate Pledge sets a significant precedent for industry-wide sustainability endeavors.
3. Alibaba's Green Logistics:
    - *Description*: Alibaba Group's Green Logistics initiative focuses on optimizing logistics and delivery to minimize emissions, including route optimization, energy-efficient vehicles, and innovative packaging.
    - *Impact*: Implementing Green Logistics aims to reduce Alibaba's e-commerce environmental impact.
4. *Zalando's do.MORE Sustainable Program*:
    - *Description*: Zalando's do.MORE Sustainable Program commits to reducing carbon emissions, increasing sustainable material usage, and enhancing supply chain transparency.
    - *Impact*: Zalando aims to drive positive change in the fashion industry by promoting sustainability.
5. *E-commerce Carbon Footprint Calculator by Shopify*:
    - *Description*: Shopify's E-commerce Carbon Footprint Calculator measures and offsets online businesses' carbon emissions, including shipping.
    - *Impact*: This tool aids e-commerce in understanding and reducing its environmental footprint.
6. *E-commerce Packaging Waste Reduction Efforts*:
    - *Description*: E-commerce companies launch initiatives to minimize packaging waste, including eco-friendly materials usage and responsible disposal promotion.
    - *Impact*: Addressing packaging waste improves e-commerce sustainability.
7. *eBay's Greener Choices*:
    - *Description*: eBay's Greener Choices program encourages sellers to adopt eco-friendly packaging, renewable energy, and waste reduction.
    - *Impact*: Empowering sellers to make sustainable choices enhances eBay's e-commerce sustainability efforts.

10.8.3 Potential Benefits and Challenges of These Initiatives

10.8.3.1 Benefits

1. *Environmental Impact*: These initiatives can significantly reduce carbon emissions and waste, contributing to a cleaner environment.
2. *Consumer Appeal*: Environmentally conscious consumers are more likely to support businesses that take sustainable actions, leading to increased brand loyalty.

3. *Cost Savings*: Energy-efficient practices and waste reduction measures can result in long-term cost savings.
4. *Compliance*: Companies that embrace these initiatives are more likely to remain compliant with evolving environmental regulations.

10.8.3.2 Challenges

1. *Initial Costs*: Implementing green initiatives often requires upfront investments, which can be a challenge for some businesses.
2. *Logistical Challenges*: Optimizing last-mile delivery and sustainable sourcing can be complex and require significant logistical adjustments.
3. *Consumer Behavior*: Changing consumer behavior to favor sustainable products or packaging can be challenging.
4. *Competitive Pressures*: Companies may face competitive pressures in an industry that does not universally adopt green practices.
5. *Regulatory Uncertainty*: Regulatory requirements can change, creating uncertainty for businesses striving for sustainability.

Balancing the benefits of these initiatives with the challenges is a complex but essential task for e-commerce businesses to reduce their environmental impact and ensure long-term sustainability.

10.9 CONCLUSION

In this chapter, we delved into the intersection of e-commerce and environmental sustainability, particularly focusing on the environmental impact of carbon emissions and waste associated with the e-commerce industry. We also explored the potential of AI in addressing these challenges. Here's a summary of the main points:

1. *E-commerce Growth*: The exponential rise of e-commerce has revolutionized retail, providing global consumers with unparalleled convenience.
2. *Environmental Impact*: However, this growth has led to increased carbon emissions, particularly in last-mile deliveries, alongside a surge in packaging waste and energy use.
3. *Sustainability Initiatives*: E-commerce firms are responding with sustainability measures like green packaging and energy-efficient warehouses.
4. *AI's Role*: AI is pivotal in reducing e-commerce's carbon footprint, optimizing delivery, and forecasting demand.
5. *Waste Reduction*: AI-driven solutions minimize overstock, optimize packaging, and suggest eco-friendly materials, aiding waste reduction.
6. *Supply Chain Efficiency*: AI optimizes supply chains, lowering transportation and logistics' environmental impact.
7. *Consumer Behavior*: By analyzing consumer actions, AI aids in accurate product stocking, reducing waste.
8. *Carbon Accounting*: AI-driven software monitors real-time carbon emissions, guiding decisions to cut footprints.
9. *Promoting Sustainability*: This chapter advocates for responsible e-commerce practices, balancing convenience with environmental stewardship.

In conclusion, as e-commerce continues to thrive, the responsible integration of AI-driven solutions and sustainability initiatives is critical for mitigating the industry's environmental impact. By embracing these technologies and practices, the e-commerce sector can evolve into a more environmentally responsible and sustainable digital marketplace, contributing to a greener and more efficient future.

10.9.1 Implications for Future Research and Business Practice

This chapter underscores several important implications for future research and business practices in the realm of e-commerce and environmental sustainability:

1. *AI Integration*: Future research should focus on the deep integration of AI solutions within e-commerce operations. This includes refining AI algorithms for last-mile delivery optimization, demand forecasting, and energy efficiency, and expanding their adoption across the industry.
2. *Innovation in Sustainable Packaging*: Businesses must invest in research and development to create innovative, sustainable packaging solutions. This includes exploring materials that are biodegradable or recyclable, as well as optimizing packaging designs to minimize waste.
3. *Supply Chain Optimization*: Continued research into supply chain optimization is essential. This includes the development of AI-driven tools for real-time logistics management, which reduce carbon emissions and improve overall efficiency (Chen., et al., 2022).
4. *Consumer Behavior Analysis*: Research should delve into understanding consumer behavior further. Advanced AI-driven consumer behavior analysis can help businesses anticipate demand, optimize inventory, and encourage sustainable purchasing patterns.
5. *Carbon Accounting Standards*: Future research should lead to the establishment of standardized carbon accounting practices within the e-commerce industry. This will allow for more accurate measurement and reduction of carbon emissions.
6. *Collaboration and Knowledge Sharing*: E-commerce companies should collaborate and share best practices in sustainability. Such collaborations can lead to industry-wide initiatives that drive positive change.
7. *Regulatory Compliance*: Research should address the evolving regulatory landscape. Businesses need to stay informed about changing environmental regulations and invest in practices that ensure compliance while reducing their carbon footprint.
8. *Business Model Evolution*: Companies should investigate how they can shift their business models to be more environmentally sustainable. This may include exploring circular economy principles, product life extension, and sustainable sourcing strategies.
9. *Consumer Education*: Future research should also focus on educating consumers about sustainable e-commerce practices. Informed consumers can drive demand for eco-friendly products and encourage businesses to adopt green practices.

10. *Economic Benefits*: Research should explore the economic benefits of sustainability. This includes investigating cost savings, increased consumer loyalty, and long-term profitability associated with environmentally responsible business practices.
11. The implications mentioned here highlight the need for ongoing research and evolving business practices in e-commerce to promote environmental sustainability. By addressing these implications, e-commerce can move toward a future that is both economically and environmentally sustainable.

10.9.2 Call to Action

1. *Environmental Responsibility*: Businesses have a critical role to play in reducing their carbon footprint and minimizing waste. Adopting AI in e-commerce supply chains demonstrates a commitment to environmental responsibility.
2. *Efficiency and Cost Savings*: AI-driven solutions not only help reduce environmental impact but also improve supply chain efficiency. This can result in cost savings and increased profitability for businesses.
3. *Competitive Advantage*: Companies that adopt AI early gain a competitive advantage in the market. Consumers are increasingly conscious of sustainability, and businesses that prioritize it can attract environmentally conscious customers.
4. *Regulatory Compliance*: Many regions are introducing environmental regulations. By proactively adopting AI for sustainability, businesses can stay ahead of regulatory requirements.
5. *Global Impact*: The collective adoption of AI in e-commerce supply chains can have a significant positive impact on global carbon emissions and waste reduction. It's a small step for each business but a giant leap for the environment.
6. *Innovation and Growth*: Embracing AI is not just about reducing negative impacts; it's also about fostering innovation and growth. AI can lead to better customer experiences, more efficient operations, and expanded business opportunities.

In conclusion, the call to action is a plea for businesses to recognize the environmental and business benefits of AI adoption in e-commerce supply chains. It's a strategic move that not only benefits the planet but also the bottom line and the long-term growth prospects of companies.

REFERENCES

Andrews, J. (2023, October 2). Revolutionizing Supply Chains: The Power of Predictive Analytics for Inventory Optimization. www.linkedin.com. Retrieved February 26, 2024, from https://www.linkedin.com/pulse/revolutionizing-supply-chains-power-predictive-justin-andrews/

Breeur Solutions. (n.d.). Data-Driven Methods from Zara. Retrieved February 26, 2024, from https://www.breeur.com/knowledge-insights/6-data-driven-methods-you-can-learn-from-zara

Brewer, M. K. (2019, October 9). Slow Fashion in a Fast Fashion World: Promoting Sustainability and Responsibility. *Laws*, *8*(4), 24. https://doi.org/10.3390/laws8040024

Chen, Y., Li, M., Song, J., Ma, X., Jiang, Y., Wu, S., & Chen, G. L. (2022, December 14). A Study of B2C Research Trends: Based on Knowledge Mapping and Literature Analysis. *Frontiers in Psychology*, *13*. https://doi.org/10.3389/fpsyg.2022.1009216

Hardesty, L. (2022, July 13). *Amazon, MIT team up to add driver know-how to delivery-routing models – Amazon Science*. Amazon Science. https://www.amazon.science/blog/amazon-mit-team-up-to-add-driver-know-how-to-delivery-routing-models

Huang, L., Xu, X., Shao, Q., & Liu, J. (2014). Improving Carbon Mitigation Potential through Grassland Ecosystem Restoration under Climatic Change in Northeastern Tibetan Plateau. *Advances in Meteorology*, *2014*, 1–11. https://doi.org/10.1155/2014/379306

Kalkandha, L. (2023). Carbon Footprints and Sustainable Practices. *Journal of Environmental Studies*, *12*(1), 45–60.

Kalkandha, T. (2024, January 23). *How Green AI transforms business sustainability?* Appventurez. https://www.appventurez.com/blog/green-ai-for-sustainable-future

Rehman, A. (2019, December 20). Barriers Affecting the Diffusion of Business-to-Consumer Online Retailing Acceptance in Kingdom of Saudi Arabia. *Indian Journal of Science and Technology*, *12*(47), 1–9. https://doi.org/10.17485/ijst/2019/v12i47/148946

Simonchuk, V. D., & Romasheva, N. V. (2021). Public Perception of Carbon Capture (Utilization) and Storage Projects: World Experience and the Situation in Russia. *E3S Web of Conferences*, *266*, 06008. https://doi.org/10.1051/e3sconf/202126606008

Steady Pace - Market Research. (2023, September 3). The Secret of Zara's Success: Data-Driven Decisions. www.linkedin.com. Retrieved October 13, 2023, from https://www.linkedin.com/pulse/secret-zaras-success-data-driven-decisions-steadypace-sa/

United States Environmental Protection Agency (US EPA). (2023, July 14). *Greenhouse Gases | US EPA*. US EPA. Retrieved October 8, 2023, from https://www.epa.gov/report-environment/greenhouse-gases

van Gelder, K. (2023, September 26). *Topic: E-commerce worldwide*. Statista. Retrieved October 8, 2023, from https://www.statista.com/topics/871/online-shopping/

11 Exploring the Role of Artificial Intelligence in Educational Technology

Girja Shankar Patel and Surendra Kumar Gautam

11.1 INTRODUCTION

With the aid of the Internet of Things, big data, cloud computing, RFID technology, artificial intelligence (AI), virtual reality, and other new technologies, smart libraries have realized the physical space intelligence, information resource (knowledge) organization intelligence, service mode intelligence, and management method intelligence of libraries during the past ten years under the influence of the new technological revolution. A smart library creates a more varied knowledge-sharing area, puts together a more beautiful environment for knowledge interconnection, and offers users additional expert and superior services (Chaturvedi & Khandelwal 2021). Various applications of smart libraries have reached a mature state, such as self-borrowing and returning at any time of day, intelligent record/position, transportable phone/network self-renewal, intellectual seat reluctance, 3D/AR/VR navigation, etc. (Lu Tingting, 2017). However, in order to boost reader services and expand reader pleasure, smart information centers—an advanced development phase of digital information centers—must bring in current systematic and scientific earnings. AI will be a new driving force for the development of smart libraries, as depending solely on the Internet of Things, RFID, and other technologies has not been able to fully meet the technical needs of smart libraries (Aithal, 2016). This essay will examine how AI technology is being used in smart libraries, provide an overview of the field's current state of application, highlight current issues, and speculate on future developments in this area.

11.2 ARTIFICIAL INTELLIGENCE AND SMART LIBRARY

11.2.1 The Smart Library

The term "smart library" was first used in 2003 by Finnish library expert Markus Attala, who defined it as a visible, space-limited mobile library service (Aittola et al., 2003). Its fundamental traits are.

DOI: 10.1201/9781003638506-11

11.2.2 All-Around Perception

To accomplish the natural integration of readers and libraries, the smart library makes use of AI technologies such as RFID, Internet of Things, "Internet +," picture recognition, speech recognition, PDA, and others.

11.2.3 People-Oriented

Readers may engage with the smart library on the same platform, and it tracks and gathers information according to their needs, offering genuine, comprehensive, and personalized services.

11.2.4 Low Cost

By developing and utilizing resources in the library logically, the smart library may lower the cost of labor and material resources.

11.2.5 Upholding the Idea of "Green Development"

The smart library may accomplish environmental preservation and energy savings, make efficient use of a range of natural resources, and encourage the coordinated growth of ecological and cultural construction (Younis, 2012).

11.3 ARTIFICIAL INTELLIGENCE

The industry currently refers to the rapidly developing AI technology as the fourth industrial revolution (Park, 2018). In 1956, the official proposal of AI was made at an academic meeting in Dartmouth, USA. Additionally, this symposium was acknowledged as a milestone in the development of AI worldwide (McCarthy et al., 1955). AI encompasses a wide range of fields, including computer science, information science, control science, cognitive science, neuroscience, neurophysiology, psychology, linguistics, and brain science. Its main goals are to imitate human intelligence activities, advance the science of human intelligence, and research the creation of intelligent computers or systems. Three categories of AI can be distinguished: behaviorism, connectionism, and representation.

In order to mimic intelligent human behavior, representation is an intelligent reproduction system built on logical reasoning. The learning algorithm and connection mechanism between one neural network and another are the fundamental ideas of connectionism. Perceptual-action control and cybernetic theory underpin behaviorism. Problem solving, natural language processing, artificial neural networks, genetic algorithms, expert systems, knowledge engineering, artificial life, deep learning, intelligent control, etc., are now the most popular technical domains in AI research (Liu Haibin, 2016). There are four distinct phases in the evolution of AI from 1956 to the present (Zhang & Zhang, 2017). Expert systems, like flight tracking and medical diagnostic systems; natural language processing, like speech recognition and automatic speech output; neural networks, like pattern recognition systems, appearance acknowledgment, character acknowledgment, and handwriting acknowledgment; and robots, like industrial and consulting robots, are the main areas of AI research. Additionally, some academics categorize AI according to the fields they study, which include machine learning (ML), computer vision, natural language processing, robotics, cognition and reasoning, and game ethics (Warwick, 2011).

There are numerous classification schemes and a fairly broad definition of AI. AI can be classified into three categories based on its overall level: weak AI, which is only good in certain areas, strong AI, which is comparable to human intelligence, and super AI, which is superior to human intelligence overall. Humans have mastered weak AI, but strong AI has not yet been achieved, according to the total level of AI progress (Hongwei, 2018).

The construction of a smart library's technological base was made possible by the use of AI expertise. In order to fully grow into a smart library and achieve the accuracy, individualization, and initiative of smart library services, more appropriate AI technology can be applied to the conversion and upgrading of smart libraries.

11.4 OBJECTIVES OF THE STUDY

1. Determine whether using AI technologies in libraries to improve knowledge management is indeed feasible.
2. Outlining how AI applications relate to the development of technological and administrative procedures in knowledge management libraries.
3. Being aware of the difficulties libraries face while implementing AI applications.

11.5 RESEARCH QUESTIONS?

1. Is it really possible to improve knowledge management in libraries by using AI applications?
2. How do AI applications relate to the development of technological and administrative procedures in knowledge management libraries?
3. What obstacles do libraries face when implementing AI applications?

11.6 TERMINOLOGY OF THE STUDY

11.6.1 Artificial Intelligence

Artificial intelligence is the science that enables machines to mimic human intelligence in behavior. They are a collection of computers that have been designed and evolved to think like people. They can learn from their errors, complete tasks quickly, and are extremely talented.

11.6.2 Libraries

Establishments that offer services to all types of students, researchers, or readers under the direction of individuals with formal training in library and administrative support. It contains all of the resources that enable the recipients to gain from them.

11.6.3 Knowledge Management

The act of applying knowledge from past decision-making experiences selectively to present and future decision-making endeavors with the explicit aim of enhancing the efficacy of the organization is known as knowledge management.

11.7 METHODOLOGY

Several considerations influenced this research. AI and ML technologies have been used in a variety of library applications. The growing interest in AI and ML in the library sector has influenced the development of research procedures in this field. This tendency has emphasized the importance of conducting a thorough study of the literature to summarize the ML and AI methodologies utilized for the many application areas of libraries. The rapid development of this new trend, in our opinion, necessitates a complete review that will trace the evolution of this subject as well as explain the uses of AI and ML in various sub-domains of libraries.

The study sought to uncover the reality of using AI applications in libraries, as well as the relationship between AI apps and their potential to develop technical and administrative procedures in academic libraries. As a result, the study employed the Descriptive technique through a content analysis of a review of the most important and relevant published Arabic and international literature. This research was actually concerned with the interaction between AI applications and their potential to establish technical and administrative processes in Knowledge Management libraries.

11.8 THE APPLICATION STATUS OF ARTIFICIAL INTELLIGENCE IN SMART LIBRARY

Almost all of the Smart Library's commercial activities are covered by AI. The three application areas are outlined through case analysis and a systematic assessment of a large number of domestic and foreign literature and practical applications: intelligent resource management, intelligent management (including intelligent warehouse management and intelligent security management), and intelligent services (including intelligent application services, intelligent consultancy services, and intelligent knowledge services).

11.8.1 Intelligent Resource System

The intelligent resource procurement system, with the advancement of big data and AI technology, can automatically collect and integrate all users' individualized demand information as well as various sorts of document resource information via a deep learning mechanism. As a result, an intelligent document resource procurement decision system can be built. Two critical points must be considered while developing an intelligent procurement system.

1. It is vital to identify the affecting factors in a scientific and reasonable manner. By combining comprehensive factors such as user group characteristics (such as gender, age, educational background, occupation, etc.) and user personalized information (such as in colleges and universities, the number of teachers and students of various majors, subject setting, subject status ranking, school key construction disciplines, teachers and students hobbies, school opening Course name), the library can establish a scientific and objective decision-making model (Bian & Chen, 2015).
2. To collect and evaluate open resources in depth. The intelligent procurement system can give reference for procurement librarians to make decisions by intelligently collecting and analyzing open materials.

11.8.2 Intelligent Management

11.8.2.1 Intelligent Warehouse Management

Intelligent warehousing management has numerous distinguishing features:

1. Implement self-service management of the book library with the goal of automatic book circulation and paper document management;
2. The books can be stored randomly on the bookshelf, with no need for the book number, reducing the multifarious bookshelf arrangement; and
3. Introduce a robot system to manage the management of automatic and unmanned counting, checking, and sorting of book storage. There are many successful cases of the library intelligent warehouse management system.

To realize the automatic library book counting function, the ultra-high-frequency RFID technology intelligent book inventory robot of Nanjing University Library in China primarily uses automatic identification technology and RF phase technology in RFID technology, as well as machine automation technology. And the library book inventory may be realized precisely and rapidly, allowing the library administrator to find and manage the books conveniently and quickly, reducing the reader's time spent looking for books (Shen et al., 2016; Ni Jie, 2017).

This system stores up to two million goods on high-density automated shelves and delivers any item within five minutes of clicking on the online catalog. Books take up only one-ninth of the space on typical bookshelves, changing the library from a storage facility to a vibrant learning and collaborating area. Each book and item is scanned as it is borrowed or returned from the system, allowing the library's online catalog to monitor all data at any moment.

Each book and object is scanned as it is borrowed or returned from the system, allowing the library's web catalog to track all data at any time. The Johns Hopkins University Library's Work Robotics Project (CAPM) in the United States mechanically collects books from shelves and transports them to a scanning station outside the bookshelf. CAPM provides real-time increased browsing and search capabilities by combining robotics, automated systems, and software technologies to find books on shelves via the web (Choudhury et al., 2019).

11.8.3 Intelligent Security Management

Seat management, loan management, identity management, and other security management are among the library's everyday services, and face recognition, fingerprint recognition, and other AI technology can help solve the library's security management (Fu Ping, 2018). Face recognition technology, for example, specially built by AI technology, is used to collect student face information and link it to student information. After binding, students no longer need to carry student identification information and can enter and exit the library by brushing their faces. Figure 11.1 depicts the smart library identity authentication flow chart.

As a technical support, the identity authentication module employs face recognition technology. Previous face recognition solutions have largely relied on basic statistical techniques such as AdBoost and Principal Component Analysis (PCA). Deep learning algorithms such as Convolution Neural Networks (CNN) and Region CNN (RCNN) have emerged as a result of the deep development of AI. The recognition accuracy and speed of such algorithms have been significantly improved. Face recognition technology now has algorithm support in the building of smart libraries, thanks to the advancement of these basic algorithms. Face recognition technology is divided into four sections: image capture and detection, image preprocessing, feature extraction, matching, and recognition (Li et al., 2018).

11.9 INTELLIGENT SERVICES

11.9.1 Intelligent Application Service

At the moment, library self-service application service technology is reasonably advanced, and the forms and contents of services are likewise extensive and diverse. Self-service seat management system, self-service library ATM, self-service print copy management, lecture training appointment management system, and so on are the primary representatives. In comparison to typical application services, self-service applications have the following advantages:

- Use AI to break through the space–time barrier and provide quick service in no-show situations;
- Extend the service form of library services and broaden the scope of service targets, decreasing library logistics and labor expenses;

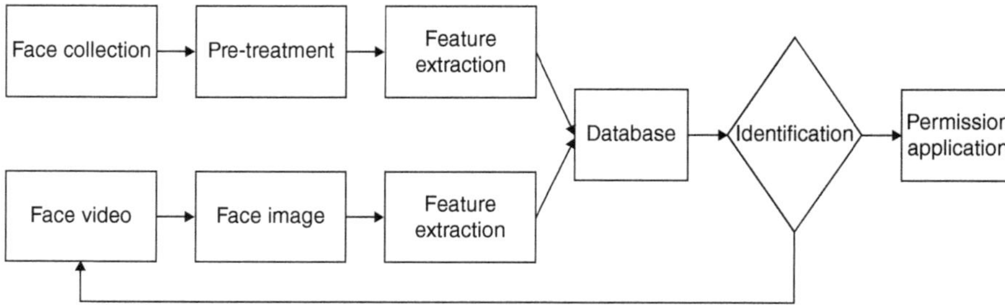

FIGURE 11.1 Smart library identity authentication flowchart.

- Increase user willingness to participate and preserve readers' service application privacy. Encourage the sensible deployment of service resources and reduce the likelihood of service mistakes caused by manual services. The clever application services mentioned above are visible in the general smart library.

11.9.2 Intelligent Consulting Service

Consulting services are an essential component of library services (Figure 11.2). Traditional consulting services are inherently insufficient due to factors such as a limited number of consulting librarians, the low efficiency of manual consultation, and the time limit for consulting work, among others. The advent of intelligent consulting services may effectively meet users' consultation needs, compensate for the deficiencies mentioned above, and achieve the library's independent, instant, convenient, and all-weather intelligent consulting services.

In the meantime, intelligent consultation services, such as "Xiao Tu" of Tsinghua University library, "Xiao Jiao" of Shanghai Jiaotong University library, and the WeChat automatic response robot of Harbin Institute of Technology (Yao et al., 2011), play an important function of a "consulting librarian" in many libraries. These are also the library's intelligent service application's shining spot and defining feature.

11.9.3 Intelligent Knowledge Service

The core of library service is knowledge service, and intelligent knowledge service is the new positioning of library service innovation, with high vitality and extensive potential. The rapid development of AI technologies, such as cross-media awareness, large data management, deep autonomous learning, virtual bionic functions, and simulation language interaction, makes it possible to intelligently and specifically specialize knowledge services. Intelligent knowledge service patterns and deep knowledge mining processes are primarily embodied in intelligent analysis of user

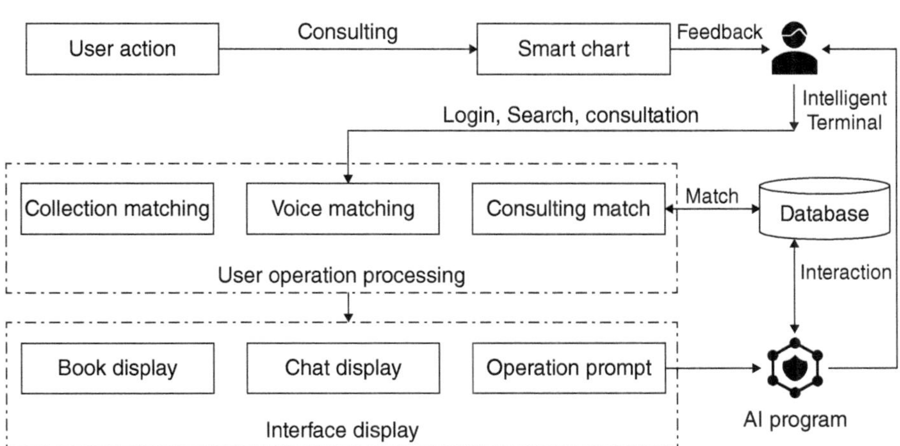

FIGURE 11.2 The working mode and process of the intelligent consulting service.

behavior, intelligent management of information data, and intelligent operation of service business, among other things, and are realized through knowledge analysis tools, knowledge presentation methods, research conceptual models, and analytical research methods (Zhang & Chen, 2016). As an example, consider the following:

1. Intelligent user behavior analysis. From the user's perspective, artificial intelligence analyzes the user's application behavior, and the required knowledge is actively recommended to meet the individualized needs of the user and improve the utilization of knowledge resources;
2. Intelligent management of information data. Use literature, patents, research, and personal data to conduct intelligent analysis and forecasting, create knowledge-related networks, and serve as a resource for knowledge services.
3. The efficient running of the service industry. To improve the core competitiveness of knowledge services through the service business and management process, On the one hand, improving service efficiency can be achieved by optimizing the knowledge service process. It can, on the other hand, provide decision-making and strategy planning for intelligent knowledge services (Xia et al., 2015).

11.10 THE APPLICATION PROSPECTS OF ARTIFICIAL INTELLIGENCE IN MART LIBRARIES

According to the 2016 IFLA movement declaration, AI can now both augment and replace existing library functions. Libraries must create modern accepted wisdom. AI is identified as one of four technical developments in the IFLA movement statement's literature analysis, and it is anticipated that the influence of AI on libraries in the future will primarily comprise three aspects:

1. Beyond keyword search and semantic analysis of web content,
2. The next generation of browsers will support real-time multi-language translation through integrated speech recognition, machine translation, and speech synthesis; and
3. Cloud services for the translation and identification of diverse and complex web content (IFLA, 2016).

Although AI has been used in a variety of ways in the intelligent library, the majority of its applications are still in the theoretical stage, which is more or less constrained and cannot be fully implemented. The following are the explanations behind this situation: Library AI hardware equipment research investment is insufficient, large data collection and data mining are challenging, the library AI talent team is scarce, and AI thinking in the library business is lacking.

Looking ahead to the future application of AI technology in the building of smart libraries, the following three features stand out:

- Intelligent guidance service for hall space. Users can use mobile phones, touch screens in the museum, wearable devices, and other mobile terminals to enjoy intelligent voice service, intelligent seat reservation, accurate positioning of books, intelligent navigation in the museum, intelligent machine consultation (which can be combined with virtual reality technology), and other intelligent guidance services based on the intelligent sensing space construction.

- The library's new generation information retrieval system employs deep learning and neural network models. Unlike the standard learning ranking model, which relies on ML techniques on manually annotated information retrieval features, the neural network model may learn language representations from actual text materials, bridging the gap between query and document vocabulary.
- Provide precise information intelligence services. Users can obtain useful information, such as reading habits, research interests, teaching content, professional fields, research directions, research teams, educational backgrounds, and related groups through information collection and behavior analysis based on big data, Internet of Things, and AI technology. According to specific scenarios, the smart library provides users with accurate, personalized, high-quality information resources and knowledge push, such as recommended borrowing books, specialized literature information in a specific field, cutting-edge research hotspots, teaching reference courseware and materials, and so on.

11.11 AREAS OF ARTIFICIAL INTELLIGENCE

11.11.1 Artificial Intelligence

The focus of AI is on symbolic, non-algorithmic problem-solving strategies. Intelligence agencies rely on their capacity to manipulate symbols. Despite being a relatively new discipline, AI has had far-reaching effects on civilization. The purpose of its subfields, which include expert systems, natural language processing, pattern recognition, and robotics, is to simulate human intelligence using computers. Some of the most current computational approaches and topics that have emerged in the advancement of AI are mentioned here.

11.11.1.1 Expert System

Expert systems are skill-based computerized systems that serve as an intelligence interface or gateway to access databases and collect relevant information. A computer software that provides expert advice, determination, or proposed solutions for a particular condition is known as an expert system. Expert system modules include a knowledge base, an inference engine, and a user interface.

11.11.1.2 Natural Language Processing

AI scientists do well in a structure-approved language interface to a large coverage, utilizing partial language and language rules. Natural Language Processing enables a computer to comprehend the core language concept contained within a question or solution. Its goal is to design and construct computers that analyze, appreciate, and create natural human language (Kumar, 2004). Speech synthesis, speech identification, machine translation, linguistic methods, and information recovery and extraction are the various modules of standard language processing.

11.11.1.3 Pattern Recognition

It is the process of determining the closest match between new data and previously stored data patterns. This process occurs continuously throughout the lifetimes of all living organisms. Pattern classification is investigated in a variety of domains, including psychology, ethnology, cognitive science, and computer science.

Pattern recognition relies on either prior knowledge or statistical data gathered from patterns. The classification patterns are a regulatory group of capacity or observation defining points in a correct multi-dimensional breathing space. Data gathering, pre-processing, feature extraction, model selection and training, and evaluation are the components of pattern recognition.

11.11.2 Robotics

The topic of robotics is frequently referred to as a subfield of AI that deals with perceptual and motor problems. A robot is a motorized piece of equipment that performs computerized tasks under direct human supervision or a pre-defined agenda or set of standard norms, employing AI skills.

11.11.3 Machine Wisdom

Arthur Samuel, an American pioneer in supercomputer betting and AI, coined the term "machine wisdom" in 1959 and defined it as "the ability of a computer to learn without explicit programming." Machine-learning applications are classified into four major groups based on the behavior of the "signal" or "responses" to the learning system, namely

- Supervised learning
- Unsupervised learning
- Reinforcement learning
- Semi-supervised learning.

11.12 APPLICATION OF AI IN LIBRARY SCIENCE

AI applications allow libraries to modify the emphasis and attention. The way we traverse the facts is constantly evolving. AI provides a highly effective and simple technique to utilize this knowledge and achieve better results. Libraries are positioned to benefit from the use of cognizable computing in general and AI in particular for their potential utility as a tool for filtering the quality of library services. The following are some ways that AI could be used in academic library services.

11.13 APPLICATIONS OF EXPERT SYSTEMS IN REFERENCE SERVICES

One of the most significant services provided by any library is reference service, and the Expert System must be used to replace the reference librarian in the following ways:

11.13.1 Research

It is a system that is designed to provide clients with standard sources to search for specific questions. This is a strategy for practicing reference skills or using technological aids for reference librarians and information specialists.

11.13.2 Pointer

It is also known as a knowledge-based method, although it functions as a computer-assisted reference agenda. It refers customers to reference materials.

11.13.3 Online Reference support (ORA)

This system is designed to encourage the use of educational reference services. Librarian for low- and medium-level questions utilizing several technologies: examples include a videotext database, a computer-aided instruction component, and a knowledge-based system. ORA is made up of directed agreements such as library locations, services, and regulations.

11.13.4 Answerman

It is a skill-based system that assists users with agricultural reference inquiries. It features a series of choices that narrow down the subject of the inquiries and the tool category required. It can function as a consultation system or as an interface to external databases and CD-ROM reference resources.

11.13.5 Plexus

This is a metric used in public libraries. It includes information on the reference technique, data retrieval for certain subject fields, reference origin, and library users. All of the systems mentioned above are advisory systems for locating specific reference books and factual material.

11.14 APPLICATION OF PROFESSIONAL ORGANISM IN CATALOGUING

Cataloguing is recognized as an old-fashioned library craft. Because it is rule-based (AACR2), current attempts to automate cataloguing using specialty systems have focused on explanatory cataloguing.

- A human–machine interaction, in which the intelligence attempt is split between the intermediate and the support system.
- An expert system with full cataloguing competence coupled to an electronic publishing system, so that as a text is written online, it may be transferred through skill-based systems and cataloguing processes can be completed without the need for any intellectual input from an intermediary.

11.15 APPLICATION OF EXPERT SYSTEMS IN CLASSIFICATION

The fundamental activity in the institute of knowledge is classification. It is well known in all systems for organizing knowledge in libraries and information hubs for this purpose. The following are examples of how professional methods are used in library classifications:

- ***Coal SORT*:** It is a conceptual browser designed to also function as a research or indexing tool. Coal SORT is essentially comprised of a frame-based semantic network and the software required to allow users to view portions of it and navigate the conceptual structure. The semantic network contains practically all of the system's expert knowledge.
- ***EP-X*:** The Environmental Pollution Expert (EP-X) and coal SORT have a focus on strengthening interfaces through the use of some knowledge-based techniques. EP-X's knowledge base is made up of a hierarchical frame-based semantic network of concepts and a set of templates that transmit patterns known as realistic relationships between concepts. Conceptual information refers to these patterns.
- ***BIOSIS*:** BIOSIS uses knowledge-based devices, which include a significant amount of procedural information, to automatically align documents to various categories. It is plotted to help indexers. BIOSIS leverages the information in biological document titles to assign as many categories as feasible to those supplied by human indexers. The indexing languages are well-organized and practical representations of information that can be utilized to benefit AI systems.

11.16 APPLICATION OF EXPERT SYSTEM IN INDEXING

Another area where expert systems are being developed is the indexing of journals. The process of indexing a monthly article begins with the identification of concepts, which is followed by the selection and assignment of restricted vocabulary terms that are conceptually equivalent to verbal descriptions. The reason for automating the cognitive parts of indexing is to improve indexing consistency and quality. The systems can arrive at appropriate preferred terms to automatically assign acceptable subdivisions based on the information provided by the indexer. The system can draw conclusions and take appropriate action based on those conclusions. "Med Index" is the best example of an indexing system used in library indexing. Few library customers have access to knowledge-based systems. Users have typically had little experience with these systems because the majority of them are not suitable for usage by the average library patron.

11.17 APPLICATION OF EXPERT SYSTEMS IN ACQUISITION

Another essential component of the library is the collection development area. The librarian or information officer is the most important individual in this enterprise. Library users play a vital part in the development of electronic collections, and their assistance and opinion should be sought early in the process. A number of systems have been included. Monograph Selection Advisor is a pioneering attempt to apply this burgeoning technology in another field of library science, namely collection construction. The job modelled is the item-by-item judgment reached by a topic bibliographer when picking monographic details. The knowledge base must be comprehensive, and the interface must be simple enough for the library to obtain the necessary information from the machine.

11.18 APPLICATIONS OF NATURAL LANGUAGE PROCESSING IN LIBRARY ACTIVITIES

When we hear the term NLPL, the first thing that comes to mind is being able to speak or write a whole phrase and have a computer action the request and speak. NLPL can be used in a variety of ways. This could be applied to the field of library and information science, namely in the discovery of databases such as Online Public Access Catalogues (OPAC). Indexing is the foundation of document recovery. "The aim of indexing is to increase precision, the portion of the retrieved documents that are relevant; and recall, the proportion of relevant documents that are retrieved."

11.19 APPLICATION OF PATTERN RECOGNITION IN LIBRARY ACTIVITIES

In this age of the Internet and information distribution, multimedia computing, new and rising types of information system applications have infiltrated office workers' lives. New applications ranging from digital libraries to multimedia systems, geographic information systems, and collaborative computing, as well as electronic commerce, have created enormous prospects for information researchers and practitioners.

11.20 APPLICATIONS OF ROBOTICS IN LIBRARY ACTIVITIES

A robot is defined as "an automatically controlled, reprogrammable, multi-purpose manipulator with three or more axes that can be fixed or mobile for use in automation applications." The Comprehensive Access to Printed Materials (CAPM) project's goal is to create a robotic on-demand and batch scanning system that will allow for real-time browsing of printed content via a web interface. The user will control the CAPM system, which will launch a robot to retrieve the specified item. The robot will transport this object to another robotic device, which will open it and automatically roll the pages.

11.21 CHALLENGES OF IMPLEMENTING ARTIFICIAL INTELLIGENCE IN LIBRARIES

Most libraries nowadays do not have functional AI systems. The following are some of the drawbacks of deploying AI systems in libraries:

- A lack of technical know-how among library workers to use and run AI systems.
- A lack of sufficient resources to expand or acquire AI systems in libraries. Because hardware and software budgets are typically modest, the type of system the library may acquire or design is always limited.
- High system development and maintenance costs for libraries' artificial brainpower systems.
- Inconsistent power supply for AI systems in libraries, particularly in underdeveloped nations.

- The inherent challenges of developing expert/AI systems.
- Natural language capabilities are limited. Intelligent system lacks that universal base of human being information, severely constraining the type of function that they can complete.
- The amount of work and technical expertise required to develop AI systems for libraries. The level and environment of effort required to expand an intellectual documentation system are inextricably linked to the system's power and complexity. This implies that the smarter the system, the more effort must be invested in it. Currently, the requisite qualified professionals with high-cost development tools or methodologies required to construct advanced intelligent systems in libraries are few or prohibitively expensive, resulting in the absence of such systems in libraries.

11.22 ADVANTAGES AND DISADVANTAGES OF ARTIFICIAL INTELLIGENCE (AI)

One of the key benefits of AI is that its conclusions are based on evidence rather than emotions. Even with our best attempts, it is a well-known fact that human decisions are constantly negatively influenced by our emotions.

11.22.1 Advantages of Artificial Intelligence (AI)

The remuneration for AI is incredible; what this field may offer us is the opportunity to progress definitively and go on to the history of artificial robots. The following are the primary benefits of AI.

- Completes tasks faster than a human being,
- Difficult and demanding work readily accomplished,
- Complex project completed in a short period of time,
- Multiple functions can be performed at the same time,
- A high success rate,
- Fewer flaws in tasks and errors,
- Greater efficiency in a shorter period of time,
- Smaller size, fewer space,
- Long-term and difficult scenario manipulation and
- Explore uncharted territory; specifically, distant space.

11.22.2 Disadvantages of Artificial Intelligence (AI)

The following are some of the major drawbacks of AI in our daily lives:

- Human beings' job is exaggerated,
- Increase in unemployment,
- Quality of AI is dependent on the programmer,
- Absence of human touch,

- New generation becomes lazy,
- Need a group of moment in occasion and currency and
- Increased advanced technology dominance.

11.23 CONCLUSION

When the occasion arises, AI will sanction a new capacity to focus on the knowledge prerequisites of library users. Documentation centers can leverage AI technology to provide not just immediate information but also substantial intelligence via Insight as a Service (IAAS). It is necessary to evaluate the benefits and drawbacks of AI and ML in order to improve user experience and their implementation in libraries and information centers. AI is changing the way information is processed and searched for, and information professionals will be able to use these exciting, updated technologies to improve their services and assist users in discovering and gaining access to specific information more easily and quickly. The construction of an efficient expert system for technical services, as well as information processing and management, will help libraries and information hubs.

The education system will change as a result of the interaction of AI. The status of the documentation will be the same as a social education, learning center, knowledge center, and communication center, and the library will have a larger development area. The introduction of AI knowledge in libraries is not intended to replace librarians, but rather to enhance and increase knowledge transfer and interpersonal connections. As a result, libraries should adapt their minds to act as AI. They should adopt a more constructive attitude toward AI and contribute to the activation of library communication functions and service efficiency.

REFERENCES

Aithal, P. S. (2016). Smart Library Model for Future Generations. *Social Science Electronic Publishing*, 1(1):693–703.

Aittola, M., Ryhänen, T., and Ojala, T. (2003). SmartLibrary – Location-Aware Mobile Library Service. In: Chittaro, L. (eds) *Human-Computer Interaction with Mobile Devices and Services. Mobile HCI 2003*. Lecture Notes in Computer Science, vol 2795. Berlin, Heidelberg: Springer.

Bian, L. and Chen, F. (2015). Analysis of Book Ordering Strategy Based on Artificial Intelligence. *Library Journal*, 34(8):39–43.

Chaturvedi, A. and Khandelwal, A.R. (2021). Impact of Virtual Learning Environment on Effective Education: A Guide to Effective Online Teaching. In *Virtual and Classroom Learning in Higher Education*. Sharjah, U.A.E.: Bentham Science, 188–198.

Choudhury, S., Lorie, M., Fitzpatrick, E., et al. (2019). Comprehensive access to printed materials (CAPM). *Journal or Proceedings*, 1(24)

Choudhury, G. S., Lorie, M., Fitzpatrick, E., Hobbs, B., Chirikjian, G., Okamura, A., & Flores, N. E. (2001, January). Comprehensive access to printed materials (CAPM). In *Proceedings of the 1st ACM/IEEE-CS joint conference on Digital libraries* (pp. 174–175).

Liu, H., (2016). *Artificial Intelligence and Its Evolution*. Beijing: Science Press, 5–8.

Hongwei, M. (2018). Thinking on the Ethical Problems of Strong Artificial Intelligence and Weak Artificial Intelligence. *Science and Society*, *8*(1).

IFLA. (2016). *IFLA trend report update 2016*. International Federation of Library Associations and Institutions. https://repository.ifla.org/handle/20.500.14598/3510

Kumar, P.S.G. (2004). *Information Technology: Applications*. New Delhi: BRPC. 401–425.

Kushins, J. (2014, March 11). *Let BookBot bring you any of this library's two million titles. Gizmodo*. https://gizmodo.com/let-bookbot-bring-you-any-of-this-librarys-two-million-1541444173

Li, P., Xie, J., Xu, C., & Shanshan, L. (2018).Application and Development *of* Artificial Intelligence in University Wisdom Library- Based on the Application of Face Recognition Technology and Its Algorithm Implementation. *Library Research and Work*, (07):27–30.

McCarthy, J., Minsky, M. L., Rochester, N., & Shannon, C. E. (1955). *A proposal for the Dartmouth summer research project on artificial intelligence*. Dartmouth College, 1–13

Ni, J. (2017). Design and Development of RFID Book Inventory Intelligent Robot. *New Century Library*, 37(2):69–72.

Park, S. (2018). The Fourth Industrial Revolution and Implications of or Innovative Cluster Policies. *AI & Society*, *33*(3), 433–445.

Ping, F. (2018). New Trends in Library Technology Development. *New Century Library*, 9(2):15–18.

Shen, K., Shao, B. and Chen, L. (2016). Design and Implementation of Book Inventory Robot Based on UHF RFID. *Library Science Research*, 7(1):24–28.

Tingting, Lu (2017). From Smart Library to Intelligent Library: The Turn of Library Development in the Age of Artificial Intelligence. *Library and Information*, 2 (3): 98–101, 140.

Warwick, K. (2011). *Artificial Intelligence*. Hoboken: Taylor & Francis.13–59.

Xia, L., Bai, Y. and Chenglong, L. (2015). Research on Wisdom Self-service Library Service System Based on So Lo Mo. *Library and Information Service*, 59(4):32–36.

Yao, F., Ji, L. and Zhang, C. (2011). A reattempt of Real-time Virtual Reference Service—Intelligent Chat Robot in Tsinghua University Library. *Modern Library and Information Technology*, (4):77–81.

Younis, M. I. (2012). SLMS: A Smart Library Management System Based on an RFID Technology. *International Journal of Reasoning-based Intelligent Systems*, 4(4):186–191.

Zhang, Q. and Chen, M. (2016). Research on Information Science Innovation in Web 4.0 Era. *Journal of the China Society for Scientific and Tech*, 35(10):1048–1061.

Zhang, K. and Zhang, J. (2017). New District, Misunderstanding, Blind Zone and Forbidden Zone in the Application and Research of Artificial Intelligence Education. *Journal of Distance Education*, (5):54–63.

12 A Comprehensive Bibliometric Analysis of Artificial Intelligence *Interventions in Agricultural Development (2019–2023)*

Abhineet Saxena, Riyanshi Khandelwal, Garima Sharma, and Priti Saxena

12.1 INTRODUCTION

It is the era in which technology is being deployed in every field of human endeavor to bring in efficiency. Even a traditional sector like agriculture does not remain untouched, as evidenced by its large employment in numerous use cases across various segments of agriculture (Tripathi et al., 2022). Agriculture, being a crucial part of the economy, is a major source of revenue for any country. The role of the agriculture sector in achieving the world development goals has been widely recognized. The World Bank (World Bank, 2023) informs that agriculture accounts for 4% of the world's Gross Domestic Product (GDP), and in some least-developed countries, 25% or more. Thus, the significance of research with respect to agriculture, its prospects and challenges, and the use of emerging technology to mitigate associated risks cannot be overstated.

The ever-expanding global population, coupled with escalating demands for food and employment, has outstripped the capabilities of traditional farming methods. The resultant shortfall has spurred the adoption of automation in agriculture, presenting a significant and burgeoning global concern (Talaviya et al., 2020). The deployment of artificial intelligence (AI) is expected to support farmers in their fight against soil contamination and reduce the use of harmful pesticides. AI has been instrumental in bringing ground-breaking changes in the domain of agriculture (Kumar et al., 2023). Ensuring the well-being of crops is a pivotal element intertwining various facets of agriculture, necessitating a thorough assessment of crop health. The application of AI and automation tools in determining the optimal times and techniques for pesticide use in large-scale farming marks a transformative shift from traditional to smart agriculture (Katiyar, 2022). Although AI and machine learning (ML) have found

DOI: 10.1201/9781003638506-12

extensive use in predictive applications within agricultural science, the opacity of the underlying reasons behind predictions remains a prevalent challenge, particularly concerning data generated by certain models. To address this problem (Ryo, 2022), a new sub-domain of AI, Explainable AI (XAI), and a related toolkit, Interpretable Machine Learning, were introduced.

Jha (2019) shed light on the usage of AI in crop yield and price forecasting (allowing better decision making), smart spraying (resulting in cost savings), agricultural robots (overcoming labor challenges), crop and soil monitoring, predictive insights, and disease diagnosis. Using AI can help improve crop production and food safety. Image acquisition followed by preprocessing using a segmentation process are several techniques for plant disease detection (Singh et al., 2020). Additionally, the adoption of AI and other related technologies has become important as part of the digital transformation in the agricultural sector. Deep learning-based computer vision and AI enable the automated execution of certain agricultural activities with the highest accuracy, enabling smart agriculture (Dhanya et al., 2022).

It is evident from various literature that about 70% of water is used for irrigation, often untreated, with adverse effects on human health. Smart farming technologies developed by IT help farmers manage water resources, reduce water waste, and detect water quality (Suhail et al., 2021). Additionally, usage of the Internet of Things (IoT), consisting of sensors and actuators, drones, edge devices, etc., has also been well documented (Srivastava, 2021).

The World Bank emphasizes that agricultural progress is a highly effective means of eradicating extreme poverty and promoting shared prosperity, with the goal of providing sustenance to a projected 9.7 billion people by 2050. Furthermore, growth in the agricultural sector yields two to four times greater income improvements for the poorest populations compared to other industries. The World Bank also takes cognizance of the grave risks posed to agriculture-led growth, poverty reduction, and food security in today's global context.

On the other hand, Climate Change Adaptation has become the most pressing need of the hour as it poses the gravest challenge to agriculture. Predictive AI models incorporating region-wise climate data can help adapt agricultural practices to changing conditions and mitigate potential risks. Further, AI-driven solutions may contribute to augmenting sustainable agricultural practices by reducing the use of harmful chemicals, minimizing greenhouse gas emissions, and promoting biodiversity. Most importantly, research in AI and its current and potential use in various aspects of agriculture may generate valuable data and crucial insights that can fuel further innovations and advancements in the field. This iterative process would not only generate continuous improvement in agricultural practices and technologies but also optimize resource usage (which are almost always constrained and under pressure) and minimize harmful environmental impact.

Quite a few research studies addressed that AI has the potential to revolutionize and augment various aspects of agriculture with a view to enhancing productivity, sustainability, and resulting in sounder resource management. AI's potential in assisting farmers based on its use of predictive analytics and data-driven insights in making better decisions about crop management, use of apt irrigation methods, fertilization, disease detection and pest control, and input management with a view to increase yields and overall efficiency has been widely acknowledged. Precision agriculture is also gaining momentum, thanks to the deployment of advanced technologies such as satellite imagery, drones, IOT devices, in conjunction with AI tools to get access to high-resolution data on a granular field-by-field basis.

The chapter attempts a bibliometric analysis of AI and agriculture to provide a realistic and representative picture of the current state of research in the area. This would generate evidence-based descriptions as well as aid in comparisons and visualizations of research output in the chosen area. This can also be used to explicate and describe the patterns of AI deployment and its impact on agriculture. Further, this analysis would also unravel areas of future research aimed at ameliorating the present food crisis and mitigating the risks and challenges posed to sustainable agriculture. Research in this area is critical to unlocking the full potential of AI technologies and addressing the unique challenges facing the agricultural sector.

12.2 RESEARCH QUESTIONS

The aim of this chapter is to identify the recent trends and prospects of AI and sustainable agriculture using science mapping review analysis. Further, the review scrutinizes the following research questions (RQs):

RQ1: What are the key findings of the literature review?
RQ2: Who are the most prolific authors in the field of AI in agriculture?
RQ3: Which documents have the highest impact on citations in the field of AI in agriculture?
RQ4: What are the most influential countries and affiliations in the field of AI in Agriculture?
RQ5: What are the gaps and areas for future research?

12.3 RESEARCH METHODOLOGY

The study followed a five-step structured procedure, which included identifying, selecting, classifying, analyzing, and evaluating information about agriculture and AI. As a first step, a suitable research design was determined, appropriate research questions were constructed – four in number – and accordingly methods were identified to answer the key research questions. The second step was the assortment of data, which included choosing the appropriate database and exporting the bibliographic data. In this study, SCOPUS was used as a database to identify the significant relevant documents, and exporting of data, including all the details such as citation information, bibliographical information, abstract and keywords, and other information in the form of a CSV file was executed. The third step involved the analysis of the data exported from the SCOPUS database with the help of a suitable bibliometric software. The study deployed VOS-viewer for the relevant analysis.

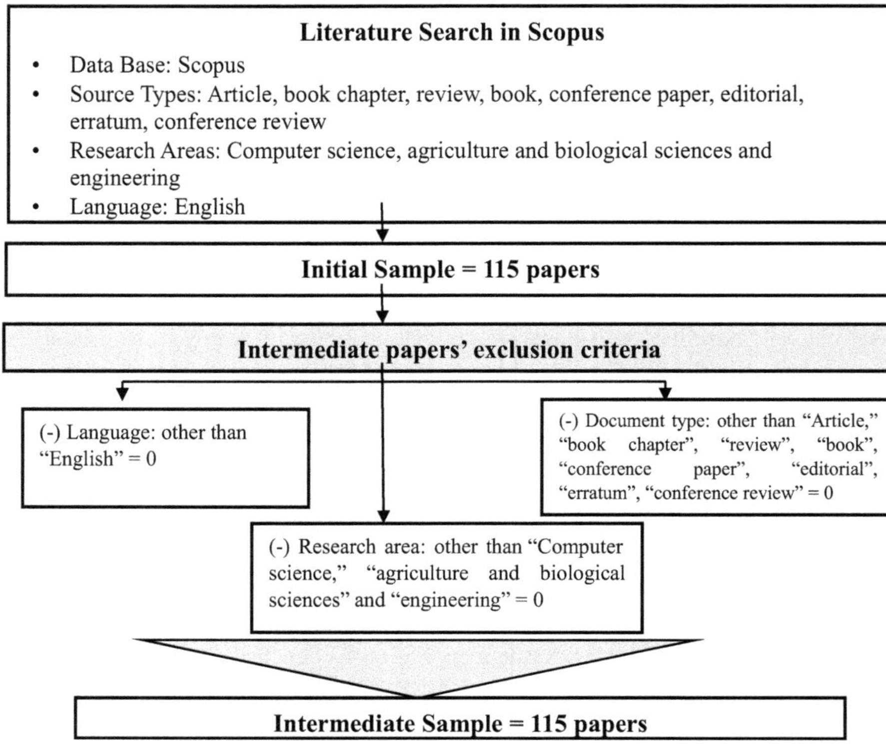

FIGURE 12.1 Sample evolution by using Scopus filters.

The fourth step entailed data visualization with proper evaluation and detection of findings. The last and final step involved the interpretation of findings from the data (Figure 12.1).

12.4 DATABASE SELECTION, DATA EXTRACTION, AND ANALYSIS

For the present study, the Scopus database was used to extract the appropriate data for the study. The reasons for choosing the Scopus database are its distinctive features of providing an all-inclusive overview of global research outputs, ease of exporting data from Scopus owing to the provision of multiple options, which include CSV, Mendeley, BibTex, etc., and the ability to provide effective search as it provides two methods for search, viz. basic and advanced search queries.

"TITLE-ABS-KEY" is used as a query string in the advanced search feature of Scopus. It helps to find out the documents that have similar keywords in the title or abstract. In this study, the keywords "Artificial Intelligence" AND "Agriculture" were used under the search source title option to extract the related documents. A total of 115 documents were found, which were then identified, arranged, evaluated, and analyzed for the bibliometric analysis.

12.5 BIBLIOMETRIC ANALYSIS

Bibliometric analysis stands out as a widely embraced and rigorous method for investigating and scrutinizing extensive sets of scientific data. Within the two primary methodologies of Bibliometric Analysis, namely performance analysis and science mapping (Cobo et al., 2011), performance analysis is notably considered a conventional and established practice (Donthu et al., 2021). The goal of performance analysis is to assess various entities in the research community, including authors/researchers, journals, countries, and departments. It involves examining the influence of their activities based on bibliographic data (Noyons et al., 1999). Science mapping, on the other hand, concentrates on the relationship between research entities. The chapter adopts Performance Analysis on account of its several merits documented in extant literature. A four-step procedure for bibliometric analysis was adopted. Step 1 involved defining the aims and scope for the bibliometric study (Mehta et al., 2023). Step 2 entailed choosing the technique for bibliometric analysis. In Step 3, relevant data were collected for further analysis. In Step 4, a bibliometric analysis was run, and the findings were reported accordingly (Donthu et al., 2021). Though there are numerous software packages available for bibliometric analysis and evaluation, such as BibExcel, Bibliometrix, etc. Pajek and CiteSpace are some software tools for the construction and visualization of bibliometric networks; for the purpose of the present study, CSV Excel has been exploited to conduct further bibliometric and statistical analyses. VOSviewer software is used to prepare for further network visualization. Kirby (2023) illustrated the effective utilization of such software for conducting initial investigations and influencing subsequent research that employs more structured methodologies.

12.5.1 Distribution of Publication by Year

Around 115 documents were extracted as the final sample data. According to the Scopus database, the first research paper with the keywords "Artificial Intelligence" AND "Agriculture" was published in 2019. Figure 12.2 shows a substantial increase in the number of publications between 2019 and 2021. The earliest papers were published in 2019 with just three documents, which rose to 50 documents in 2021, with the highest number of publications to date. By April 2023, eight documents were published. With the rise in technology around the world, the agriculture sector is not excluded from tech development. Therefore, there has been a significant growth of AI in agriculture to accelerate high production. Figure 12.2. represents growing awareness of AI in the agriculture sector, leading to an escalation of research publications. Overall, the findings represent that AI in agriculture has accumulated considerable attention from researchers.

12.5.2 Country Analysis

Table 12.1 and Figure 12.3 represent the top 10 countries in the field of "AI in agriculture" among the total 115 documents. The average publication count per country is 10.8, which means only two countries are higher than the average publication count: India and the United States. Furthermore, India is the highest

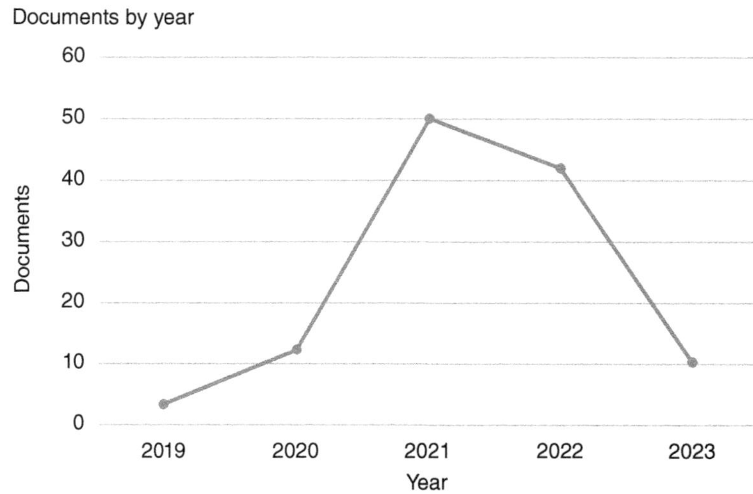

FIGURE 12.2 Year-wise analysis of AI in Agriculture from the Scopus database.

TABLE 12.1
Compiled Country and Publication Data from Scopus Database

Country	Publication Count	Percentage
India	56	51.85
United States	15	13.89
China	10	9.26
Italy	5	4.63
Nigeria	5	4.63
Ethiopia	4	3.70
Saudi Arabia	4	3.70
Australia	3	2.78
Brazil	3	2.78
France	3	2.78
Total	108	100.00

contributor in the research field with 56 publications, followed by the United States with 15 publications. The remaining eight countries have less than average publications per country. Table 12.1 shows that even though being a developing country, India is leading in this research field, leaving some of the developed countries like the United States, Italy, Australia, and France behind. Another interesting observation is that there is a noteworthy contribution of research output from the developing countries cohort as well (India, Nigeria, Ethiopia, and Brazil), though the developed countries outnumber the developing ones in the list. Perhaps, this is a testament to the importance of research in agriculture and the exploration of advanced and emerging technologies in finding solutions to some of the problems it is facing.

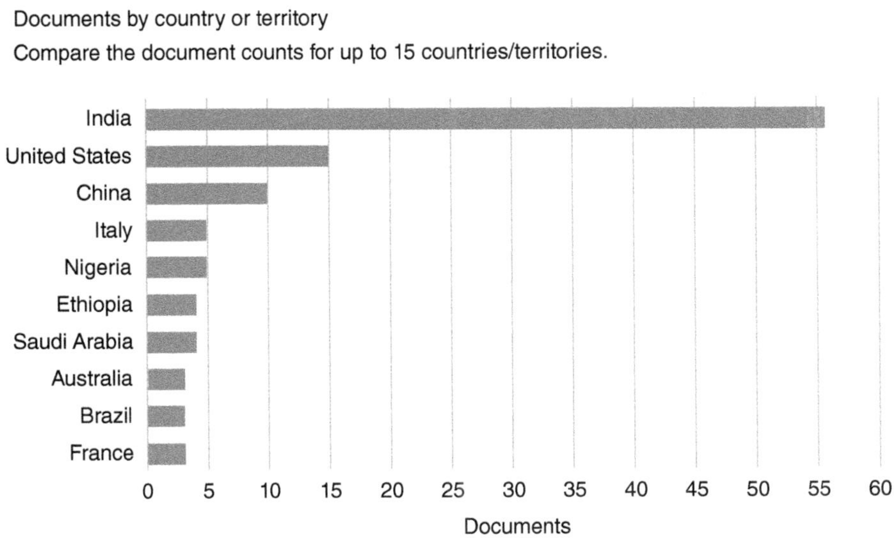

FIGURE 12.3 Top 10 leading countries in the field of AI in agriculture from the Scopus database.

12.5.3 Citation Analysis

Citation analysis indicates the number of times any article and/or author has been cited. Table 12.2 represents the top 10 highest cited articles out of 115 documents, including other columns which are document title, publication year, source title, first author, and citation score. The highest cited document for AI in agriculture is the one authored by Jha et al. (2019) with the title "A comprehensive review on automation in agriculture using artificial intelligence" published under the source title "Artificial Intelligence in Agriculture" in 2019 with 358 citations. This chapter focuses on different automation practices like IOT, Wireless communication, ML, AI, and Deep Learning. It also talks about the current implementation of automation in agriculture.

12.5.4 Co-occurrence of Keywords

The co-occurrence of keywords serves as a focal point in delineating the research landscape within the field of AI in agriculture. The analysis of the keyword table shows that out of 319 words, only ten meet the threshold of a minimum of four occurrences of keywords.

Table 12.3 and Figure 12.4 represent different clusters and network analysis of the co-occurrence of keywords.

Figure 12.5 represents density visualization which is retrieved using VOSviewer. The keywords like deep learning, ML, AI, and agriculture show the highest density.

12.5.5 Author Analysis

Figure 12.6 represents the top 10 most influential/prolific authors in the selected field of study.

TABLE 12.2
Compiled Citation Data from Scopus Database

Document Title	Publication Year	First Author	Citation Score
"A comprehensive review on automation in agriculture using artificial intelligence"	2019	Jha, K	358
"Implementation of artificial intelligence in agriculture for optimisation of irrigation and application of pesticides and herbicides"	2020	Talaviya, T	207
"A review of imaging techniques for plant disease detection"	2020	Singh, V.	66
"Plant disease detection using hybrid model based on convolutional autoencoder and convolutional neural network"	2021	Bedi, P	58
"Artificial cognition for applications in smart agriculture: A comprehensive review"	2020	Pathan, M	58
"Sunflower leaf diseases detection using image segmentation based on particle swarm optimization"	2019	Singh, V.	55
"Deep learning approach for recognition and classification of yield affecting paddy crop stresses using field images"	2020	Anami, B.S	44
"Transfer Learning for Multi-Crop Leaf Disease Image Classification using Convolutional Neural Network VGG"	2022	Paymode, A.S	37
"Review of agricultural IoT technology"	2022	Xu, J	31
"Modeling and optimization of *Terminalia catappa L.* kernel oil extraction using response surface methodology and artificial neural network"	2020	Agu, C.M	28

TABLE 12.3
Compiled Affiliation Analysis Data from Scopus Database

Affiliation	Documents	Percentage (%)
Delhi Technological University	6	14.63
ABES Institute of Technology	6	14.63
Indian Institute of Technology Kharagpur	4	9.76
ICAR – Central Institute of Agricultural Engineering, Bhopal	4	9.76
Indian Council of Agricultural Research	4	9.76
Gautam Buddha University	4	9.76
University of Bisha	4	9.76
University of Johannesburg	3	7.32
Salesforce. Comm Inc.	3	7.32
Dr. A.P.J. Abdul Kalam Technical University	3	7.32
Total	**41**	**100.00**

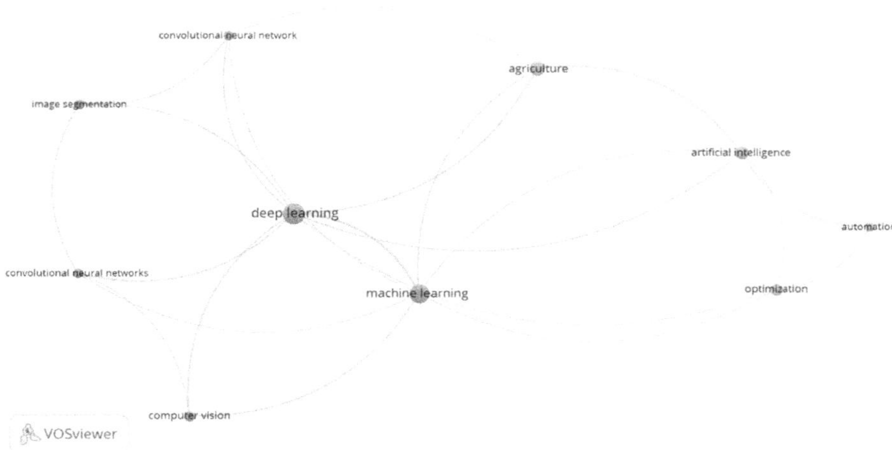

FIGURE 12.4 Keyword cluster analysis from VOSviewer.

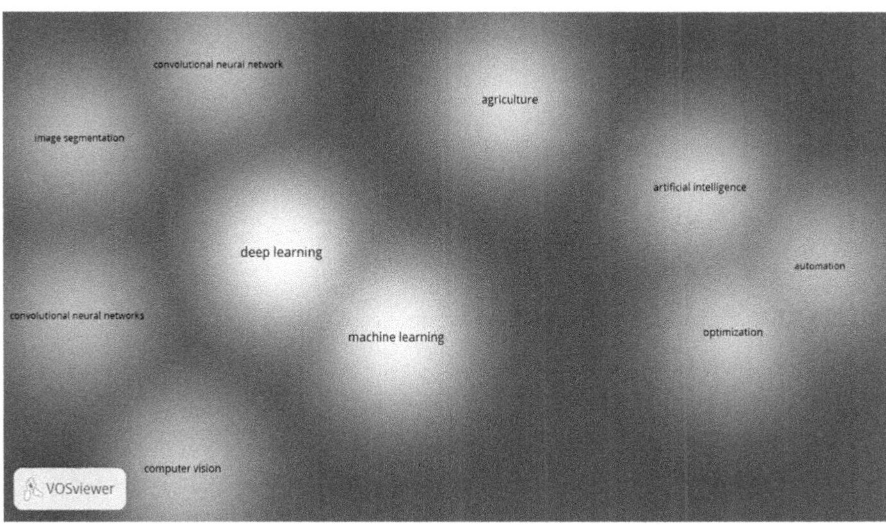

FIGURE 12.5 Keywords density visualization from VOSviewer.

FIGURE 12.6 WordCloud from RStudio Biblioshiny.

12.5.6 Affiliation Analysis

Figure 12.6 illuminates the details of documents published by the notable affiliation organization in the subject area of AI in Agriculture. A total of 41 documents were fetched from the top 10 affiliating organizations with an average of 4.1 documents per organization. Only two affiliating organizations have more than the average number of documents, and the remaining eight have less than average number of documents. Delhi Technological University and ABES Institute of Technology are the most prominent organizations in the field of AI in Agriculture with six documents each (Figures 12.7 and 12.8).

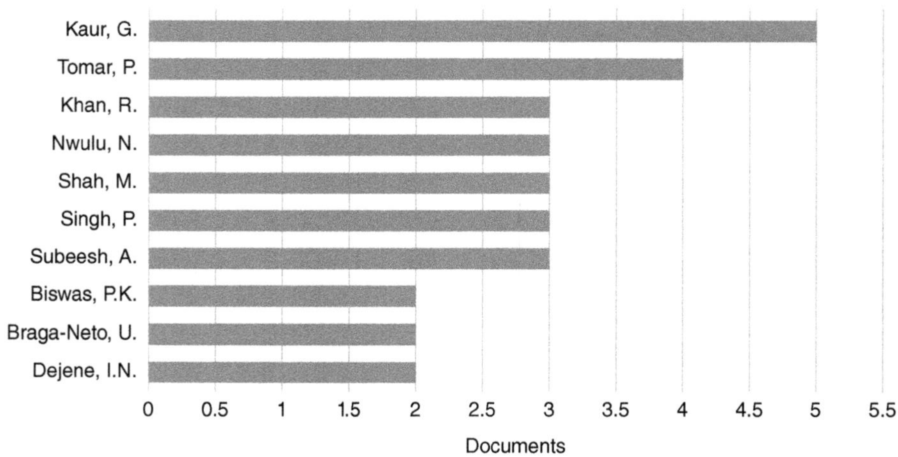

FIGURE 12.7 Documents by author from Scopus.

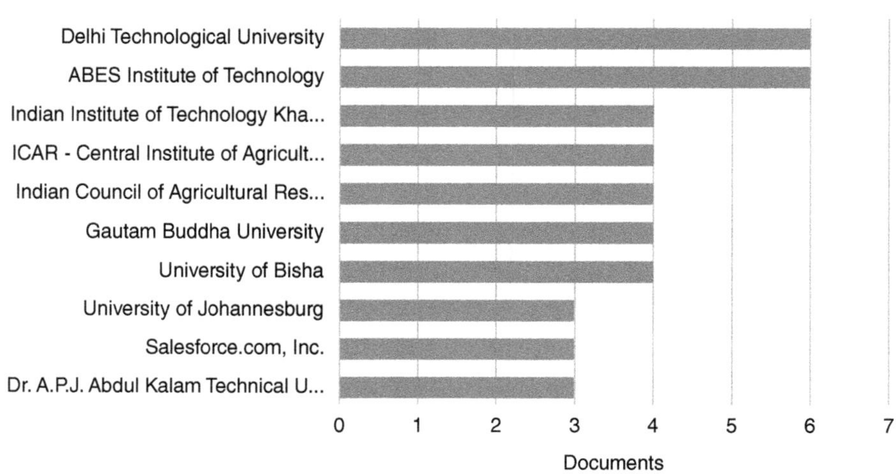

FIGURE 12.8 Documents by affiliation from Scopus.

12.6 RESEARCH GAPS IN EXISTING LITERATURE AND DIRECTION FOR FUTURE RESEARCH

After doing a systematic research literature review of selected papers, we have identified major research gaps that are impeding the current research. Hence, we propose several domains worthy of exploration within the ongoing research landscape of AI in agriculture.

12.6.1 Predomination of Quantitative Research and Lack of Qualitative Studies

Researchers have investigated AI in Agriculture using quantitative techniques in many articles. To analyze primary data, ML and deep learning approaches are used mostly. Numerous statistical and ML models are utilized in single-environment trials, presuming that the environment has no impact on quantitative traits. However, it is imperative to incorporate both genomic and environmental data in the development of new varieties, as reliance on these stringent assumptions may result in the oversight of optimal lines tailored to specific environments (Jubair et al., 2023). Thus, future researchers must address this issue of dominance of quantitative research and give due importance to the qualitative aspect of research as well.

12.6.2 AI for Aquaculture and Livestock Management

While AI applications in crop agriculture, disease prediction in crops, and food security have been extensively studied, there is relatively scarcity of research on AI's role in aquaculture and livestock management. It is a crucial part of agriculture, along with yield detection, soil protection, food security, improved irrigation facilities, etc. There are very few researchers to date who have explored the use of AI in aquaculture and livestock management. Some of them are Gomes et al. (2023), Mishra & Sharma (2023), Shao et al. (2023), and Yang et al. (2023).

Therefore, future research should focus and examine the potential of AI in refining aquaculture operations, disease detection in fish and livestock, and enhancing animal welfare.

12.6.3 Integration of Multimodal Data

The agriculture sector produces a vast amount of data from a variety of sources, including satellite imagery, sensors, historical records, and weather forecasts. In order to develop AI models that provide a holistic insight and prediction for farmers, there is a need to examine and explore different effective methods for integrating multi-modal data.

12.6.4 Lack of AI Applications for Small-Scale Farming

The majority of the existing research focuses on AI applications for large-scale farmers. There exists an unexplored research gap regarding the customization of

AI-based solutions to address the distinct requirements and obstacles encountered by small-scale farmers. Future research could explore affordable and accessible AI solutions for small-scale farmers to improve their livelihood and productivity.

12.6.5 Enhancing Farm Robotic Automation Experience

Although studies on AI-driven agricultural robots have been conducted, there are still research gaps in the development of adaptable and robust robots that can carry out a variety of tasks in constantly changing outdoor environment situations. Future research could concentrate on boosting the autonomy, mobility, and intelligence of agricultural robots so that they can handle complex farm operations.

12.6.6 Practical Applications of AI

AI methods are commonly integrated into existing Farm Management Information Systems or, in some instances, embedded within separate Decision Support Systems. Alternatively, AI may be an integral component of machine or robot operations. Presently, AI methods predominantly find application in machine vision scenarios, with practical implementations utilizing methods beyond image data remaining limited (Backman et al., 2023). Therefore, future research can explore different reasons as to why practical applications of AI in the field are so scarce.

12.7 CONCLUSION

Agriculture stands as the backbone of developing countries, and the incorporation of emerging technologies such as AI, ML, and the IoT aims to bolster its productivity. By providing precise insights into seeds, soil, weather, diseases, and other factors influencing farming, these technologies contribute to informed decision-making. AI leverages predictive analysis, image analysis, learning techniques, and pattern analysis to recommend cost-effective and high-yield strategies for agriculturists (Awasthi, 2020). The widespread application of AI is increasingly recognized as a solution to address the diverse challenges faced by farmers, particularly in meeting the escalating demand for crops driven by the exponential increase in population. Few fields like soil management, weed management, and IoT are of great potential in the future. Considering the growing importance of AI in agriculture, this study performs a bibliometric analysis of 115 articles on AI and agriculture from the Scopus database with the help of RStudio and VOS-viewer. This chapter gives a structured review and bibliometric analysis of influential authors, countries, and emerging clusters in the field of AI in agriculture. The major findings of the study include a growing trend in the quantum of research published in the area in the past five years rising from three in 2019, peaked to 50 in the year 2021, and with eight publications already in place in the first quarter of 2023. India emerges as the largest contributor followed by the United States in the country-wise analysis of research publications in the area. In the evaluation of citation metrics, Indian authors Jha et al. emerge as winners with the maximum citations. Affiliation analysis reveals seven of the top 10 institutes are from India, and the remaining three are from Saudi Arabia, South Africa, and the

United States, respectively. A systematic review of research papers also revealed that there are a few challenges in the application of AI. Three major challenges are uneven distribution of mechanization, the ability of algorithms to process large sets of data accurately and quickly, and the security and privacy of data, as well as the devices (Zha, 2020). Currently, in the era of Agriculture 5.0, precision agriculture principles using AI, ML, and IoT are being used widely. But unfortunately, India still relies on the traditional farming system. Educational level, inadequate training, and indigent farmers put India at a disadvantage and way behind the world in applying technology in agriculture. Technology helps in sustaining agriculture, proving to be beneficial for the environment. Hence, ML techniques, AI, and IoT can be of great advantage and can possibly cater to various agricultural challenges. Future research studies can utilize other bibliometric tools such as BibExcel, HistCite, CiteSpace, etc., for better visualization and comprehensive review. Further, the study reveals the gaps in terms of the lack of literature on qualitative aspects, as well as those covering allied areas of aquaculture, livestock, and small-scale farming. Future research would do well to examine these aspects, which have been overlooked in extant literature but are of great relevance to both developed and developing countries.

REFERENCES

Agu, C. M., Menkiti, M. C., Ekwe, E. B., & Agulanna, A. C. (2020). Modeling and optimization of Terminalia catappa L. kernel oil extraction using response surface methodology and artificial neural network. *Artificial Intelligence in Agriculture*, *4*, 1–11.

Anami, B. S., Malvade, N. N., & Palaiah, S. (2020). Deep learning approach for recognition and classification of yield affecting paddy crop stresses using field images. *Artificial intelligence in agriculture*, *4*, 12–20.

Awasthi, Y. (2020). Press "a" for artificial intelligence in agriculture: A review. *International Journal on Informatics Visualization*, *4*(3), 112–116. https://doi.org/10.30630/JOIV.4.3.387

Backman, J., Koistinen, M., & Ronkainen, A. (2023). Agricultural process data as a source for knowledge: Perspective on artificial intelligence. *Smart Agricultural Technology*, *5*. https://doi.org/10.1016/J.ATECH.2023.100254

Bedi, P., & Gole, P. (2021). Plant disease detection using hybrid model based on convolutional autoencoder and convolutional neural network. *Artificial Intelligence in Agriculture*, *5*, 90–101. https://doi.org/10.1016/j.aiia.2021.05.002

Cobo, M. J., López-Herrera, A. G., Herrera-Viedma, E., & Herrera, F. (2011). An approach for detecting, quantifying, and visualizing the evolution of a research field: A practical application to the Fuzzy Sets Theory field. *Journal of informetrics*, *5*(1), 146–166.

Dhanya, V. G., Subeesh, A., Kushwaha, N. L., Vishwakarma, D. K., Nagesh Kumar, T., Ritika, G., & Singh, A. N. (2022). Deep learning based computer vision approaches for smart agricultural applications. *Artificial Intelligence in Agriculture*, *6*, 211–229. https://doi.org/10.1016/J.AIIA.2022.09.007

Donthu, N., Kumar, S., Mukherjee, D., Pandey, N., & Lim, W. M. (2021). How to conduct a bibliometric analysis: An overview and guidelines. *Journal of Business Research*, *133*, 285–296. https://doi.org/10.1016/J.JBUSRES.2021.04.070

Gomes, J., Esteves, I., Neto, V. V. G., David, J. M. N., Braga, R., Arbex, W., Kassab, M., & de Oliveira, R. F. (2023). A scientific software ecosystem architecture for the livestock domain. *Information and Software Technology*, *160*. https://doi.org/10.1016/J.INFSOF.2023.107240

Jha, K., Doshi, A., Patel, P., & Shah, M. (2019). A comprehensive review on automation in agriculture using artificial intelligence. *Artificial Intelligence in Agriculture*, *2*, 1–12. https://doi.org/10.1016/j.aiia.2019.05.004

Jubair, S., Tremblay-Savard, O., & Domaratzki, M. (2023). GxENet: Novel fully connected neural network based approaches to incorporate GxE for predicting wheat yield. *Artificial Intelligence in Agriculture*, *8*, 60–76. https://doi.org/10.1016/J.AIIA.2023.05.001

Katiyar, S. (2022). The use of pesticide management using artificial intelligence. *Artificial Intelligence Applications in Agriculture and Food Quality Improvement*, 74–94. https://doi.org/10.4018/978-1-6684-5141-0.CH005

Kirby, A. (2023). Exploratory bibliometrics: Using VOSviewer as a preliminary research tool. *Publications*, *11*(1), 10.

Kumar, A. S., Sharma, P., Kaur, S., Saleh, O. S., Chennamma, H. R., & Chaturvedi, A. (2023). AI-Equipped IoT Applications in High-Tech Agriculture Using Machine Learning. In *Handbook of Research on AI-Equipped IoT Applications in High-Tech Agriculture* (pp. 38–64). IGI Global.

Mehta, M., Pancholi, G., & Saxena, D. A. (2023). Metaverse changing realm of the business world: a bibliometric snapshot. *Journal of Management Development.*, *42*(5), 373–387.

Mishra, S., & Sharma, S. K. (2023). Advanced contribution of IoT in agricultural production for the development of smart livestock environments. *Internet of Things (Netherlands)*, *22*. https://doi.org/10.1016/J.IOT.2023.100724

Noyons, E. C., Moed, H. F., & Luwel, M. (1999). Combining mapping and citation analysis for evaluative bibliometric purposes: A bibliometric study. *Journal of the American society for Information Science*, *50*(2), 115–131.

Pathan, M., Patel, N., Yagnik, H., & Shah, M. (2020). Artificial cognition for applications in smart agriculture: A comprehensive review. *Artificial Intelligence in Agriculture*, *4*, 81–95. https://doi.org/10.1016/j.aiia.2020.06.001

Paymode, A. S., & Malode, V. B. (2022). Transfer learning for multi-crop leaf disease image classification using convolutional neural network VGG. *Artificial Intelligence in Agriculture*, *6*, 23–33.

Ryo, M. (2022). Explainable artificial intelligence and interpretable machine learning for agricultural data analysis. *Artificial Intelligence in Agriculture*, *6*, 257–265. https://doi.org/10.1016/J.AIIA.2022.11.003

Shao, D., He, Z., Fan, H., & Sun, K. (2023). Detection of cattle key parts based on the improved Yolov5 algorithm. *Agriculture*, *13*(6), 1110. https://doi.org/10.3390/AGRICULTURE13061110

Singh, V. (2019). Sunflower leaf diseases detection using image segmentation based on particle swarm optimization. *Artificial Intelligence in Agriculture*, *3*, 62–68.

Singh, V., Sharma, N., & Singh, S. (2020). A review of imaging techniques for plant disease detection. *Artificial Intelligence in Agriculture*, *4*, 229–242. https://doi.org/10.1016/J.AIIA.2020.10.002

Srivastava, K. (2021). Farming automation. *Artificial Intelligence and IoT-Based Technologies for Sustainable Farming and Smart Agriculture*, 68–82. https://doi.org/10.4018/978-1-7998-1722-2.CH005

Suhail, A., Hasteer, N., Sharma, A., & Singh, S. (2021). Use of smart farming techniques to mitigate water scarcity. *Artificial Intelligence and IoT-Based Technologies for Sustainable Farming and Smart Agriculture*, 140–150. https://doi.org/10.4018/978-1-7998-1722-2.CH009

Talaviya, T., Shah, D., Patel, N., Yagnik, H., & Shah, M. (2020). Implementation of artificial intelligence in agriculture for optimisation of irrigation and application of pesticides and herbicides. *Artificial Intelligence in Agriculture*, *4*, 58–73. https://doi.org/10.1016/J.AIIA.2020.04.002

Tripathi, P., Kumar, N., Rai, M., & Khan, A. (2022). Applications of deep learning in agriculture. *Artificial Intelligence Applications in Agriculture and Food Quality Improvement*, 17–28. https://doi.org/10.4018/978-1-6684-5141-0.CH002

World Bank. (2023, March 31). Agriculture and food. In https://www.worldbank.org/en/topic/agriculture/overview. Retrieved August 16, 2023, from https://www.worldbank.org/en/topic/agriculture/overview

Xu, J., Gu, B., & Tian, G. (2022). Review of agricultural IoT technology. *Artificial Intelligence in Agriculture*, 6, 10–22.

Yang, X., Bist, R., Subedi, S., & Chai, L. (2023). A deep learning method for monitoring spatial distribution of cage-free hens. *Artificial Intelligence in Agriculture*, *8*, 20–29. https://doi.org/10.1016/J.AIIA.2023.03.003

Zha, J. (2020). Artificial intelligence in agriculture. *Journal of Physics: Conference Series*, *1693*(1), 012058. https://doi.org/10.1088/1742-6596/1693/1/012058

13 The Research on AI and Sustainable Social Development

A Bibliometric Analysis of Three Decades

Nisha Kumari and Mukesh Kondala

13.1 INTRODUCTION

The ever-expanding frontiers of commerce, corporate conduct, and government policies owe their rapid expansion to advancements in artificial intelligence (AI). AI presents a wealth of opportunities for enhancing the well-being of individuals and fostering economic and societal growth (Thiebes et al., 2021). However, it also gives rise to a host of ethical, legal, social, and technological challenges that necessitate thoughtful consideration and well-crafted responses (Awan et al., 2018; Toniolo et al., 2020). Deep learning–capable machines and intelligent robots have wielded a transformative influence, disrupting and facilitating changes in the realms of business, politics, and society, while also leaving a substantial imprint on global sustainability trends (Di Vaio et al., 2020).

The exploration of AI within the context of promoting sustainable social development is a dynamic and rapidly evolving domain with considerable potential to address contemporary societal dilemmas. This interdisciplinary field aims to leverage AI to foster sustainable development, enrich people's lives, and confront universal challenges in areas like healthcare (Chui et al., 2017), education (Nguyen et al., 2020), environmental preservation (Ferreira et al., 2023), and poverty alleviation (Florey et al., 2020). Social sustainability, also referred to as social sustainable development, is an integral facet of the broader concept of sustainable development (Di Vaio et al., 2020). It places paramount emphasis on fostering social inclusion, equity, and overall well-being, ensuring the fulfillment of both current and future generations' needs (Awan et al., 2018). At its core, social sustainable development aspires to create societies where individuals, regardless of their background or circumstances, can enjoy a decent standard of living and have equitable access to essential services (Kondala et al., 2023a). Social equity lies at the heart of social sustainability, working to bridge gaps in opportunity, resources, and fundamental services like healthcare and education (Zimmer et al., 2016). This entails creating

DOI: 10.1201/9781003638506-13

environments that empower all members of society to flourish and fully participate, with inclusivity being equally vital in nurturing diverse communities that promote a sense of identity and social cohesion (Rajaram et al., 2014). Key components of promoting health and well-being include ensuring access to high-quality healthcare and support for both physical and mental health. Equally vital is the provision of quality education with an emphasis on accessibility, equity, and empowerment. Furthermore, stable employment and livelihoods are fundamental for economic well-being, offering employment opportunities and fair working conditions (Toniolo et al., 2020).

The impact of AI on social sustainability is both substantial and multifaceted. AI-driven telemedicine and educational platforms expand access to top-tier education and healthcare, fostering social sustainability. Upholding social equity mandates that AI systems operate with impartiality and fairness. AI plays a crucial role in emergency response and community engagement, encouraging participation and resilience. Additionally, AI can surmount language barriers, making services accessible to speakers of all languages, thus promoting inclusivity (Ferreira et al., 2023). AI also contributes to poverty reduction and mental health support, further enhancing social sustainability (Thiebes et al., 2021). Sustainable social development emerged in recognition of the fact that focusing solely on economic growth could not secure the well-being of both humans and the environment (Awan et al., 2018). It entails the integration of social, economic, and environmental dimensions to achieve long-term prosperity while addressing issues such as social inequality and environmental degradation (Kondala et al., 2023a). The increasing urgency to find sustainable solutions for poverty, inequality, and climate change has led to its significant prominence in recent years (Nguyen et al., 2020). Governments, businesses, and civil society are placing greater emphasis on social sustainability to construct a more equitable, just, and environmentally conscious future (Kondala et al., 2023b). Addressing critical issues like gender equality, healthcare access, education, and poverty is crucial for a society's resilience and well-being, and thus, these are central priorities for sustainable social development. The growing understanding of the interdependence of environmental and social well-being has propelled the momentum of the social sustainability trend.

Bibliometric analysis, which counts academic works like books, articles, and conference papers, has become increasingly popular (Saravanan et al., 2023). Researchers can evaluate the impact, influence, and trends in particular academic fields or across disciplines by methodically analyzing citation data (Omar et al., 2020). The popularity of bibliometric analysis is due to several factors. The availability of digital data has increased interest in bibliometric analysis by making studies more practical (Chawla & Goyal, 2022). It is frequently employed for determining the effectiveness of research, allocating funding, and rating academics, organizations, and journals. In the era of interdisciplinary research, it assists researchers and institutions in increasing the dissemination and visibility of their research while also helping discover intricate citation networks and cross-disciplinary collaborations. Policymakers also use bibliometrics to allocate resources, support industry–academia collaboration, and inform science and innovation policies (Butcher & Jeffrey, 2005). Researchers use bibliometrics to find new topics, gaps, and interdisciplinary intersections as research trends become more dynamic and global. It directs decision-making and resource allocation in academia and beyond in the ever-changing knowledge landscape (Lochab et al., 2018).

Bibliometric analysis in the realm of social sustainability and AI presents a unique and specialized lens to understand the evolving landscape at the intersection of these two fields. Understanding these collaborative networks is vital for recognizing where the most influential and impactful research is originating, as well as for fostering international collaboration and knowledge exchange (Kumari & Kondala, 2023). Research in the context of social sustainability and AI has significant real-world implications for policy, practice, and societal well-being. It offers a structured method to gauge which studies and methodologies are gaining the most traction and exerting the greatest influence within the field. Furthermore, it provides valuable insights into the evolution of ethical and policy discussions in the context of AI and social sustainability. This type of analysis can also reveal how diverse disciplines collaborate to tackle complex challenges holistically, fostering a more comprehensive and effective approach to issues at the intersection of social sustainability and AI.

13.2 METHODOLOGY

The methodology employed for this research utilized bibliometric analysis to explore the intersection of sustainable social development and AI within the Scopus database. Alan Pritchard is credited with developing bibliometrics, or bibliometric analysis, in 1969, which was a crucial turning point for the field of academic research evaluation and quantitative analysis of bibliographic data (Tierno et al., 2017). Bibliometric methods constitute an indispensable approach for the comprehensive examination of academic research within a specific domain or scholarly journal (Raan, 1996). These methodologies facilitate the extraction of in-depth insights into various facets of scholarly output, encompassing the latest trends in publications, the prominence of keywords, affiliations, author contributions, and the extent of citations (Sarkar et al., 2022). The specific steps followed during the methodology are visually presented in Figure 13.1, illustrating the systematic approach undertaken to identify and analyze the most relevant and credible sources in this research field.

The initial dataset was obtained by conducting a comprehensive search using the query: “Sustainable” AND “Social” AND “Development” AND “AI” OR “Artificial Intelligence” within the “titles, abstracts, and keywords” fields, resulting in 1,318 documents. To narrow down the focus and ensure relevance, subject-based exclusion criteria were applied, limiting the study to documents categorized under Social Sciences, Business, Management and Accounting, and Economics, Econometrics, and Finance. This step reduced the dataset to 492 documents. Further refinement was carried out by filtering based on document type and publication stage. Articles, conference proceedings, and book chapters that were published and deemed as credible sources were retained, while document types such as retracted documents, conference reviews, editorials, notes, short surveys, and reviews were excluded to avoid including grey literature. This rigorous selection process resulted in a final

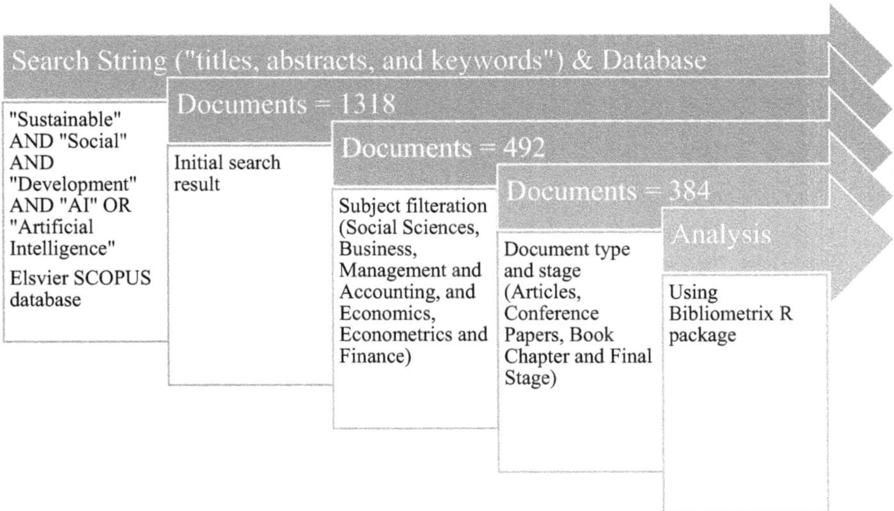

FIGURE 13.1 Data processing steps.

dataset comprising 399 documents for in-depth analysis. The analysis of this dataset was performed using the Bibliometrix R package, allowing for a comprehensive exploration of the research landscape at the intersection of sustainable social development and AI.

13.3 RESULTS

The descriptive statistics of the literature that was retrieved show important developments in the field of sustainable social development and AI (see Table 13.1). Spanning from 1997 to 2023, it portrays a thriving and steadily expanding field, involving 218 distinct sources contributing to 384 documents from 517 affiliations and 52 countries, with an annual growth rate of 15.07%. The average document age of 3.09 years signifies a focus on recent developments, and the high average citations per document (13.55) underscores its academic impact. This research domain involves 1,197 authors, with 67 single-authored documents and an average of 3.28 co-authors per document, emphasizing collaboration. Notably, 27.08% of collaborations are international, highlighting the global nature of the research on the intersection of sustainable social development and AI. Within this scholarly domain, an array of document types is evident, encompassing 257 articles, 55 book chapters, and 72 conference papers. This diversity in document types, coupled with the substantial growth rate and international collaborations, collectively underscores the significance of this interdisciplinary field. It highlights the pivotal role it plays in examining the application of AI in the context of sustainable social development, while also shedding light on the opportunities and challenges arising from the utilization of AI for the broader goals of sustainable development.

TABLE 13.1
Descriptive Data of Extracted Data

Description	Results
Main Information About Data	
Timespan	1997–2023
Sources (Journals, Books, etc.)	218
Documents	384
Annual Growth Rate %	15.07
Document Average Age	3.09
Average citations per doc	13.55
References	20,425
Document Contents	
Keywords Plus (ID)	2,226
Author's Keywords (DE)	1,508
AUTHORS	
Authors	1,197
Authors of single-authored docs	67
Authors Collaboration	
Single-authored docs	69
Co-Authors per Doc	3.28
International co-authorships %	27.08
Document Types	
Article	257
Book Chapter	55
Conference Paper	72
Affiliations	517
Countries	52

13.3.1 Number of Publications and Average Citation per Year

The graph illustrates the evolution of research in the field from 1997 to 2023 (see Figure 13.2). In the early years, minimal articles led to negligible citations. Significant activity started in 2006, with one article driving increased average citations. From 2011 onward, a notable surge in articles occurred, peaking in 2022 and 2023 with 81 and 77 publications, respectively. This increase corresponded with a rise in average annual citations. A remarkable peak in 2016, with an average of 11.34 citations, signified heightened research impact. The data signifies a growing interest in the field, particularly in recent years, with research gaining recognition and relevance in addressing issues related to sustainable social development and AI.

13.3.2 Most Relevant Sources and Authors

Table 13.2 presents information about significant sources (journals) and authors in the field of sustainable social development and AI. To identify the most impactful

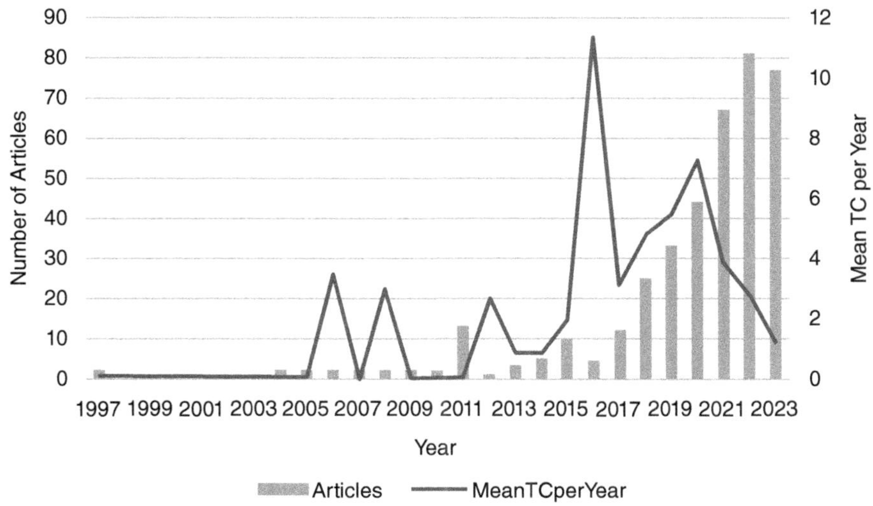

FIGURE 13.2 Annual scientific production and average TC per year.

sources and authors, we can focus on h-index (a measure of an entity's impact based on both the number of publications [NPs] and their citations), total citations (TCs), NPs, and the starting year of their publication history (PY_start).

Among the journals, "Journal of Cleaner Production" stands out as the most significant source, with an h-index of 19 and a substantial TC count of 929. It began its publication history in 2015, which indicates a relatively recent prominence. "Sustainability (Switzerland)" also holds considerable influence, with an h-index of 15 and 847 TCs. Similarly, "Sustainable Cities and Society" is notable with an h-index of 7 and 235 TCs. Regarding the authors, "Gupta S" emerges as a prominent contributor with an h-index of 3 and 47 TCs, commencing their publication journey in 2020. "Rhyner J" follows closely with an h-index of 3 and 57 TCs, also starting in 2020. "Guo Y" is another noteworthy author with an h-index of 2 and 128 TCs, beginning their publication history in 2019.

13.3.3 Most Relevant Affiliations and Countries

Table 13.3 offers a comprehensive overview of significant affiliations (institutions) and their corresponding countries within the research domain of sustainable social development and AI. To assess the prominence of these affiliations and countries, key metrics such as the number of articles published (Articles) are taken into consideration, along with two pivotal metrics: Single-country publications (SCPs) and multiple-country publications (MCPs). These metrics collectively enable us to discern the most influential affiliations and countries in this research field. Among the affiliations, "Beijing" emerges as the most significant with 13 articles. Following closely is "Universitat de València" with 11 articles, and "Beijing Technology and Business University" with 10 articles.

TABLE 13.2
Most Significant Source and Authors

Element	h_index	TC	NP	PY_start	Element	h_index	TC	NP	PY_start
Journal of Cleaner Production	19	929	33	2015	Gupta S	3	47	4	2020
Sustainability (Switzerland)	15	847	72	2016	Rhyner J	3	57	3	2020
Sustainable Cities and Society	7	235	8	2018	Frangopol DM	2	51	2	2017
International Journal of Production Research	6	511	6	2016	Guo Y	2	128	3	2019
Technology in Society	6	230	10	2013	Jang D	2	30	2	2016
Technological Forecasting and Social Change	5	248	7	2015	Kim J	2	30	2	2016
IEEE Technology and Society Magazine	3	31	3	2021	Li Y	2	35	4	2022
Journal of Urban Planning and Development	3	53	3	2012	Liao H-T	2	7	2	2020
Smart and Sustainable Built Environment	3	43	3	2015	Liu Y	2	8	2	2023
Building and Environment	2	51	2	2020	Lombardi P	2	19	2	2015
Contributions to Management Science	2	40	2	2020	Osanloo M	2	59	2	2017
Environmental Impact Assessment Review	2	16	2	2021	Pan C-L	2	9	2	2021
Industrial Management and Data Systems	2	53	2	2017	Park S	2	30	2	2016

TABLE 13.3
Most Significant Affiliations and Countries

Affiliation	Articles	Country	Articles	SCP	MCP
Beijing	13	China	55	41	14
Universitat De València	11	Italy	20	13	7
Beijing Technology and Business University	10	Germany	18	13	5
Indian Council of Agricultural Research – Central Marine Fisheries Research Institute (ICAR-CMFRI)	10	USA	17	15	2
The Bucharest University of Economic Studies	10	Spain	15	11	4
Wuhan University	10	United Kingdom	15	10	5
University of Michigan	9	Portugal	12	8	4
Universiti Teknologi Malaysia	8	Australia	8	2	6
Jiangsu University	7	India	8	6	2
Kaunas University of Technology	7	Iran	7	4	3
Korea University	7	Korea	7	6	1
Northwest A and F University	7	Romania	7	6	1
Sun Yat-Sen University	7	Brazil	6	1	5

In terms of countries, "China" holds the top position with 55 articles, signifying its substantial contribution to research in this field. "Italy" follows with 20 articles, and "Germany" is third with 18 articles. When considering the SCP metric, "China" remains the most prominent, indicating a strong domestic research presence with 41 articles, followed by the United States with 15 articles. "Italy" and "Germany" are the third most significant countries in terms of SCP, with 13 articles each. Upon examining the MCP metric, again China was the most prominent one with 14 articles, followed by Italy and Australia with seven and six articles, respectively.

13.3.4 Keyword Analysis

Keywords serve as the fundamental unit of analysis as they serve as descriptors of academic research. The objective of employing a word cloud for the discussed papers is to assess the prevalence of terms signifying the primary focus of the analysis (Patil et al., 2023). Figure 13.1 highlights the prominent keywords within the literature at the intersection of AI and sustainable social development as revealed through word cloud analysis. Notably, "artificial intelligence" stands out as the most frequently occurring term, emphasizing its central role in this research area. "Sustainability" and "sustainable development" are also prominent, underlining the emphasis on environmentally and socially responsible practices, while "sustainable development goals" (SDGs) align research with global sustainability targets. Key technological and data-related terms such as "big data," "decision support system," "digital transformation," and "machine learning," underscore the technological dimensions of AI in sustainable development. The presence of terms like "ethics," "climate change," and "social sustainability" signals a growing awareness of ethical and environmental considerations. Emerging trends in virtual and social dimensions are reflected in terms like "metaverse" and "social network analysis." Additionally, the focus on "renewable energy," "supply chain management," and "business model innovation" highlights the multifaceted aspects of sustainable development under study. In summary, this analysis of keywords illustrates the interdisciplinary nature of research in this field, encompassing technology, ethics, sustainability, and societal well-being.

Further, keyword co-occurrence analysis was performed. In co-occurrence analysis, the distances between the links represent the frequency of associations, while node sizes indicate their prevalence (Kumar et al., 2023). Figure 13.3 offers insights into the linkage between the keywords in the context of co-occurrence analysis within the literature related to AI and sustainable social development. Figure 13.3 illustrates the clusters of keywords, denoted as nodes to differentiate clusters, which are indicative of common themes and trends in the research landscape. Cluster 1 appears to revolve around terms like "economy" and "environment," signifying a focus on economic and environmental aspects within the field. These keywords have relatively low betweenness centrality, indicating their limited influence as connecting terms in the broader network. Cluster 2 encompasses a set of terms, including "sustainability," "decision support system," "industry 4.0," "climate change," and others. "Sustainability" holds a significant betweenness centrality, suggesting its pivotal role in connecting various topics within this cluster. This cluster appears to be characterized by a diverse range of keywords related to decision support, technology, and environmental concerns. Cluster 3 includes keywords such as "big data," "smart cities," "IoT," "machine learning," and "AI." "IoT" and "big data" exhibit

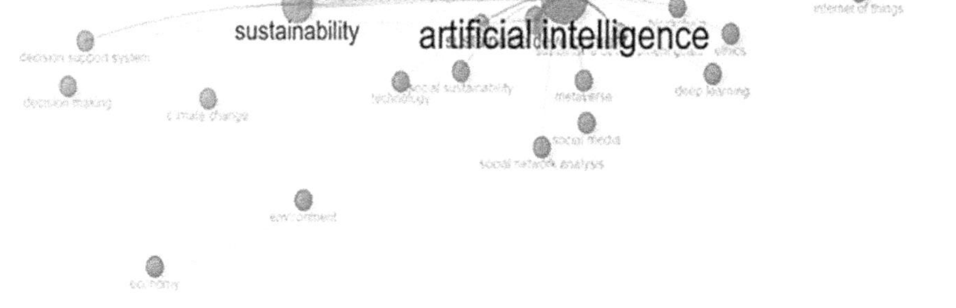

FIGURE 13.3 Keyword co-occurrence analysis.

notable betweenness centrality, indicating their roles as key connectors within this technological cluster. This cluster likely represents the technological dimensions of AI in the context of sustainable development, with an emphasis on data-driven and smart city solutions. Cluster 4 is the largest and most diverse, featuring central keywords like "artificial intelligence," "sustainable development," "digital transformation," "ESG," and "SDGs." "Artificial intelligence" has the highest betweenness centrality, underscoring its critical role in connecting various facets of research in sustainable development. This cluster encompasses a wide spectrum of themes, including ethical considerations, sustainability goals, and technology adoption. Lastly, Cluster 5 associated with "cloud computing," appears to be somewhat isolated, with limited connections to other clusters. This suggests that cloud computing, while relevant, maybe a relatively distinct subdomain within the broader landscape of AI and sustainable social development research.

13.3.5 Thematic Analysis

Thematic analysis, a vital method in scholarly research, involves identifying and categorizing prevalent themes or subjects within a dataset. This process offers valuable insights into prevailing research trends and themes, facilitating further exploration and knowledge development (Kumari & Kondala, 2023). Through thematic analysis, keywords can be categorized into distinct themes, reflecting the diverse landscape of research in this field (See Figure 13.4). Central to the discourse, AI-related terms such as "artificial intelligence," "big data," "machine learning," and

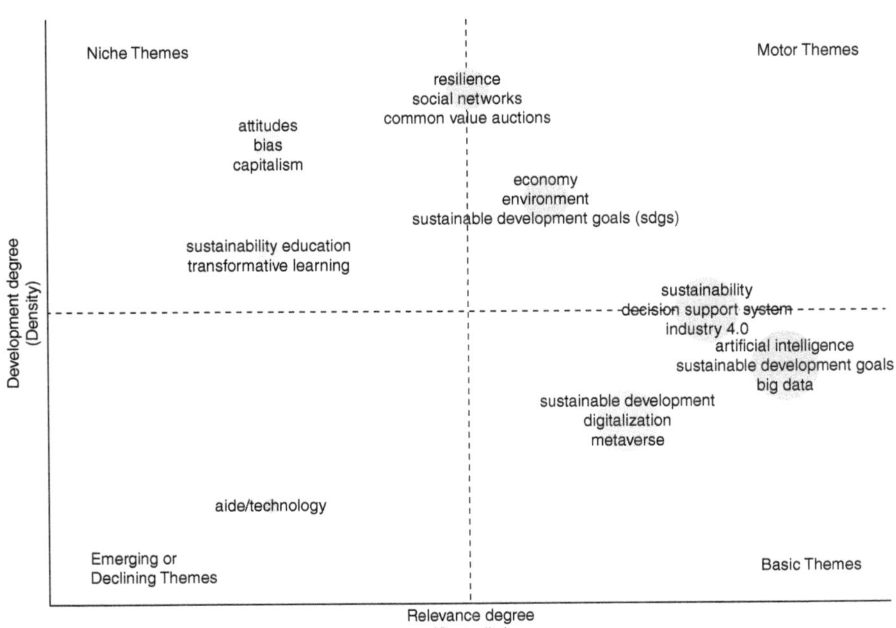

FIGURE 13.4 Thematic analysis of keywords.

"digital transformation" emerge as motor themes. These terms signify the driving force behind research, highlighting the pivotal role of advanced technology in sustainable social development efforts. Another prominent motor theme encompasses words like "sustainability," "sustainable development goals (SDGs)," and "sustainable development." These terms underscore the core focus of the literature, emphasizing the need for sustainable practices in social and economic development.

Within the niche themes, keywords like "ethics," "social sustainability," "corporate social responsibility (CSR)," and "social challenges" indicate a growing emphasis on ethical considerations and social responsibility in AI applications. These themes reflect the evolving discourse around the ethical implications of AI technologies. Terms like "IoT," "blockchain," "cloud computing," and "5G" represent niche themes related to technological advancements. These keywords highlight the exploration of cutting-edge technologies and their integration into sustainable social development strategies. Basic themes like "decision support system," "decision-making," and "framework" underline foundational concepts in the literature. These terms signify the fundamental tools and methodologies employed in decision-making processes within sustainable development initiatives.

Emerging themes include novel concepts like "metaverse," "chatbot," "smart tourism," and "social networks." These keywords indicate emerging areas of interest within the intersection of AI and sustainable social development, potentially pointing toward future research directions. On the other hand, declining themes like "transhumanism," "virtual reality," and "digital economy" suggest a decreasing focus on certain topics in the literature. This decline might signify shifts in research priorities or the maturation of these areas of study.

13.4 DISCUSSION

The research at the nexus of AI and Sustainable Social Development epitomizes a dynamic and burgeoning landscape, underscoring technology's pivotal role in addressing intricate societal dilemmas. Spanning the years from 1997 to 2023, this field has demonstrated a noteworthy annual growth rate of 15.07%, attesting to the mounting recognition of AI's significance in the realm of sustainable development and its pertinence in confronting multifaceted global challenges. The research's focal emphasis on recent developments, exemplified by an average document age of 3.09 years, underscores its ever-evolving nature. Furthermore, the robust average citations per document, amounting to 13.55, attests to the scholarly impact of this work, underscoring its potential to shape academic discourse and extend its influence beyond the academic sphere. Collaboration is a hallmark of research within this domain, with an average of 3.28 co-authors per document, serving to underscore the interdisciplinary character of this research endeavor and the imperative need for a diverse spectrum of expertise to effectively address the intricate challenges inherent in sustainable social development. Notably, 27.08% of these collaborative efforts are of international origin, underscoring the global dimension of this research and signifying the collective dedication of researchers from a multitude of countries to address overarching global issues through the analytical lens of AI and its contributions to sustainable development.

The inaugural work in this domain was published in 1997 under the title "The Nature of Virtual Organizations and Their Anticipated Social and Psychological Impacts" by Taylor et al. (1997). This research explored the prospective proliferation of virtual organizations in the future, leveraging advanced technologies. The authors posited that virtual organizations and their agents represent a natural progression in Western society's historical evolution toward efficient commercial and communicative structures. With the advent of high-speed communication and intricate interconnectivity, a widespread dispersion of commerce and education is anticipated. The researchers argued that technology will attain a level of sophistication wherein traditional educational institutions, with their substantial physical assets, will become unsustainable. Consequently, these institutions are projected to be supplanted by virtual organizations, which will deliver education and training through innovative means, minimizing the dependence on physical infrastructure. However, Zimmer et al.'s (2016) work is the one that has received the most citations in this field. The title of their research is "Sustainable Supplier Management: A Review of Models Supporting Sustainable Supplier Selection, Monitoring, and Development." In their research, a comprehensive content analysis methodology was employed to scrutinize the body of academic literature pertaining to sustainable supplier management. Specifically, the study concentrates on formal models designed to facilitate decision-making processes concerning sustainable supplier selection, ongoing monitoring, and progressive development.

Further, the three prominent authors "Gupta S, Rhyner J, and Guo Y" have performed research work related to technology and sustainable development. Gupta S and Rhyner J's work covers digitalization, AI, SDGs, and disaster risk. They mentioned that digitalization with the globally acknowledged SDGs is crucial for inclusive and holistic sustainable development in the digital era and introduced a new "Digitainability Assessment Framework (DAF)" for context-aware practical assessment of the impact of digitalization intervention on the SDGs and proposed an "Digitalization-Sustainability Matrix (DSM)" approach for the mutual relationship between digitalization and AI and SDGs. They also mentioned in their research that the emerging vulnerabilities are due to the dependency on digital services and identified novel semi-human groups and their vulnerabilities to disaster risks (Fekete & Rhyner, 2020; Gupta et al., 2020; Gupta & Rhyner, 2022; Nasir et al., 2023). Whereas Guo Y proposed a collaborative decision-making framework for sustainable supply chain planning, tested "continuous adoption of artificial intelligence technology (CAAIT)" for sustainable agricultural development, and also illustrated computational intelligence for the development of ecological architecture, protecting environment material implementation, and the natural environment to provide designers with certain new concepts of the art design using the "Hybrid Conformal Prediction Algorithm Framework (HCPAF)" (Allaoui et al., 2019; Guo et al., 2023; Xu et al., 2021).

Furthermore, it is noteworthy to mention that influential journals such as the "Journal of Cleaner Production," "Sustainability (Switzerland)," and "Sustainable Cities and Society" have emerged as significant contributors to this scholarly domain, as demonstrated by their notable h-indices. Furthermore, the prominence of specific affiliations and countries within this research field can be delineated through an analysis based on the NPs of affiliations and the countries of origin. In this regard, affiliations such as "Beijing," "Universitat de València," and "Beijing Technology and Business University" have notably advanced the scholarly literature. Similarly, countries such as "China," "Italy," and "Germany" have played a pivotal role in driving the research discourse forward, underscoring their substantial contributions to this evolving field of study.

The intersection of AI and Sustainable Social Development constitutes a multifaceted and dynamic research field, as evidenced by the analysis of prominent keywords and their co-occurrence in the literature. "Artificial intelligence" takes center stage, emphasizing its fundamental role in addressing the complex challenges of sustainability. This underscores AI's significance as a driving force in sustainable social development. Notably, the prevalence of keywords like "sustainability," "sustainable development," and "sustainable development goals" underscores a resolute commitment to environmentally and socially responsible practices, aligning research with global sustainability objectives. These keywords collectively stress the imperative of integrating sustainability principles into various facets of societal advancement. Furthermore, the inclusion of terms such as "ethics," "climate change," and "social sustainability" highlights the increasing recognition of ethical and environmental dimensions within AI applications. This signifies a holistic approach that addresses not only the technical aspects but also the moral and ecological imperatives of sustainable development. Additionally, the prominence of technological and data-related terms like "big data," "decision support system," "digital transformation," and "machine learning" underscores the pivotal role of cutting-edge technologies in advancing sustainable development, emphasizing the adoption of data-driven solutions to achieve sustainability goals.

The co-occurrence analysis unveils distinct clusters of interconnected keywords that provide valuable insights into the thematic structure of this research field. The central cluster, prominently featuring "artificial intelligence," "sustainable development," "digital transformation," "ESG," and "Sustainable Development Goals," serves as the focal point of research activities. This cluster underscores the interdisciplinary nature of the field, encompassing ethical considerations, sustainability objectives, and technological adoption, highlighting the necessity of addressing these aspects collectively. Additionally, niche themes, represented by keywords like "IoT," "blockchain," "cloud computing," and "5G," demonstrate a keen interest in exploring and integrating innovative technologies into the sustainable development agenda, signifying the ongoing pursuit of leveraging technological advancements for societal betterment. The identification of emerging themes, such as "metaverse," "chatbot," "smart tourism," and "social networks," hints at forward-looking perspectives and potential future research directions. Conversely, the waning emphasis on themes like "transhumanism," "virtual reality," and the "digital economy" may suggest a shift in research priorities or the maturation of these specific areas.

13.4.1 Implications

This study, which employs bibliometric analysis to explore the intersection of AI and Sustainable Social Development, offers both theoretical and practical implications in the context of academic research and real-world applications.

This study contributes to the theoretical framework of interdisciplinary collaboration by showcasing the intersection of AI and Sustainable Social Development. It reveals how these traditionally distinct fields are increasingly converging, emphasizing the broader role of technology in addressing complex societal challenges. Additionally, the research identifies emerging themes and areas of interest within the AI and sustainable development landscape, such as "metaverse," "chatbot," and "smart tourism." These findings offer theoretical insights for future research directions in these evolving domains. Moreover, the study underscores the growing emphasis on ethical and environmental considerations in AI applications for sustainable development, emphasizing the imperative of addressing not only the technological aspects but also the moral and ecological dimensions of sustainable development.

The insights from this research also hold practical value for policymakers, governments, organizations, businesses, and technology developers. They can inform strategic decisions related to technology adoption and sustainability initiatives. Understanding the prominence of specific keywords and thematic clusters can guide the development of policies and practical approaches that promote sustainable technology solutions. Furthermore, the significant international collaboration observed in this study offers practical implications for fostering global partnerships in addressing sustainable development challenges. Researchers, organizations, and academic institutions can draw on the successful collaborative models highlighted in this research for practical implementation in international projects.

13.5 CONCLUSION

The research at the intersection of AI and Sustainable Social Development signifies a dynamic and burgeoning field with remarkable scholarly impact. This interdisciplinary landscape, driven by the core tenets of technology, ethics, and sustainability, serves as a testament to the collective commitment of researchers and institutions worldwide. The substantial international collaboration observed underscores the global dimension of this research, emphasizing the united effort to address pressing global challenges through the lens of AI and its role in sustainable development. This commitment to collaboration mirrors the interconnected nature of the research itself, where AI not only plays a pivotal role but also intersects with environmental and ethical considerations.

In summary, the comprehensive analysis of the research literature related to AI and sustainable social development underscores the interdisciplinary essence of this field. The co-occurrence analysis illuminates intricate connections between diverse keywords, showcasing the multidisciplinary nature of research in this domain. Central themes like "sustainability" and "artificial intelligence" serve as pivotal points linking various concepts, while specialized areas, particularly ethical considerations and technological advancements, represent focused avenues of exploration. This dynamic landscape reveals ongoing research efforts driven by core themes such as AI and sustainability, reflecting the field's continuous evolution. The emergence of niche areas and the evolving research interests highlight the adaptability of scholars in response to the ever-changing challenges and opportunities within AI and sustainable social development.

13.5.1 Limitations and Scope for Future Work

Limitations of this research stem from the data and analysis methodologies employed. The bibliometric analysis primarily relies on the data available within the Scopus database, which might not encompass all relevant sources in the rapidly evolving field of AI and sustainable social development. While the data has provided valuable insights, it cannot capture the entirety of interdisciplinary research occurring outside traditional academic publication channels, such as industry reports, policy documents, or real-world AI applications. Moreover, while the study identifies emerging themes, it does not delve deeply into the qualitative aspects of research in these areas, leaving ample room for more in-depth investigations into the specific content and impact of research within these niche topics.

Future work in this field can expand on these limitations and explore emerging areas of interest in more detail. Future research in this field should delve deeper into emerging niche areas like "metaverse," "chatbot," and "smart tourism" by conducting qualitative content analyses to understand specific challenges and opportunities. Longitudinal studies beyond 2023 are needed to track the evolving landscape. Comparative studies across regions can highlight regional variations, calling for tailored policies. Additionally, exploring practical AI applications through case studies can offer valuable insights for policymakers and organizations striving to address complex societal issues while upholding ethical and environmental principles.

REFERENCES

Allaoui, H., Guo, Y., & Sarkis, J. (2019). Decision support for collaboration planning in sustainable supply chains. *Journal of Cleaner Production*, *229*, 761–774.

Awan, U., Kraslawski, A., & Huiskonen, J. (2018). Governing interfirm relationships for social sustainability: The relationship between governance mechanisms, sustainable collaboration, and cultural intelligence. *Sustainability*, *10*(12), 4473.

Butcher, J., & Jeffrey, P. (2005). The use of bibliometric indicators to explore industry–academia collaboration trends over time in the field of membrane use for water treatment. *Technovation*, *25*(11), 1273–1280.

Chawla, R. N., & Goyal, P. (2022). Emerging trends in digital transformation: A bibliometric analysis. *Benchmarking: An International Journal*, *29*(4), 1069–1112.

Chui, K. T., Alhalabi, W., Pang, S. S. H., Pablos, P. O. D., Liu, R. W., & Zhao, M. (2017). Disease diagnosis in smart healthcare: Innovation, technologies and applications. *Sustainability*, *9*(12), 2309.

Di Vaio, A., Palladino, R., Hassan, R., & Escobar, O. (2020). Artificial intelligence and business models in the sustainable development goals perspective: A systematic literature review. *Journal of Business Research*, *121*, 283–314.

Fekete, A., & Rhyner, J. (2020). Sustainable digital transformation of disaster risk—Integrating new types of digital social vulnerability and interdependencies with critical infrastructure. *Sustainability*, *12*(22), 9324.

Ferreira, J. J., Lopes, J. M., Gomes, S., & Rammal, H. G. (2023). Industry 4.0 implementation: Environmental and social sustainability in manufacturing multinational enterprises. *Journal of Cleaner Production*, *404*, 136841.

Florey, C., Hellin, J., & Balié, J. (2020). Digital agriculture and pathways out of poverty: The need for appropriate design, targeting, and scaling. *Enterprise Development & Microfinance*, (2), 126–140.

Guo, Y., Dong, Y., Wei, X., & Dong, Y. (2023). Effects of continuous adoption of artificial intelligence technology on the behavior of holders' farmland quality protection: The role of social norms and green cognition. *Sustainability*, *15*(14), 10760.

Gupta, S., & Rhyner, J. (2022). Mindful application of digitalization for sustainable development: The Digitainability Assessment Framework. *Sustainability*, *14*(5), 3114.

Gupta, S., Motlagh, M., & Rhyner, J. (2020). The digitalization sustainability matrix: A participatory research tool for investigating digitainability. *Sustainability*, *12*(21), 9283.

Kondala, M., Nudurupati, S. S., & Pappu, R. P. (2023a). The challenges in adoption of circular economy in SMEs–a research agenda and way forward. *Benchmarking: An International Journal*, *31*(5), 1667–1699.

Kondala, M., Nudurupati, S. S., Riwayadi, E., Chavan, A., Asif, S. R., & Gupta, N. (2023b). Original research article moving towards a sustainable world with the circular economy practices concerning the SMEs in Visakhapatnam's ice-cream industry. *Journal of Autonomous Intelligence*, *6*(1), 1–14.

Kumar, S., Khan, A., Lochab, A., Gupta, V. P., & Arora, A. K. (2023). Boundaryless career: A bibliometric analysis. *Prabandhan: Indian Journal of Management*, *16*(8), 24–44.

Kumari, N., & Kondala, M. (2023). Role of Soft Power in Tourism Development: A Bibliometric Analysis of the Past Decade. In *Global Perspectives on Soft Power Management in Business* (pp. 245–260). IGI Global.

Lochab, A., Kumar, S., & Tomar, H. (2018). Impact of human resource analytics on organizational performance: A review of literature using R-software. *International Journal of Management, Technology And Engineering*, *8*, 1252–1261.

Nasir, O., Javed, R. T., Gupta, S., Vinuesa, R., & Qadir, J. (2023). Artificial intelligence and sustainable development goals nexus via four vantage points. *Technology in Society*, *72*, 102171.

Nguyen, H. D., Mai, L. T., & Anh Do, D. (2020). Innovations in creative education for tertiary sector in Australia: Present and future challenges. *Educational Philosophy and Theory*, *52*(11), 1149–1161.

Omar, W. M. W., Ismail, M. N., Ismail, M., Ahmad, G. O., & Ya, M. S. (2020). Visualizing research literature in halal supply chain: A bibliometric analysis. *Malaysian Journal of Consumer and Family Economics*, *24*(S2, 7), 5–92.

Patil, R. R., Kumar, S., Rani, R., Agrawal, P., & Pippal, S. K. (2023). A bibliometric and word cloud analysis on the role of the internet of things in agricultural plant disease detection. *Applied System Innovation*, *6*(1), 27.

van Raan, A. (1996). Advanced bibliometric methods as quantitative core of peer review based evaluation and foresight exercises. *Scientometrics*, *36*(3), 397–420.

Rajaram, S. S., Grimm, B., Giroux, J., Peck, M., & Ramos, A. (2014). Partnering with American Indian communities in health using methods of strategic collaboration. *Progress in Community Health Partnerships: Research, Education, and Action*, *8*(3), 387–395.

Roig-Tierno, N., Gonzalez-Cruz, T. F., & Llopis-Martinez, J. (2017). An overview of qualitative comparative analysis: A bibliometric analysis. *Journal of Innovation & Knowledge*, *2*(1), 15–23.

Saravanan, J., Thomas, V., & Ashikho, A. (2023). Mapping the research trends on political communication in Asia: A bibliometric analysis using R package and VOS. *Asian Journal of Comparative Politics*, *9*(2), 254–273.

Sarkar, A., Wang, H., Rahman, A., Memon, W. H., & Qian, L. (2022). A bibliometric analysis of sustainable agriculture: Based on the Web of Science (WOS) platform. *Environmental Science and Pollution Research*, *29*(26), 38928–38949.

Taylor, R. G., Peltsverger, B. W., & Vasu, M. L. (1997). The nature of virtual organizations and their anticipated social and psychological impacts. *Education and Information Technologies*, *2*, 347–360.

Thiebes, S., Lins, S., & Sunyaev, A. (2021). Trustworthy artificial intelligence. *Electronic Markets*, *31*, 447–464.

Toniolo, K., Masiero, E., Massaro, M., & Bagnoli, C. (2020). Sustainable business models and artificial intelligence: Opportunities and challenges. In F. Matos, V. Vairinhos, I. Salavisa, L. Edvinsson, & M. Massaro (Eds.), *Knowledge, People, and Digital Transformation: Approaches for a Sustainable Future*, 103–117. Contributions to Management Science. Cham: Springer. https://doi.org/10.1007/978-3-030-40390-4_8

Xu, Y., Guo, Y., Jumani, A. K., & Khatib, S. F. (2021). Application of ecological ideas in indoor environmental art design based on hybrid conformal prediction algorithm framework. *Environmental Impact Assessment Review*, *86*, 106494.

Zimmer, K., Fröhling, M., & Schultmann, F. (2016). Sustainable supplier management – A review of models supporting sustainable supplier selection, monitoring and development. *International Journal of Production Research*, *54*(5), 1412–1442.

14 Artificial Intelligence and Sustainable E-Commerce

A Study on Rising Human and Market-Related Issues

Rajani H. Pillai, Deeksha S, and Santosh N.

14.1 INTRODUCTION

The onset of the COVID-19 pandemic has had a profound impact on the global economy, resulting in significant disruptions and adversely affecting the livelihoods of billions of individuals worldwide. The current outbreak serves as a unique and significant reminder of the existing disparities and deficiencies that the 2030 Agenda for Sustainable Development and the Paris Climate Agreement aim to rectify. Through strategic utilization of a crisis, characterized by the disruption of established policies and cultural norms, valiant endeavors can contribute to the restoration of the Sustainable Development Goals (SDGs) on a global scale. The SDGs play a crucial role in fostering sustainable societies that are characterized by inclusivity, resilience, and robustness. The advancements in computing technology, processing speed, Internet technology, and the accessibility of vast amounts of data have triggered a significant transformation in the ability to replicate or potentially exceed human intellectual capabilities. The current revolution has been instigated due to recent developments in technology. The current impact of artificial intelligence (AI)-driven forces is resulting in disruption across various industries worldwide, encompassing communication, transportation, healthcare, manufacturing, and finance, among other sectors. According to projections made in 2017, the market value of AI technologies was estimated to be $16.06 billion. It is expected that this value will experience significant growth and reach $190.61 billion by the year 2025. In recent years, there has been a notable rise in the popularity and significance of AI systems, including digital personal assistants like Apple's Siri and Amazon's Alexa, as well as robots and other AI gadgets. These technologies have become integral to the daily lives of individuals. AI is widely recognized as a beneficial tool among scholars and professionals in various industries. This is mostly attributed to its notable attributes, including its rapid processing capabilities, enhanced efficiency, vast storage capacity, and its capability for autonomous learning. According to statements made by Sundar Pichai, the Chief Executive Officer of Google, AI is regarded as a crucial area of focus for humanity. It possesses a greater level of relevance in comparison to both electricity and fire.

DOI: 10.1201/9781003638506-14

This chapter examines the presence of AI-enabled products or services in consumer markets and explores the ways in which firms might adopt socially responsible practices to effectively handle the ethical challenges associated with AI. The broader population recognizes the enhanced capabilities of AI but expresses caution on the possible threats to their privacy and safety associated with such advanced technologies. The concept of sustainable development plays a crucial role in the operation of an economy that relies on internet commerce. The notion of sustainable development has garnered significant attention in the domain of business operations, eliciting interest from both customers and businesses alike. Consequently, sustainable development is currently recognized as an effective approach for maintaining competitiveness and attracting a larger client base in online markets. Sustainable development models exert significant influence on various aspects within e-commerce firms, including economic, societal, and environmental dimensions. Nevertheless, the integration of sustainable development principles in the realm of electronic commerce (e-commerce), along with the careful management of many variables, has the potential to positively impact an organization's total efficiency and productivity.

This chapter will focus on the adverse ramifications that have emerged due to advancements in information technology. This chapter aims to elucidate a diverse range of challenges associated with the implementation of AI, encompassing behavioral, cultural, ethical, social, and economic dimensions. From a theoretical perspective, we will discuss a diverse range of issues that have emerged due to advancements in technology. The aforementioned issues can be categorized into two main domains: human factors encompassing behavioral, cultural, ethical, and social aspects, and market factors involving responsibility, legal considerations, security, and privacy concerns. It is important to note that these categories are intricately linked to one another.

When considering the practical implications, the primary focal points of AI systems should be around the preservation of individuals' rights to personal privacy and safety. It is imperative for individuals to prioritize their endeavors toward ensuring the ethical integrity of the system to the greatest extent feasible, while maintaining personal agency and direct influence. The present study is a valuable contribution to the field of AI, encompassing both the human and market perspectives. This research aims to examine the various individual concerns associated with AI, including behavioral, psychological, and ethical considerations, and their impact on human existence. Additionally, this study will ascertain the security and privacy risks associated with AI in electronic markets. Furthermore, this discourse acknowledges the legal and accountability issues associated with AI in online markets and underscores the imperative for government intervention in safeguarding the ethical and secure operation of such systems for the benefit of the wider populace.

To optimize the utilization of sophisticated technology, it is imperative to delve into the respective roles assumed by individuals and society in the adoption of novel and cutting-edge technologies. The comprehension of businesses regarding the necessity for the establishment of ethical norms and laws in electronic markets, aimed at ensuring customer safety, may be enhanced by an extensive analysis that deliberates on the influence of AI on human behavior and society. To gain the trust of consumers and organizations, it was imperative for enterprises and marketing agencies to develop comprehensive rules, ethical guidelines, laws, and regulations to enhance the security of AI systems.

AI systems possess both advantageous and detrimental effects on the promotion of sustainable development inside electronic markets. This research aims to enhance theoretical and practical understanding by examining several aspects of AI and encompassing behavioral, psychological, ethical, social, and cultural dimensions. Additionally, this study will propose potential remedies to improve the current state of affairs. Nevertheless, it is imperative to acknowledge that the paramount aspect of utilizing AI technology is the careful examination of social and ethical implications. The advent of AI has had detrimental effects on the market and economy, characterized by a significant increase in unemployment rates and a consequential shift in the job market that has created an atmosphere of uncertainty. Moreover, the progression of AI is giving rise to apprehensions regarding security and privacy, particularly in light of the emergence of social media platforms. The attainment of AI security will remain elusive unless all individuals possessing the ability to compromise AI security adhere to ethical principles. Developing a framework that prioritizes the well-being of persons and fosters a harmonious coexistence between humans and AI could be a feasible approach to creating a mutually beneficial environment for both parties. Enhancing the human-like characteristics of robots will result in a convergence between machines and human psychology, thereby exerting a positive influence on society.

AI has undeniably had a profound impact on human existence, with both advantageous and detrimental outcomes. Nevertheless, the principal objective of AI ought to be its utilization in the service of humanity's overarching concerns. The rapid expansion of AI throughout various domains of individuals' personal and occupational spheres necessitates a careful approach. This prudence stems from historical evidence indicating that, regardless of a tool's potency, it remains susceptible to acts of sabotage. It is imperative to enhance the capabilities of AI while concurrently upholding our ability to exercise control over it. The development of AI should be conducted within a controlled environment, employing meticulous and well-aligned data collection methods, in order to effectively achieve our objectives.

14.2 REVIEW OF LITERATURE

14.2.1 Review of Literature on AI in Various Fields

Shukla and Vijay (2013) in their study elucidate the characteristics of AI, its developmental trajectory, and a selection of its principal applications. John McCarthy, a prominent figure in the establishment of AI research, provided a concise definition of the discipline as the endeavor to enable computers to do tasks that are typically associated with human intelligence. The purpose of the definition was that he was entirely at ease with continuing his investigation without the necessity of initially justifying any certain philosophical perspective, regarding the concept of "intelligence." The origins of AI can be traced back to the realms of philosophy, fiction, and imagination. The advent of early technological advancements in fields such as electronics, engineering, and various other disciplines has exerted a significant influence on the development of AI. Several significant achievements in the field of AI can be identified, particularly in the area of problem-solving. These accomplishments encompassed fundamental advancements in learning, knowledge representation, and inference. Additionally, notable progress was made through the development of demonstration programs that showcased the capabilities of language understanding, translation, theorem proving, associative memory, and knowledge-based systems.

According to Haleem et al. (2019), AI is poised to revolutionize numerous domains within the medical industry in the present context. There is a requirement to examine the research conducted in this particular technology and ascertain its various uses within the medical domain. Numerous scholarly studies and research publications have extensively examined the subject matter of "artificial intelligence" and its specific use within the medical sector. These investigations have successfully identified and elucidated various noteworthy implementations of AI technology. This chapter examines the utilization of AI in addressing complex issues within the medical domain, supported by comprehensive research and development efforts. The present study has successfully identified and delineated five noteworthy technologies that are now being employed in the domain of AI within the medical area. Additionally, the study has also examined the intricate procedure involved in the implementation of AI in this context. This chapter presents a comprehensive analysis of 10 key applications of AI within the medical domain, accompanied by concise explanations. AI offers a valuable clinical decision-making tool that has the potential to enhance patient outcomes. Various technologies are being embraced and experimented with in order to enhance automation within the medical domain. In contemporary times, AI is being implemented within the realm of medicine to facilitate the digitization of medical records and enable patient checkups through the utilization of intelligent technologies. The technology offers solutions, particularly in the realm of targeted treatments, novel drug compositions, and individualized therapy. AI is a cutting-edge technology that assists surgeons in the process of medicine administration, treatment planning, and surgical procedures. The primary utilization of this technology is to enhance decision-making processes in complex scenarios. Furthermore, it has the potential to facilitate the monitoring, identification, examination, and management of infections inside the hospital setting. This technology is designed to enhance and optimize internet platforms for scheduling patient appointments. In the future, the application of medical advancements will prove beneficial across all domains, ultimately contributing to the betterment of humanity.

According to Patel et al. (2021), in the forthcoming times, it is anticipated that intelligent machines will assume the role of either substituting or augmenting human talents across several domains. AI refers to the manifestation of intelligence in robots or software systems. This particular domain is within the discipline of computer science (CS). The discipline of AI is gaining prominence within the realm of CS due to its significant contributions in enhancing several aspects of human life. Over the past two decades, there has been significant advancement in the capabilities of AI, leading to notable enhancements in the performance of several sectors such as industry, services, and education. The field of AI has contributed to the emergence of a fast-expanding technology referred to as expert systems. The application domains of AI have significantly influenced several spheres of human life. Expert systems, in particular, are extensively employed to address intricate challenges in industries such as education, engineering, business, medicine, and weather forecasting. The domains that have implemented AI technology have experienced a notable enhancement in both the caliber and effectiveness of their operations. This chapter provides a comprehensive overview of AI technology and its use in various domains, with a particular focus on its utilization in the field of education. It encompasses an examination of the concept, search methodologies, notable advancements, and future prospects of this technology.

According to Jha et al. (2019) The automation of agriculture is a prominent and burgeoning topic of interest across nations. The global population is experiencing rapid growth, resulting in a corresponding increase in the need for food. The conventional agricultural practices employed by farmers have proven inadequate in meeting the growing demand, necessitating the greater application of toxic pesticides, hence compromising soil health. This phenomenon significantly impacts agricultural practices, ultimately resulting in land infertility and barrenness. This chapter discusses many automation approaches, including Internet of Things (IoT), wireless communications, machine learning, AI, and deep learning. Several factors are contributing to challenges in the agricultural sector, including crop diseases, inadequate storage management, insufficient pesticide control, weed management, and inadequate irrigation and water management. These issues can be effectively addressed by the implementation of various approaches as indicated above. Currently, there exists a pressing demand to unravel the complexities surrounding the utilization of detrimental pesticides, regulated irrigation practices, pollution control measures, and the impacts of the environment on agricultural activities. The implementation of automated farming procedures has demonstrated a positive impact on both crop yield and soil fertility. This study provides a comprehensive examination of the research conducted by numerous scholars, aiming to offer a concise summary of the present state of automation in the agricultural sector. The research additionally examines a proposed system that might be deployed within a botanical farm to facilitate the identification of flowers and leaves, as well as automate the process of watering, by leveraging IoT technology.

According to Oana et al. (2017), it is possible for individuals to recollect the significance of corporate information technology solutions from a few years back. In the current market, a distinct segmentation may be observed among various software solutions, such as Enterprise Resource Planning (ERP), Customer Relationship Management (CRM), Document Management, Business Intelligence (BI), and other related systems. The process of segmentation has progressively incorporated additional components, such as Analytics and, more recently, Big Data, along with its prominent counterpart, Cognitive Computing. The emergence of technological advancements, particularly in the realm of business, has led to a diminishing significance of segmentation in the market for business solutions. This shift may be attributed to the increasing demand for timely and practical information derived from expanding data sets. The establishment of distinct demarcations. Currently, it is not feasible to discuss a comprehensive solution that encompasses all the sub-segments within this industry. An integrated system of this nature would likely incur significant costs and potentially exhibit inflexibility in accommodating the diverse requirements of various market participants.

According to Surden (2019), there has been a significant amount of scholarly discourse in recent times regarding the intersection between AI and the field of law. The inquiry pertains to the definition and connection between AI and its application within the legal field, encompassing both legal practice and administration. This chapter aims to elucidate the aforementioned inquiries by offering a comprehensive overview of AI and its application in the field of law. The debate strives to be complex but yet intelligible to individuals without a technical background. In pursuit of this objective, I initially delve into a comprehensive examination of AI. Subsequently, attention is directed toward the use of AI within the legal profession, encompassing its application by legal practitioners, individuals, and organizations subject to legal regulations, as well as governmental authorities responsible for the enforcement and implementation of legal frameworks. The primary objective of this essay is to present an objective and accurate perspective on AI, devoid of any misconceptions or exaggerations, by focusing on its genuine potential. This is intended to juxtapose discussions pertaining to the futuristic aspects of AI and law.

According to Jamankulova et al. (2022), the area of informatics, which encompasses intellectual information technology, is often regarded as a rapidly advancing scientific and practical domain with promising prospects for the future. Computers play a crucial role in various scientific and technological domains, exerting a significant impact on society. The practical outcomes derived from their utilization are of utmost importance, as they have the potential to bring about substantial changes in numerous fields. The chapter examines the intellectual aspects of scientific technologies in the context of the foundational components of AI.

According to Stone et al., (2022), the inaugural report of the "One Hundred Year Study on Artificial Intelligence" project (AI100) at Stanford University was released in September 2016. This study marked the initial step in the project's comprehensive and ongoing evaluation of AI and its societal implications. The composition of the authorship of this work consists of a group of 17 individuals who possess extensive expertise in the field of AI. Notably, the panel was led by Peter Stone, an esteemed researcher affiliated with the University of Texas at Austin. The report titled "Artificial Intelligence and Life in 2030" investigates the potential influence of AI on eight domains commonly found in urban environments. These domains include transportation, home and service robots, healthcare, education, public safety and security, low-resource communities, employment and workplace, and entertainment. The report explores the expected impact of AI on these domains in the upcoming years. The primary objective of this initiative is to offer the general populace an accurate depiction of the present status and potential of AI from a scientific and technological standpoint. Additionally, it seeks to assist industry and governmental entities in making informed decisions, while also providing valuable insights for research and development endeavors within the field. The AI100 Standing Committee, led by Barbara Grosz from Harvard University, assigned the responsibility of producing this report to the panel.

14.2.2 Review of Literature on AI in E-Commerce

According to Bawack et al. (2022), the chapter presents a synthesis of existing research on AI within the context of e-commerce. Additionally, it puts out a set of suggestions for information systems (IS) research to make valuable contributions to this particular area of study. In order to achieve this objective, the research employed a novel methodology that involved integrating bibliometric analysis with a comprehensive literature evaluation. A total of 4,335 papers were subjected to bibliometric analysis, while a subset of 229 publications published in prominent IS journals was carefully analyzed. The bibliometric analysis unveiled that the predominant research focus in the field of AI in e-commerce pertains to recommender systems. The core research themes that were found include sentiment analysis, trust, customization, and optimization. Furthermore, this phenomenon positions institutions based in China as frontrunners in this particular field of research. Furthermore, it is worth noting that a significant majority of research articles pertaining to AI in the field of e-commerce have been published in academic venues related to CS, AI, business, and management. The literature review is an overview of the primary study subjects, methodologies, and recurring motifs that have captured the attention of scholars in the field of ISs. Based on the findings presented, proposals for future research are formulated. This chapter introduces a novel study that endeavors to amalgamate existing research on AI within the domain of e-commerce. For researchers, this study provides valuable insights and suggestions for future directions in this particular field of inquiry. For professionals in the field, this resource offers a structured compilation of knowledge regarding the ways in which AI might be utilized to enhance e-commerce initiatives.

According to Khrais (2020), the introduction and integration of technology in several industries have significantly transformed operational processes. Significantly, key technological advancements in the field of e-commerce are designed to exert an influence on consumer behavior, with the objective of promoting certain items and brands. AI has emerged as a crucial and innovative technique for achieving personalization and customization of products in order to cater to unique requests. This study reveals that the ethical implications of AI systems in the field of e-commerce are a subject of debate, particularly with regard to the notion of explainability. The research included word cloud analysis, Voyance analysis, and Concordance analysis as methodologies to comprehensively examine the concept of explainability as it has been employed by scholars within the field of AI. This study is driven by an examination of a corpus and aims to provide a consistent framework, thereby making a valuable contribution to the advancement of Explainable Artificial Intelligence (XAI) models. XAI is a specialized domain within the study of machine learning that aims to scrutinize and comprehend the underlying models and processes employed in the decision-making mechanisms of AI systems sometimes referred to as black boxes. By doing so, XAI endeavors to offer valuable insights into the pivotal decision junctures, factors, and data utilized in generating recommendations. This study posited that in order to implement explainable XAI systems, it is necessary to enhance ML models to render them interpretable and intelligible.

According to Song et al. (2019), the increasing prevalence of AI in conjunction with the quick advancements in science, technology, and the economic landscape has resulted in a significant impact on both our professional endeavors and daily routines. AI technology has been effectively implemented and yielded favorable outcomes in the domain of e-commerce. AI has emerged as a significant catalyst for the advancement of e-commerce. This paper provides a comprehensive description of the current state and future prospects of AI technology in the field of e-commerce. It analyzes the current application of AI technology in e-commerce, focusing on the areas of AI assistants, intelligent logistics, recommendation engines, and optimal pricing. The research is primarily based on a case study of Baidu takeaway, which exemplifies the intelligent operation of e-commerce. The paper explores the significant impact and importance of AI on the development of e-commerce.

According to Vanneschi et al. (2018), the escalating volume of e-commerce transactions necessitates heightened risk management measures to mitigate the occurrence of payment defaults. Payment default refers to the inability of a client to fulfill their financial obligation by settling a bill within a period of 90 days from the date of receipt. Credit scoring (CS) is commonly utilized to assess the likelihood of clients defaulting on their obligations. The field of CS has undergone extensive research, resulting in the development of numerous computational techniques. The main objective of this study is to create a CS model that can replace the pre-risk check function of the e-commerce risk management system known as Risk Solution Services (RSS). Currently, RSS is widely utilized for assessing the likelihood of clients defaulting. The pre-risk check procedure utilizes data obtained from the order process and incorporates exclusion rules along with a generic customer service model. The proposed model is intended to serve as a comprehensive replacement for the pre-risk check process, functioning effectively both alone and in conjunction with the RSS primary risk check. This chapter presents an application of genetic programming (GP) in the field of CS. The model was constructed using a real-world dataset obtained from a reputable German financial solutions provider. The collection comprises order requests that have been handled by the RSS system. The findings indicate that the performance of the GP model surpasses that of the general CS model in terms of both classification accuracy and profit in the context of pre-risk check. The performance of GP in terms of classificatory accuracy was comparable to that of many advanced machine learning techniques, including logistic regression, support vector machines, and boosted trees. Moreover, the utilization of the GP model in conjunction with the RSS primary risk check can yield a model that possesses enhanced discriminatory capability.

According to Pallathadka et al. (2023), AI has been implemented in the e-commerce and financial sectors to enhance customer experience, optimize supply chain management, increase operational efficiency, and minimize material waste. The primary objective is to develop standardized and dependable methods for product quality control, as well as explore novel approaches to customer acquisition and service delivery while keeping costs at a minimum. Machine learning and deep learning are often employed as AI methodologies. These models are employed by individuals, organizations, and government agencies to forecast and get insights from data. Currently, there is ongoing development of machine learning models that aim to address the intricacy and variability of data within the food business. This chapter examines the various uses of machine learning and AI in the domains of e-commerce, corporate management, and finance. Some of the primary applications are sales growth, profit maximization, sales forecasting, inventory management, security measures, fraud detection, and portfolio management.

According to Fedorko et al. (2022), the popularity of AI is on the rise due to advancements in information and communication technology. The primary objective of enterprises operating in the contemporary e-commerce landscape is to exert influence on customer behavior in order to promote specific items and brands. The utilization of AI as a novel instrument in the realm of e-commerce may be perceived as a favorable advancement. This chapter centers on elucidating the fundamental nature of e-commerce and AI, as well as exploring their respective advantages. The objective is to assess the significance of AI and its use within the realm of e-commerce, drawing upon existing research on this subject matter.

According to Micu et al. (2021), the objective of this study is to determine the instruments utilized in e-commerce that have the capability to enhance the effectiveness of marketing efforts. The optimization of managerial and marketing processes utilizing AI has been recognized in the existing body of knowledge. Consequently, a quantitative research approach has been employed to develop a questionnaire. The research sample comprises 201 individuals holding managerial roles in the field of e-commerce. These individuals are affiliated with enterprises that were operational in the year 2020 and employed at least one staff member. This chapter elucidates the managerial tools employed for the promotion of items in the online environment, as well as the business procedures that aim to enhance efficiency through the utilization of AI. Simultaneously, the quantitative study has examined three hypotheses in order to ascertain the underlying motivations for online purchasing, as well as the communication strategies employed by online store administrators. The study's flaws arise from its exclusive focus on the management perspective, neglecting the viewpoint of the end consumer, hence potentially giving rise to ethical problems. The future research will focus on optimizing the flow of stocks and logistical operations, since it has been identified as the primary problem for management based on quantitative research findings.

According to Areiqat et al. (2021), the e-commerce business is currently employing AI to provide chatbot services, analyze user feedback, and deliver personalized assistance to online customers. A recent impartial assessment conducted in 2019 revealed that approximately 20% of customers express anxiety while considering the purchase of items or services from a chatbot. Additionally, it was determined that 40% of online clients actively seek out exceptional discounts and shopping deals through chatbot interactions. According to Gartner's projections, the total value of web-based business transactions is expected to reach $4.8 billion annually by 2021. Furthermore, Gartner forecasts that around 80% of customer interactions will be conducted by computer-based intelligent technologies, without the need for human intervention, by the end of 2020.

14.2.3 Literature Review on Artificial Intelligence and Sustainable E-Commerce

According to Deng & Guo (2023), the objective of this essay is to elucidate the determinants that impact the sustainable development of mobile e-commerce, including perspectives from both users and operational aspects. This study addresses the existing gap in qualitative research pertaining to the sustainable development of AI technology in the context of mobile e-commerce, utilizing the grounded theory approach. This study offers significant insights and serves as a source of inspiration for sustainable development in the respective sector. Moreover, it establishes a solid theoretical framework and serves as a helpful research reference for future studies. The influence of AI apps on mobile e-commerce mostly manifests itself across three key stages of the client shopping experience. There exist pre-shopping, mid-shopping, and after-shopping AI services, with each step encompassing distinct dimensions that warrant careful consideration. The study and its enduring characteristics are being examined. The findings of this study offer prospective recommendations and strategies for the development and enhancement of forthcoming e-commerce platforms. Moreover, they contribute to the sustainable progression of e-commerce and the enduring and robust expansion of the social economy. This study presents a set of sustainable development measures for the implementation of AI in the domain of mobile e-commerce. These measures encompass several aspects, including operational procedures and supervisory mechanisms. The findings of this study serve as a valuable resource for facilitating synchronized and expeditious socio-economic progress.

The objective of this study is to enhance the effectiveness of logistics distribution within the context of international trade by optimizing the distribution route of the e-commerce supply chain model (SCM). This study examines the historical context of international trade and e-commerce and identifies the challenges that now exist within sustainable SCM. Next, a research experiment is formulated with the objective of investigating sustainable SCM within the context of the Business to Consumer (B2C) E-commerce business model and subsequently doing a comparative analysis with the Consumer to Consumer (C2C) business model. The findings of this study hold significant academic merit in terms of enhancing resource efficiency within the domain of logistics. The experimental findings of the model indicate that for an order quantity of 2,000, the vehicle competition ratio (VCR) is merely 1.1 when the vehicle capacity (VC) is set at 12. In contrast, when the variable cost (VarC) is equal to 72, the variable cost ratio (VarCR) is calculated to be 1.45. Hence, it can be inferred that there exists a positive correlation between the load capacity of each individual vehicle within the logistics management system and the overall effectiveness of logistics distribution across the supply chain. The research findings include significant implications for enhancing the effectiveness of logistics distribution within the prevailing e-commerce landscape and fostering the digitalization and electronic advancement of international trade logistics (Qi et al., 2023).

The assessment of credit risk holds significant importance for e-commerce platforms as a result of the inherent uncertainty and transaction risk that accompany both customers and sellers in these contexts. Furthermore, it serves as a crucial component for the advancement of the e-commerce ecosystem and the long-term viability of the financial sector. The primary aim of this study is to devise a method that is both efficient and user-friendly for the evaluation of seller credit risk. The study examined three hybrid AI models: (1) decision tree—artificial neural network (ANN), (2) decision tree—logistic regression, and (3) decision tree—dynamic Bayesian network. The models utilized in this study were trained using a dataset consisting of sellers' credit instances obtained from Taobao. The dataset had a total of 609 cases, each of which contained 23 attributes encompassing both category and numerical variables. The findings indicate that the integration of a decision tree and an ANN yields the highest level of accuracy. This outcome has the potential to enhance the efficiency and effectiveness of transactions between buyers and sellers on various platforms. The aforementioned model is widely recognized as a robust instrument that enables the construction of a sophisticated credit risk assessment system. Moreover, it fulfills the criteria of the platform transaction mode by being adaptable and capable of autonomous learning. Ultimately, the implementation of this model is expected to foster the long-term growth and stability of the e-commerce ecosystem. The empirical findings can be utilized as a point of reference for e-commerce platforms in their efforts to enhance the e-commerce transaction environment by supporting an optimal credit risk evaluation model. Additionally, these results can also aid buyers and investors in making informed decisions (Xu et al., 2019).

The future of e-commerce will be influenced by various factors such as personalization, mobility, same-day deliveries, and maybe AI. In order to attain success and effectively respond to evolving customer preferences and requirements, it is imperative to deliberate upon the features and standards that online shops must adhere to. This includes accommodating the customers' heightened awareness of environmental concerns, such as sustainable development and their perception of the surrounding environment. Therefore, it is vital to establish a defined set of characteristics that will impact the determination to utilize the services provided by a particular e-commerce platform. Despite extensive efforts, numerous internet retailers experience failure due to their inability to match client expectations. Simultaneously, the task of satisfying client expectations is a significant obstacle for nascent e-commerce enterprises. Numerous scholarly investigations have been conducted on the subject of sustainable development within the realm of e-commerce. However, a notable gap exists in terms of a distinct approach for the design of e-shops, particularly with regard to the integration of sustainable development principles. The proposed methodology by the authors uses Kano's model and customer satisfaction as a basis for investigating both explicit and implicit consumer demands. These needs are subsequently categorized into distinct groups, each having varying effects on overall customer happiness. This chapter presents a case study that examines the factors contributing to customer satisfaction in a recently launched e-commerce platform specializing in organic products. The study focuses on the Central European market and incorporates specific principles for sustainable growth. The study was conducted as an author-initiated, comprehensive survey that possesses the potential for application in analogous research endeavors. The analysis considered a total of 1,069 surveys that were completed accurately. The participants in the study provided feedback on 16 essential qualities that an e-commerce platform must possess in order to enhance customer experience and maximize the benefits derived from its services. Additionally, they identified 11 specific features that are crucial in determining the level of consumer satisfaction. Within the realm of essential attributes, there existed a subset pertaining to sustainable development, hence signifying the discernible environmental consciousness exhibited by prospective clientele. The results acquired were presented to the management of the research e-shop and were incorporated into the design of its operational processes. The outcomes of the e-shop's operations will be assessed after a period of one to two years (Ingaldi & Ulewicz, 2019).

Many clients have utilized conventional e-commerce platforms, which lack guarantees of product quality, comprehensive feature displays, image search capabilities, virtual chat services, product recommendations, and tracking functionalities. As a result of these drawbacks, clients have transitioned to utilizing items offered by Alibaba and Tencent, thereby maintaining their presence within the respective ecosystems. The present study utilized the Quo Bias hypothesis as a framework to examine consumer behavior in relation to their retention on an e-commerce platform. A total of 28 comprehensive interviews were undertaken to extract the variables and develop a model grounded in risk theory and the Consumer Relationship Continuity Behavior (CRCB) framework. A survey was sent to a total of 649 e-commerce consumers, ensuring that only valid responses were included in the analysis. The collected data were then evaluated and subjected to structural equation modelling (SEM) for further analysis. The findings indicate that the CRCB was impacted by factors such as switching costs and comparative attractiveness. Furthermore, it can be shown that negative attitudes have a mediating role in the association between risk perception and consumer resistance to brand co-creation (CRCB), ultimately leading to a favorable influence on negative word-of-mouth (WoM) behavior. The findings of this study provide valuable insights for managers and policymakers, enabling them to develop and implement more effective policies that enhance customer service (Rashidin et al., 2021).

The objective of this study is to establish a complete analytical framework for evaluating the sustainability of the e-commerce business model. This framework involves the amalgamation of various key performance indicators into a unified measure of business model sustainability. The proposed vector methodology enables the prediction and assessment of the impacts of various measures, facilitating the identification and implementation of the most efficacious strategies for sustainable e-commerce business growth. The methodology employed in this study is grounded in correlation, cluster, and regression analysis techniques. The primary scientific contribution of this study lies in the introduction of a novel methodological approach. This approach enables the analysis of business model sustainability and facilitates the comparison of companies operating within a competitive environment. By employing this approach, it becomes possible to identify the key priorities that companies must address in order to attain leadership positions while promoting sustainable development. The findings from the correlation study demonstrate that in contemporary circumstances, the economic and environmental factors play a crucial role in determining the effectiveness of the business model in the realm of e-commerce. The examination of the e-commerce enterprises under study presented an advantageous prospect to consider the distinctive characteristics of these organizations and categorize them based on comparable performance measures.

This facilitated the development of more precise regression models for each cluster. In this particular scenario, a discernible relationship exists between the sustainability dimension of a company's business model and its categorization into a specific cluster. The modeling and assessment of business model sustainability were carried out to establish a correlation between this sustainability and the success of e-commerce enterprises within the economic, environmental, and social domains. Simultaneously, the findings indicate that enhancing the sustainability dimension leads a corporation toward achieving the business sustainability standard (Sun et al., 2021).

14.3 RESEARCH METHODS

The goal of the research is to see how AI is utilized in e-commerce and issues related to the sustainable development of e-commerce. As a result, the research utilizes a systemic literature review approach (Dziopa & Ahern, 2011), which involves reviewing secondary data and materials, as well as previous investigations. Indeed, some researchers claimed that a systematic or semi-systematic literature review (Nightingale, 2009), as well as a study of secondary data, might aid researchers in gaining a better understanding of the phenomena being studied. Because only research, including meta-analyses, that has been done on the subject matter assists the identification, analysis, comprehension, and synthesis of the ways in which AI has altered and impacted e-commerce and its results, this method guarantees that the study is based on empirical or empirically supported evidence (Lame, 2019).

The research utilized a rigorous literature review approach to assess both comprehensive reviews and individual studies about the application of AI in the domain of e-commerce (Okoli, 2015). A comprehensive selection process was conducted to identify and include a total of 64 studies in the review. The inclusion criteria specified that only English research published within the last ten years was included. A detailed evaluation of the quality of literature was conducted by selecting a total of 64 pieces of literature. The whole texts of the respective study works were acquired for this purpose. Six papers were removed from the analysis as a result of iterative procedures and concerns pertaining to their quality. The current investigation is based on an extensive examination of 58 pertinent academic references related to the topic at hand.

This study employed a systematic literature review approach to analyze previous scholarly works that were pertinent to the research inquiries under investigation (Xiao & Watson, 2019). The articles were sourced from reputed journals and were scrutinized to determine the level of quality exhibited by each study. Elsevier database, Routledge, and CRC Press Taylor and Francis database, Emerald Group Publishing database, Springer Nature database, and Sage database. Several supplementary articles were acquired from reputable academic databases such as Wiley, Academia, JSTOR, and Guildford Press.

14.4 RESULTS AND DISCUSSION

14.4.1 Aspects of Human Behavior, Psychology, and Culture

Human labor is adapting to AI in new ways. As machines continue to take over previously human-only tasks, it becomes increasingly important to identify emerging areas of work that can best be performed by humans. Creativity, empathy, critical thinking, and the ability to solve complicated problems are all examples of what make humans special (Livingston & Risse, 2019). As AI takes over mundane chores, people will have more time to focus on creative endeavors. The introduction of AI into the workplace also calls for a shift in perspective and ongoing education to meet the ever-changing requirements of the labor market. The adoption of AI technology can result in a more efficient and effective workforce in which humans and machines collaborate effectively (Zhang, 2021).

AI increases societal stress, induces behavioral changes, and adds emotional strain to people's already difficult jobs. However, the potential advantages of AI should not be overlooked (Liu, 2021). Automating routine chores frees up people to focus on what they do best: think critically and creatively. This move frees people to make use of their innate capacities for creativity, problem solving, and emotional intelligence, all of which are incomparable to anything a machine might do (Shaikh & Cruz, 2023). Despite any difficulties at the outset, individuals who are open-minded and committed to lifelong learning stand a much better chance of succeeding in today's rapidly changing labor market. Humans and AI can work together to improve productivity and efficiency in the workplace, which will have positive effects on the economy and society as a whole (Park et al., 2019).

In the future, AI might potentially drag the entire globe into an AI war that could inflict huge setbacks because AI systems do not express any human-like emotions (Choung et al., 2023). While it's true that AI systems don't have feelings, it's also crucial to remember that the humans responsible for creating and deploying these systems are ultimately responsible for their activities and the societal impacts they may have (Liu, 2021). Organizations and people must put ethical concerns and responsible behavior first when using AI. We can avoid the dangers and maximize the opportunities of this revolutionary technology by regulating its usage for the common good. So, reducing the likelihood of AI conflicts and paving the way for a more cooperative and prosperous future (Flavián et al., 2022).

Data fuel AI systems, but as more and more information about our demographics is collected, our privacy is at risk. Because of the profound effects that computer interaction has had on human behavior and social dynamics, it presents a formidable challenge to modern society. Guidelines and laws that preserve user privacy and ensure ethical usage of data are essential as AI becomes more pervasive in everyday life (Choung et al., 2023). We can ensure that AI's benefits are responsibly exploited, ushering in a day when people and robots may coexist peacefully and mutually improve civilization by addressing these concerns and increasing openness (Bao et al., 2021).

Compassionate people are more likely to be the ones who waste time on these devices. While acknowledging the potential benefits of AI technology, some experts express their reservations, noting that AI has the potential to bring widespread harm if it is not properly monitored and controlled (Flavián et al., 2022). The hazards of AI can be reduced, and its ethical usage guaranteed, if we promote a culture of responsibility and accountability. Trust in and apprehension of AI can be bolstered by making its underlying algorithms and decision-making processes public. With these safeguards in place, AI can be a useful tool for improving our lives and moving us forward (Liu, 2021).

14.4.2 Ethical and Social Issues

Many social and ethical concerns have arisen with the advent of AI technology, particularly when it comes to the management of autonomous systems. The potential for AI to displace human labor, resulting in economic instability and social division, is a serious ethical concern. There are also concerns regarding fairness and accountability when AI is used in decision-making processes, such as in the criminal justice or healthcare systems. As AI systems sift through mountains of data, they pose a threat to individuals' privacy and security. To guarantee that AI technology is used responsibly and for the greater good, these ethical and societal challenges require careful examination and regulation (Kolodinet et al., 2020).

Highlighted in the context of AI are societal challenges like the potential for widespread joblessness, diminished autonomy, and a loss of well-being. Many people are currently losing their jobs as a direct result of the development of AI technologies. With the development of new technologies, the problem is only becoming worse (Du & Xie, 2021). The development of increasingly sophisticated AI has raised concerns that people will learn to rely too heavily on machines, giving up control over their own lives and decisions. Important ethical questions are also raised by AI's propensity for bias and discrimination. Regulations are essential to deal with these problems and guarantee that AI is developed and used in a responsible and useful way, putting people's and society's best interests first (Hermann, 2022).

Facebook, Instagram, and YouTube are just a few examples of AI-powered platforms that have global user bases and have negative ethical and social impacts because they keep people glued to their screens for extended periods of time, encouraging unhealthy smartphone and social media addictions (Kaplan, 2020). The developers of these AI platforms must assume accountability and introduce safeguards to encourage positive interactions with users. For example, algorithms should give more weight to meaningful and relevant information than clickbait and sensationalism, and features should be included to encourage users to take breaks from their devices. To ensure that AI is developed and used to benefit individuals and society as a whole, there should be legislation and guidelines in place to hold these platforms liable for any harm produced by their algorithms (Gerlick & Liozu, 2020).

Addiction to digital devices is on the rise, and it's disrupting people's lives in many ways, including their productivity at work or school, their happiness, and their relationships (Hagerty & Rubinov, 2019). In order to combat this problem, it is essential for platforms to include tools that encourage good usage behaviors, such as notifications to remind users to take breaks and reduce their screen time. Additionally, initiatives and campaigns should be created to educate people on the risks of digital addiction and give help for those who need it. We can make the internet a better place for everyone by putting people's health first and striving to reduce the harm caused by internet use (Wang et al., 2023).

14.4.3 Privacy and Safety Concerns

It is unethical and perhaps dangerous to use AI systems to gain access to, gather, or distribute consumers' private information. All sorts of information about people's internet activities, purchases, and chats might be collected by such systems

(Bandara, 2020). This leads to worries about the privacy of sensitive data and its potential abuse. Moreover, as people become more reliant on AI systems, it opens up new entry points for hackers to breach security and access sensitive data, which poses a serious risk to people's safety and privacy. As a result, in this age of AI, it is more important than ever that those businesses and governments emphasize strict security measures and regulations to protect consumers' private data (Singh, 2021).

Due to the data-centric nature of AI systems, privacy has emerged as one of the world's most pressing concerns. The proliferation of AI technology has made it difficult to regulate the dissemination of information to the general public. Personal information is continuously being collected and analyzed by AI algorithms, which are embedded in everything from social media platforms to smart home devices (Jin, 2018). Identity theft, monitoring, and invasive marketing can all take advantage of the wealth of personal data now at anyone's fingertips. Privacy safeguards need to keep up with the development of AI. In order to provide people the peace of mind they deserve in today's quickly developing digital landscape, stricter restrictions and encryption mechanisms must be enacted to protect personal data (Saeed, 2023).

Because AI is not under the user's control and data sharing is not defined, data can be accessed without the user's knowledge or permission. Individuals are more vulnerable to identity theft and spying as a result of this lack of control, which increases the danger of unauthorized access to personal data (Srivastava & Jeet, 2023). Strict rules and encryption mechanisms governing the usage and sharing of data by AI systems are necessary to address this risk. By doing so, we can provide people the peace of mind they need to confidently adapt to the ever-evolving digital environment, knowing that their personal information is safe and secure (Manheim & Kaplan, 2019).

As a result, the increasing prevalence of social media, cloud data, and mobile settings, along with the increasing sophistication of AI systems and AI-enabled goods, provides a greater threat of cybercrime and hence increases the importance of investing in cybersecurity. The growing reliance on AI systems in fields like healthcare, banking, and transportation also highlights the significance of strong cybersecurity precautions. In order to prevent hostile activities and secure sensitive data from unauthorized access, certain standards and restrictions must be established. Prioritizing cybersecurity alongside the advancement of AI will allow us to fully realize the benefits of emerging technologies while protecting users and businesses from harm (Akour et al., 2022).

Data breaches have recently occurred at numerous companies and organizations, including social media platforms (such as Google, Facebook, Instagram, LinkedIn, and Yahoo), software makers (such as Adobe, where more than 150 million user passwords were compromised), retailers (where more than 40 million debit and credit cards were stolen in stores), banks (where the website of the Federal Reserve Bank of the United States was hacked), and many others. Millions of consumers' private information, including names, addresses, social security numbers, and credit card numbers, has been compromised as a result of these data breaches (Manheim & Kaplan, 2019). Because of this, many people and businesses have lost a lot of money, had their reputations ruined, and had their trust broken. Professionals in the field of cybersecurity are always toiling away at novel solutions, solid security measures, and user education in the hopes of preventing such attacks (Kesh et al., 2002).

14.4.4 Accountability and Legal Concerns

When AI advances beyond its traditional support role and makes judgments on its own, the question of who should be held responsible for those choices arises (Yildiz & Beloff, 2020). When AI is making decisions with legal ramifications, the situation becomes much more complex. Since many people may have a hand in creating and releasing an AI system, pinpointing legal responsibility for its conduct can be difficult. AI's opaque decision-making algorithms can also make understanding the reasoning behind the system's actions challenging. As AI develops further, these accountability and legal challenges will need to be addressed by politicians and legal professionals to guarantee the ethical and responsible application of the technology (Cobbe & Singh, 2021).

In the event that an AI tool makes a mistake, whose responsibility is it? AI relies entirely on data and the preprogrammed algorithms with which it is equipped. Complex and intricate algorithms make it difficult for humans to comprehend the underlying justifications for AI's actions (Liu et al., 2019). Concerns regarding bias, discrimination, or mistakes stemming from a lack of transparency exist. Frameworks and rules establishing liability and accountability in these circumstances will need to be established by policymakers and legal professionals. The promotion of responsible and ethical use of AI technology may also necessitate processes to audit and review AI systems to ensure they are making fair and unbiased decisions. (Yildiz & Beloff, 2020).

AI systems or networks cannot replace the human brain in its ability to adapt its thinking and decision-making to new circumstances. These are merely programs and are limited to the decisions they have been taught to make, but the advantages of these gadgets are that they are faster and more precise (Santoni & Mecacci, 2021). AI systems may perpetuate pre-existing biases or discriminate against particular groups due to their inability to reason like humans. That's why it's so important to have audit and evaluation processes in place. Regular assessments will help us spot unfair or biased decision-making practices and try to eliminate them. This will ensure that AI is used in a fair and ethical manner, improving conditions for people of all backgrounds (Thiebes et al., 2021).

As a result, the role of law and responsibility in AI judgments will continue to be open to debate. From these vantage points, however, it's reasonable to assume that interdisciplinary boards will shoulder blame in tricky scenarios, evaluating the evidence presented as pertinent rather than decisive (Thiebes et al., 2021). These panels will assess the strengths and weaknesses of AI algorithms and make recommendations for improving the field. Guidelines and laws will be established to make AI systems accountable for their actions and open to public scrutiny. These boards will help build an AI ecosystem that is more welcoming to marginalized groups by incorporating their insights and knowledge (Reier & Garay, 2021).

14.5 PRACTICAL AND THEORETICAL IMPLICATIONS

This study explores the potential negative outcomes of AI as a result of technological progress. Questions like "How does AI affect behavioral, psychological, ethical, social, and cultural issues in electronic markets?" are at the heart of the study. In the study, we explore the many cultural, ethical, societal, and economic concerns that arise from the use of AI. We also hoped to lessen the privacy and security concerns associated with AI software. The topic of the paper is timely and important. AI, as a technology with extraordinary transformational potential in many such fields, must nevertheless be examined in terms of both its good and negative implications on society. From a theoretical vantage point, we covered a wide range of problems caused by technological progress from the perspectives of both people (their behavior, culture, ethics, and society) and businesses (their accountability, law, security, and privacy). Concerns about employment loss, prejudice in algorithms, ethical repercussions in decision-making, and the possibility of AI exacerbating societal imbalances all fall under this category. To guarantee AI is created and implemented ethically, politicians and business leaders must confront the difficulties of governing the field. Computer scientists, ethicists, lawyers, sociologists, and psychologists will need to work together to find solutions to these societal ramifications. We can only maximize the benefits of AI while reducing its drawbacks if we take the time to carefully examine and comprehend these complex implications.

In terms of real-world applications, AI systems should prioritize protecting users' personal information and keeping them safe. The system should be as fair and within our control as feasible, and that is where our attention should be focused. The current research makes this kind of contribution to AI on the consumer and commercial levels. The study identifies the security and privacy hazards connected with AI in electronic marketplaces and explores the unique difficulties of AI and its effects on human life, including behavioral, psychological, and ethical concerns. Legal and accountability concerns related to AI in online marketplaces are also addressed, as is the government's role in ensuring the safety and integrity of such platforms. The potential economic impact of AI on online markets is being investigated, along with related topics like employment displacement, retraining, and upskilling. In addition, it highlights the significance of responsible AI deployment by investigating the function of AI in bettering customer experience and corporate operations. Overall, this study's goal is to help policymakers, organizations, and individuals better understand the AI landscape so that they can maximize the technology's benefits while minimizing its risks.

We elucidated how individuals and communities might most productively embrace these cutting-edge innovations. Such in-depth explanations of the AI impact debate on human and social behavior may help businesses see the need for adopting ethical norms and regulations in electronic markets to protect shoppers. To earn the public's and the business world's trust, organizations and marketing firms must implement sufficient policies, ethical guidelines, laws, and regulations to make AI systems more secure. This is crucial for the widespread adoption of AI technologies, as it allows businesses to gain consumers' trust and confidence. Furthermore, the creation of ethical standards and laws would address worries about privacy and data protection, guaranteeing the safety of individuals' private information. In the end, both consumers and companies profit from these reforms, since they will lead to a more open and accountable AI world.

14.6 CONCLUSION

The growth of sustainable electronic marketplaces can be affected both positively and negatively by the use of AI technologies. This study addresses the negative and positive aspects of AI, including behavioral, psychological, ethical, social, and cultural difficulties, and suggests ways to improve the current state of affairs. There's no denying that AI has had a good effect on online marketplaces. Economic growth and ecological sustainability have both benefited from the enhanced efficiency, accuracy, and productivity made possible by these systems. Nonetheless, it is essential to also address the negative effects. Displacement of workers, invasions of privacy, and unfair decision-making are all possible outcomes of AI development. Stakeholders can work together to build regulatory and ethical frameworks for AI systems that will help mitigate these risks. Workers can better use the positive effects of AI technology while mitigating the negative ones if investments are made in education and training programs to prepare them for the evolving labor environment. However, taking into account social and ethical concerns is the most crucial part of utilizing AI technology. As a result of the AI revolution, markets and economies have been damaged by unprecedented levels of joblessness and an unstable change in the job market. Moreover, as new social media platforms arise, advances in AI are generating security and privacy concerns. No amount of ethically upstanding humans will be enough to solve the problem of AI security. Establishing a shared context between robots and humans may be possible through the modeling of a people-friendly AI system and an AI-friendly environment for individuals. Improving robots' human-like qualities will help bring technology in line with human psychology. By encouraging moral action and mutual comprehension, this approach has the potential to lessen the damage caused by AI security breaches. Risks associated with social media can be reduced by putting an emphasis on privacy and data protection during the AI development process. Incorporating human psychology into AI systems allows for the protection of people's personal information and the overall improvement of society's security. However, the moral concerns raised by the risk of AI and human autonomy becoming indistinguishable through the creation of too humanlike robots must also be addressed.

Without a doubt, AI has altered people's lives, both positively and negatively. However, AI should be used primarily for the greater good of humanity. Despite the rapid expansion of AI into every sector of society, we must proceed with caution because even the most potent of tools have been compromised in the past. We need to improve AI while making sure we keep the reins on it. To achieve our objectives, AI research and development must take place in a regulated setting where accurate and standardized data can be collected. This is the only way to guarantee that AI continues to serve mankind and is in line with its values. The development of AI should also prioritize ethical concerns to avoid its misuse or unforeseen repercussions. When we put society's needs first, we can maximize AI's benefits while limiting its drawbacks.

REFERENCES

Akour, I., Alnazzawi, N., Alshurideh, M., Almaiah, M. A., Al Kurdi, B., Alfaisal, R. M., & Salloum, S. (2022). A conceptual model for investigating the effect of privacy concerns on E-commerce adoption: A study on United Arab Emirates consumers. *Electronics, 11*(22), 3648. https://doi.org/10.3390/electronics11223648

Areiqat, A. Y., Hamdan, A., Alheet, A. F., & Alareeni, B. (2021). Impact of artificial intelligence on E-commerce development. In *The Importance of New Technologies and Entrepreneurship in Business Development: In the Context of Economic Diversity in Developing Countries: The Impact of New Technologies and Entrepreneurship on Business Development* (pp. 571–578). Springer International Publishing. https://link.springer.com/chapter/10.1007/978-3-030-69221-6_43

Bandara, R., Fernando, M., & Akter, S. (2020). Privacy concerns in E-commerce: A taxonomy and a future research agenda. *Electronic Markets, 30*(3), 629–647. https://link.springer.com/article/10.1007/s12525-019-00375-6

Bao, Y., Cheng, X., De Vreede, T., & De Vreede, G. J. (2021). Investigating the relationship between AI and trust in human-AI collaboration. https://aisel.aisnet.org/hicss-54/cl/it_enabled_collaboration/3/

Bawack, R. E., Wamba, S. F., Carillo, K. D. A., & Akter, S. (2022). Artificial intelligence in E-commerce: A bibliometric study and literature review. *Electronic Markets, 32*(1), 297–338. https://link.springer.com/article/10.1007/s12525-022-00537-z

Choung, H., David, P., & Ross, A. (2023). Trust in AI and its role in the acceptance of AI technologies. *International Journal of Human–Computer Interaction, 39*(9), 1727–1739. https://arxiv.org/pdf/2203.12687

Cobbe, J., & Singh, J. (2021). Artificial intelligence as a service: Legal responsibilities, liabilities, and policy challenges. *Computer Law & Security Review, 42*, 105573. https://doi.org/10.1016/j.clsr.2021.105573

Deng, Z., & Guo, M. (2023). Research on the impact of the application of artificial intelligence technology on the sustainable development of mobile e-commerce. *Benchmarking: An International Journal.* https://doi.org/10.1108/bij-11-2022-0697

Du, S., & Xie, C. (2021). Paradoxes of artificial intelligence in consumer markets: Ethical challenges and opportunities. *Journal of Business Research, 129*, 961–974. https://doi.org/10.1016/j.jbusres.2020.08.024

Dziopa, F., & Ahern, K. (2011). A systematic literature review of the applications of Q-technique and its methodology. *Methodology.* https://doi.org/10.1027/1614-2241/a000021

Fedorko, R., Kráľ, Š., & Bačík, R. (2022, July). Artificial intelligence in e-commerce: A literature review. In *Congress on Intelligent Systems: Proceedings of CIS 2021* (Vol. 2, pp. 677–689). Singapore: Springer Nature Singapore. https://link.springer.com/chapter/10.1007/978-981-16-9113-3_50

Flavián, C., Pérez-Rueda, A., Belanche, D., & Casaló, L. V. (2022). Intention to use analytical artificial intelligence (AI) in services–the effect of technology readiness and awareness. *Journal of Service Management, 33*(2), 293–320. https://doi.org/10.1108/JOSM-10-2020-0378

Gerlick, J. A., & Liozu, S. M. (2020). Ethical and legal considerations of artificial intelligence and algorithmic decision-making in personalized pricing. *Journal of Revenue and Pricing Management, 19*, 85–98. https://link.springer.com/article/10.1057/s41272-019-00225-2

Hagerty, A., & Rubinov, I. (2019). Global AI ethics: A review of the social impacts and ethical implications of artificial intelligence. *arXiv preprint arXiv:1907.07892.*

Haleem, A., Javaid, M., & Khan, I. H. (2019). Current status and applications of Artificial Intelligence (AI) in medical field: An overview. *Current Medicine Research and Practice*, *9*(6), 231–237. https://doi.org/10.1016/j.cmrp.2019.11.005

Hermann, E. (2022). Leveraging artificial intelligence in marketing for social good—An ethical perspective. *Journal of Business Ethics*, *179*(1), 43–61. https://link.springer.com/article/10.1007/s10551-021-04843-y

Ingaldi, M., & Ulewicz, R. (2019). How to make e-commerce more successful by use of Kano's model to assess customer satisfaction in terms of sustainable development. *Sustainability*, *11*(18), 4830. https://doi.org/10.3390/su11184830

Jamankulova, F., Toshniyozova, M., & Nizomova, D. (2022). The most necessary aspects of artificial intelligence in different fields. *Mathematical Statistician and Engineering Applications*, *71*(4), 7536–7542. https://doi.org/10.17762/msea.v71i4.1372

Jha, K., Doshi, A., Patel, P., & Shah, M. (2019). A comprehensive review on automation in agriculture using artificial intelligence. *Artificial Intelligence in Agriculture*, *2*, 1–12. https://doi.org/10.1016/j.aiia.2019.05.004

Jin, G. Z. (2018). Artificial intelligence and consumer privacy. In *The Economics of Artificial Intelligence: An agenda* (pp. 439–462). University of Chicago Press. https://doi.org/10.1016/j.aiia.2019.05.004

Kaplan, A. (2020). Retailing and the ethical challenges and dilemmas behind artificial intelligence. In *Retail futures: The good, the bad and the ugly of the digital transformation* (pp. 181–191). Emerald Publishing Limited. https://doi.org/10.1108/978-1-83867-663-620201020

Kesh, S., Ramanujan, S., & Nerur, S. (2002). A framework for analyzing e-commerce security. *Information Management & Computer Security*, *10*(4), 149–158. https://doi.org/10.1108/09685220210436930

Khrais, L. T. (2020). Role of artificial intelligence in shaping consumer demand in E-commerce. *Future Internet*, *12*(12), 226. https://doi.org/10.3390/fi12120226

Kolodin, D., Telychko, O., Rekun, V., Tkalych, M., & Yamkovyi, V. (2020, March). Artificial intelligence in E-commerce: Legal aspects. In *III International Scientific Congress Society of Ambient Intelligence 2020 (ISC-SAI 2020)* (pp. 96–102). Atlantis Press. https://doi.org/10.2991/aebmr.k.200318.012

Lame, G. (2019, July). Systematic literature reviews: An introduction. In *Proceedings of the Design Society: International Conference on Engineering Design* (Vol. 1, No. 1, pp. 1633–1642). Cambridge University Press. https://doi.org/10.1017/dsi.2019.169

Liu, B. (2021). In AI we trust? Effects of agency locus and transparency on uncertainty reduction in human–AI interaction. *Journal of Computer-Mediated Communication*, *26* (6), 384–402. https://doi.org/10.1093/jcmc/zmab013

Liu, H. W., Lin, C. F., & Chen, Y. J. (2019). Beyond State v Loomis: Artificial intelligence, government algorithmization and accountability. *International journal of law and information technology*, *27*(2), 122–141. https://doi.org/10.1093/jcmc/zmab013

Livingston, S., & Risse, M. (2019). The future impact of artificial intelligence on humans and human rights. *Ethics & International Affairs*, *33*(2), 141–158. https://doi.org/10.1017/S089267941900011X

Manheim, K., & Kaplan, L. (2019). Artificial intelligence: Risks to privacy and democracy. *Yale JL & Technology*, *21*, 106. https://heinonline.org/HOL/LandingPage?handle=hein.journals/yjolt21&div=4&id=&page=

Micu, A., Micu, A. E., Geru, M., C⊠p⊠ţîn⊠, A., & Muntean, M. C. (2021). The impact of artificial intelligence use on the e-commerce in Romania. *Amfiteatru Economic*, *23*(56), 137–154. https://www.ceeol.com/search/article-detail?id=929507

Nightingale, A. (2009). A guide to systematic literature reviews. *Surgery (Oxford)*, *27*(9), 381–384. https://doi.org/10.1016/j.mpsur.2009.07.005

Oana, O., Cosmin, T., & Valentin, N. C. (2017). Artificial Intelligence-a new field of computer science which any business should consider. *Ovidius University Annals, Economic Sciences Series*, *17*(1), 356–360. https://ideas.repec.org/a/ovi/oviste/vxviiy2017i1p356-360.html

Okoli, C. (2015). A guide to conducting a standalone systematic literature review. *Communications of the Association for Information Systems*, *37*. https://doi.org/10.17705/1CAIS.03743

Pallathadka, H., Ramirez-Asis, E. H., Loli-Poma, T. P., Kaliyaperumal, K., Ventayen, R. J. M., & Naved, M. (2023). Applications of artificial intelligence in business management, e-commerce and finance. *Materials Today: Proceedings*, *80*, 2610–2613. https://doi.org/10.1016/j.matpr.2021.06.419

Park, S. Y., Kuo, P. Y., Barbarin, A., Kaziunas, E., Chow, A., Singh, K., . . .& Lasecki, W. S. (2019, November). Identifying challenges and opportunities in human-AI collaboration in healthcare. In *Conference Companion Publication of the 2019 on Computer Supported Cooperative Work and Social Computing* (pp. 506–510). https://doi.org/10.1145/3311957.3359433

Patel, A. R., Ramaiya, K. K., Bhatia, C. V., Shah, H. N., & Bhavsar, S. N. (2021). Artificial intelligence: Prospect in mechanical engineering field—A review. *Data Science and Intelligent Applications: Proceedings of ICDSIA 2020*, 267–282. https://link.springer.com/chapter/10.1007/978-981-15-4474-3_31

Qi, B., Shen, Y., & Xu, T. (2023). An artificial-intelligence-enabled sustainable supply chain model for B2C E-commerce business in the international trade. *Technological Forecasting and Social Change*, *191*, 122491. https://www.sciencedirect.com/science/article/abs/pii/S0040162523001762

Rashidin, M. S., Gang, D., Javed, S., & Hasan, M. (2021). The role of artificial intelligence in sustaining the e-commerce ecosystem: Alibaba vs. Tencent. *Journal of Global Information Management (JGIM)*, *30*(8), 1–25. https://doi.org/10.4018/JGIM.304067

ReierForradellas, R. F., & GarayGallastegui, L. M. (2021). Digital transformation and artificial intelligence applied to business: Legal regulations, economic impact and perspective. *Laws*, *10*(3), 70. https://doi.org/10.3390/laws10030070

Saeed, S. (2023). A customer-centric view of E-commerce security and privacy. *Applied Sciences*, *13*(2), 1020. https://doi.org/10.3390/app13021020

Santoni de Sio, F., & Mecacci, G. (2021). Four responsibility gaps with artificial intelligence: Why they matter and how to address them. *Philosophy & Technology*, *34*, 1057–1084. https://link.springer.com/article/10.1007/s13347-021-00450-x

Shaikh, S. J., & Cruz, I. F. (2023). AI in human teams: Effects on technology use, members' interactions, and creative performance under time scarcity. *AI & Society*, *38*(4), 1587–1600. https://link.springer.com/article/10.1007/s00146-021-01335-5

Shukla Shubhendu, S., & Vijay, J. (2013). Applicability of artificial intelligence in different fields of life. *International Journal of Scientific Engineering and Research*, *1*(1), 28–35. https://mineracaodedados.wordpress.com/wp-content/uploads/2014/12/applicability-of-artificial-intelligence-in-different-fields-of-life.pdf

Singh, R. (2021). A Study of Artificial Intelligence and E-Commerce Ecosystem–A Customer's Perspective. *International Journal of Research in Engineering, Science and Management*, *4*(2), 78–87. https://journal.ijresm.com/index.php/ijresm/article/view/507

Song, X., Yang, S., Huang, Z., & Huang, T. (2019, August). The application of artificial intelligence in electronic commerce. In *Journal of Physics: Conference Series* (Vol. 1302, No. 3, p. 032030). IOP Publishing. http://dx.doi.org/10.1088/1742-6596/1302/3/032030

Srivastava, S., & Jeet, S. (2023). E-commerce and Privacy Issues. *Russian Law Journal*, *11*(5). https://cyberleninka.ru/article/n/e-commerce-and-privacy-issues

Stone, P., Brooks, R., Brynjolfsson, E., Calo, R., Etzioni, O., Hager, G., ... & Teller, A. (2022). Artificial intelligence and life in 2030: The one hundred year study on artificial intelligence. *arXiv preprint arXiv:2211.06318.*

Sun, M., Grondys, K., Hajiyev, N., & Zhukov, P. (2021). Improving the e-commerce business model in a sustainable environment. *Sustainability*, *13*(22), 12667. https://doi.org/10.3390/su132212667

Surden, H. (2019). Artificial intelligence and law: An overview. *Georgia State University Law Review*, *35*, 19–22. https://heinonline.org/HOL/LandingPage?handle=hein.journals/gslr35&div=59&id=&page=

Thiebes, S., Lins, S., & Sunyaev, A. (2021). Trustworthy artificial intelligence. *Electronic Markets*, *31*, 447–464. https://link.springer.com/article/10.1007/S12525-020-00441-4

Vanneschi, L., Horn, D. M., Castelli, M., & Popovič, A. (2018). An artificial intelligence system for predicting customer default in e-commerce. *Expert Systems with Applications*, *104*, 1–21. https://doi.org/10.1016/j.eswa.2018.03.025

Wang, C., Ahmad, S. F., Ayassrah, A. Y. B. A., Awwad, E. M., Irshad, M., Ali, Y. A., ... & Han, H. (2023). An empirical evaluation of technology acceptance model for Artificial Intelligence in E-commerce. *Heliyon*, *9*(8). https://doi.org/10.1016/j.heliyon.2023.e18349

Xiao, Y., & Watson, M. (2019). Guidance on conducting a systematic literature review. *Journal of Planning Education and Research*, *39*(1), 93–112. https://doi.org/10.1177/0739456X17723971

Xu, Y. Z., Zhang, J. L., Hua, Y., & Wang, L. Y. (2019). Dynamic credit risk evaluation method for e-commerce sellers based on a hybrid artificial intelligence model. *Sustainability*, *11*(19), 5521. https://doi.org/10.3390/su11195521

Yildiz, Z. O., & Beloff, N. (2020, February). The emerging AI policy for e-commerce industry. In *Proceedings of the 2020 5th International Conference on Intelligent Information Technology* (pp. 66–70). https://dl.acm.org/doi/abs/10.1145/3385209.3385210

Zhang, Z., Genc, Y., Wang, D., Ahsen, M. E., & Fan, X. (2021). Effect of ai explanations on human perceptions of patient-facing AI-powered healthcare systems. *Journal of Medical Systems*, *45*(6), 64. https://doi.org/10.1007/s10916-021-01743-6

15 Metaverse Changing the Landscape of Game-Based Immersive Corporate Training

Khushi Desai and Deepali Bhatnagar

15.1 INTRODUCTION

The Metaverse, a word coined in Neal Stephenson's science fiction novel Snow Crash, is a shared cybernetic world that exists in cyberspace [Stephenson, 1992]. With advancements in virtual (VR) and augmented reality (AR) technology, the metaverse is no longer just a science fiction concept but a rapidly emerging reality. This development has significant implications for education, especially in the realm of game-based learning [Stephenson, 1992]. *"Metaverse isn't a thing a company builds. It's the next chapter of the internet overall."* – Mark Zuckerberg.

Commercial training is an indispensable aspect of any association, as it empowers employees to acquire the necessary skills and knowledge required to perform their job duties effectively. However, conventional training approaches frequently fall short of involving staff members and giving them a practical learning opportunity. This is where the metaverse comes in, providing an opportunity to create a more interactive and engaging training experience.

Game-based immersive corporate training in the metaverse enables employees to study in a safe and controlled situation while enjoying the benefits of interactive gameplay. Companies can adapt their training courses to suit their unique requirements and include gamification components to boost motivation and involvement [Abinaya et al., 2023]. Moreover, the metaverse fosters collaboration, networking, and teamwork, providing employees with a sense of community that is not possible with conventional training techniques.

Another benefit of the metaverse for game-based knowledge is that it offers the prospect for learners to collaborate as well as communicate with each other in real time. This can be particularly valuable for learners who are working on projects or solving problems together. In the metaverse, learners can also engage in virtual simulations and role-playing activities, which can help to develop their problem-solving, critical thinking, and communication skills.

The metaverse also provides a unique opportunity to gamify the learning experience. As a result, learners can receive incentives and praises for their

DOI: 10.1201/9781003638506-15

accomplishments, which may inspire and interest them. Additionally, the metaverse provides an opportunity for learners to compete against each other in virtual challenges and games, which can help to foster a sense of community and encourage friendly competition [Eliane Schlemmer, August 2014].

The use of the metaverse for corporate training is also cost-effective, as it eliminates the need for expensive travel and venue rental for in-person training programs. Thus, the metaverse has the potential to transform commercial training by offering an enjoyable, realistic, as well as economical option that augments employee abilities as well as knowledge.

Recently, there has been a lot of awareness of the concept of the metaverse as a possible platform for a range of applications, including education. The shared online virtual world known as the metaverse possesses the unique and fascinating potential to alter the field of immersive game-based business training.

Moreover, the metaverse introduces a distinctive opportunity to gamify the learning experience, offering learners incentives and commendations for their achievements, thereby fueling inspiration and sustained interest. Additionally, the competitive aspect within the metaverse enables learners to engage in virtual challenges and games, fostering a sense of community and encouraging friendly competition.

The economic advantages of utilizing the metaverse for corporate training are noteworthy, eliminating the substantial expenses associated with travel and venue rentals for traditional in-person training programs. Consequently, the metaverse emerges as a transformative force in corporate training, presenting an enjoyable, realistic, and cost-effective option that not only enriches employee abilities and knowledge but also transcends into a captivating realm with multifaceted applications, including education. This shared online virtual world, known as the metaverse, stands poised to revolutionize the landscape of immersive game-based business training, unlocking unique and fascinating potentials for diverse sectors.

15.2 EXAMPLE OF METAVERSE CHANGING THE LANDSCAPE OF GAME-BASED IMMERSIVE CORPORATE TRAINING

The use of VR training programs is one instance of how the metaverse is altering the landscape of game-based immersive business training. Companies such as Walmart have been using VR to train their employees in various aspects of their job duties, from customer service to compliance training [Chun et al., 2022].

In Walmart's VR training program, employees are immersed in a virtual store environment where they can practice scenarios such as handling a customer complaint or identifying a potential safety hazard [Makransky & Mayer, 2022]. The program incorporates gamification elements such as points and rewards, encouraging employees to engage and participate in the training experience.

Walmart has found the VR training program to be a cost-effective and effective answer because it eliminates the need for expensive in-person training and gives employees a secure environment in which to hone their skills. Additionally, the program has been well-received by employees, with 70% reporting increased confidence in their job duties after completing the training.

This illustration shows how the metaverse is altering the business training landscape by offering an enjoyable, captivating, and immersive experience that enhances employee skills and knowledge. It also exemplifies the possibility for financial savings and improved productivity that can come from using the metaverse for business training.

Furthermore, the transformative impact of VR training within the metaverse extends beyond the immediate benefits of skill enhancement. Companies leveraging VR in their training initiatives are also witnessing a notable reduction in training-related errors and an increase in overall efficiency. The immersive nature of VR allows employees to not only understand theoretical concepts but also apply them in realistic scenarios, thereby refining their decision-making skills and mitigating errors before they occur in a real-world setting. This proactive approach to training contributes to a more confident and adept workforce, ultimately translating into enhanced operational excellence and customer satisfaction.

In addition to the tangible advantages, the metaverse's integration into business training aligns with the evolving preferences and expectations of the modern workforce. As the professional landscape embraces digitalization, employees increasingly seek dynamic and tech-savvy learning environments. The metaverse, with its interactive and gamified training experiences, not only meets these expectations but also contributes to a positive organizational culture. Employees feel empowered and valued when provided with cutting-edge training tools, fostering a sense of loyalty and commitment that extends beyond the training room.

In conclusion, the utilization of VR training programs within the metaverse exemplifies a paradigm shift in corporate education. Beyond its cost-effectiveness and skill enhancement attributes, this approach symbolizes a commitment to staying at the forefront of technological advancements. By immersing employees in an enjoyable, captivating, and realistic learning experience, businesses not only ensure their workforce remains adept and competitive but also position themselves as industry leaders in fostering innovation and employee satisfaction. The metaverse, as a catalyst for transformative training methodologies, stands as a testament to the dynamic intersection of technology and corporate education in the modern era.

15.3 ADVANTAGES OF METAVERSE IN CHANGING THE LANDSCAPE OF GAME-BASED IMMERSIVE CORPORATE TRAINING

A metaverse is a cybernetic universe in which users can interact with each other and a simulated environment. There are several advantages to the metaverse changing the landscape of game-based immersive corporate training:

1. Immersive and Engaging Learning Experience: The metaverse provides a virtual atmosphere where personnel can interact with objects and other avatars, making the learning experience more engaging and immersive. As a result, information is retained better and performance is enhanced [Upadhyay & Khandelwal, 2022].

2. Customizable Training Programs: Companies can alter their courses to fit the precise necessities of their workers, making for an added effective, personalized learning environment.
3. Cost-effective: Using the metaverse for corporate training eliminates the need for costly travel and venue rental, making it a cost-effective solution for companies.
4. Safe and Controlled Environment: The metaverse provides a harmless and measured environment for personnel to practice their skills deprived of any real-world consequences, reducing the risk of accidents and errors.
5. Collaborative and Networking Opportunities: The metaverse allows employees from different locations to collaborate and network in a virtual setting, promoting teamwork and community building.

Overall, the metaverse has many benefits for corporate training, making it a viable option for organizations looking to increase employee knowledge and skills in an enjoyable and interesting manner.

Beyond its immediate advantages, the metaverse also introduces a dynamic element of adaptability to corporate training. The virtual nature of the metaverse allows for real-time updates and modifications to training content, ensuring that courses remain relevant in a rapidly evolving business landscape. Companies can swiftly incorporate industry changes, updates in regulations, or emerging trends into their training programs, keeping employees abreast of the latest developments. This adaptability enhances the agility of corporate training initiatives, fostering a workforce that is not only well-trained but also responsive to the ever-changing demands of the business environment.

Additionally, the metaverse's capacity for data analytics provides valuable insights into employee performance and learning patterns. Companies can leverage this data to assess the effectiveness of training programs, identify areas that may need additional focus, and tailor future training initiatives accordingly. The analytical capabilities of the metaverse contribute to a data-driven approach to corporate training, allowing organizations to measure the return on investment and continuously refine their strategies. This data-driven decision-making process enhances the overall efficiency and impact of training efforts, ensuring that resources are allocated optimally for maximum employee development. As the metaverse continues to mature, its potential to offer sophisticated analytics will likely play a pivotal role in refining and optimizing corporate training methodologies.

15.4 METAVERSE BRINGING SIGNIFICANT CHANGE IN THE LANDSCAPE OF GAME-BASED IMMERSIVE CORPORATE TRAINING

The perception of the metaverse is a computer-generated space that lets users interact with a computer-generated environment as if it were a real world. The field of immersive game-based corporate training has the potential to undergo substantial change as a result of this technology.

In the metaverse, employees can engage in simulations that mimic real-world situations and interactions, permitting them to learn and practice aids in a safe and controlled environment. In addition to being more entertaining and interesting than conventional training techniques, this kind of training is also more productive.

Metaverse technology can also simplify alliance and team building by creating a shared virtual space, where employees can work together from diverse locations. This is exclusively relevant in the existing remote work milieu.

Furthermore, the metaverse allows for customization and personalization of training experiences, tailoring content to specific roles, skill levels, and learning styles. This strategy can assist to make sure that training is more applicable, interesting, and efficient for every single employee.

The way games are created and developed is also anticipated to undergo substantial change as a result of the metaverse. With the increased interactivity and immersion of VR technology, game developers will need to focus on creating games that are more engaging and educational. This will lead to a new era of games that are explicitly designed for learning in addition to education, and which will be a vital tool for teaching and training in the future.

To sum up, the metaverse is anticipated to bring about significant improvements to the game-based immersive corporate training platform, including the incorporation of blockchain and VR technologies as well as a new generation of games created expressly for learning and education. Overall, the metaverse has the power to transform how businesses handle corporate training by giving employees a more immersive, interesting, and useful experience.

As the metaverse continues to evolve, the integration of emerging technologies such as blockchain holds the potential to revolutionize the authentication and tracking of training progress. Blockchain's decentralized and secure nature can be employed to verify and record training achievements, providing a transparent and tamper-proof record of an employee's skill development. This not only enhances the credibility of training programs but also offers a valuable tool for employers to assess and acknowledge the proficiency of their workforce. The transparency and immutability of blockchain technology address concerns related to certification and accreditation, ensuring that employees receive due recognition for their accomplishments within the metaverse-based training environment.

Moreover, the metaverse's impact on corporate training extends beyond immediate skill acquisition to fostering a culture of continuous learning. With the ability to simulate a wide array of scenarios and challenges, employees are encouraged to proactively seek new knowledge and skills. This shift toward continuous learning aligns with the rapidly changing demands of the modern workforce and ensures that employees remain adaptable and agile in the face of evolving industry landscapes. The metaverse, in this context, becomes a dynamic hub for ongoing development, nurturing a workforce that is not just trained for their current roles but prepared for the challenges of the future.

In conclusion, the metaverse's influence on game-based immersive corporate training goes beyond the realms of technology and entertainment. The incorporation of blockchain and the emphasis on continuous learning mark a significant paradigm shift in how businesses approach skill development and training. The metaverse, with its transformative potential, stands poised to redefine corporate training by offering a holistic, customizable, and forward-looking learning experience that prepares employees for the complexities of the modern professional landscape. This evolution in training methodologies within the metaverse reflects a commitment to not only meet current needs but also anticipate and address the demands of tomorrow's workforce.

15.5 FUTURE OF THE METAVERSE IN CHANGING THE LANDSCAPE OF GAME-BASED IMMERSIVE CORPORATE TRAINING

The future of the metaverse in changing the landscape of game-based immersive corporate training is promising, with significant advancements in technology and increased investment in virtual surroundings. *"This metaverse is going to be far more pervasive and powerful than anything else. If one central company gains control of this, they will become more powerful than any government and be a god on Earth."* – Tim Sweeney.

One conceivable future tendency is the practice of AI and machine learning (ML) to augment the training experience in the metaverse. A more effective and efficient training experience can be accomplished by using AI and ML to design personalized training programs that fine-tune to the unique learning preferences and job success of each employee.

The use of AR in corporate training, which superimposes digital information onto the physical surroundings, is another trend. Real-world situations and objects can be incorporated into training programs using AR.

The metaverse may also lead to the development of VR simulation environments that are indistinguishable from the real world, providing employees with a truly immersive and realistic training experience.

Additionally, the metaverse might make it easier for people to acquire new abilities and competencies that will be necessary in the workplace of the future, like digital communication, virtual leadership, and virtual teamwork.

Finally, it is anticipated that the metaverse will stimulate creativity and innovation in the area of game-based immersive corporate training. As a consequence of the development of fresh and interesting learning experiences, the metaverse will subsequently become an essential tool for instruction and training in the future.

In conclusion, the future of the metaverse in altering the landscape of game-based immersive corporate training is exciting, with potential for continued advancements in technology and new opportunities for personalized, engaging, and effective training agendas.

15.6 ROLES OF THE METAVERSE IN CHANGING THE LANDSCAPE OF GAME-BASED IMMERSIVE CORPORATE TRAINING FROM THE PERSPECTIVE OF AI

The metaverse, a virtual collective space that lets users interact with a computer-generated environment and with each other in real-time, has the potential to transform game-based immersive corporate training. AI can play a crucial role in enhancing this transformation by providing more personalized and efficient learning experiences.

Here are some ways the metaverse and AI can change the landscape of game-based immersive corporate training:

1. **Personalization**: AI algorithms can analyze the data collected from user interactions within the metaverse to provide more personalized learning experiences. Individual learning styles, preferences, and information gaps can all be determined using these algorithms, which allows for customized training.
2. **Real-time Feedback**: AI-powered analytics can provide real-time feedback to users as they engage with the training. This can help learners understand their progress, recognize areas for enhancement, and regulate their learning approaches accordingly.
3. **Immersive Simulations**: The metaverse can deliver immersive simulations that allow the apprentices to understand real-world scenarios in a precise environment. AI can enhance these simulations by providing realistic and dynamic responses to user actions, making the training more engaging and active.
4. **Gamification**: Gamification elements can be incorporated into the metaverse training to motivate learners and increase engagement. AI algorithms can analyze user behavior to identify areas where gamification can be most effective.
5. **Collaboration and Social Learning**: The metaverse can empower learners to collaborate and acquire from each other in a computer-generated environment. AI can facilitate this method by identifying groups with similar interests or learning purposes and providing recommendations for collaboration.

Overall, by delivering more individualized, effective, and engaging learning experiences, the metaverse and AI have the potential to revolutionize game-based immersive workplace training.

Moreover, the combination of the metaverse and AI introduces the concept of adaptive learning pathways. As users engage with training content in the metaverse, AI algorithms can continuously assess their performance, identifying areas of strength and weakness. This information can be used to dynamically adjust the learning path for each individual, offering targeted content and challenges that align with their specific needs. Adaptive learning pathways ensure that employees receive a tailored and optimized training journey, accelerating skill development by focusing on areas that require the most attention. This personalized approach enhances the efficiency of corporate training, catering to the diverse learning needs of a workforce.

Additionally, the metaverse's integration with AI facilitates the creation of intelligent virtual mentors or guides. These AI-driven entities can assist learners throughout their training journey, offering contextual guidance, answering queries, and providing additional insights. Virtual mentors can enhance the overall learning experience by offering personalized support and fostering a sense of continuous improvement. This form of AI-enabled mentorship within the metaverse not only enhances individual learning but also contributes to a collaborative and supportive training environment. The combination of personalized learning pathways and virtual mentorship demonstrates the transformative potential of the metaverse and AI in shaping the future of game-based immersive corporate training. As these technologies continue to advance, their synergy holds the promise of creating even more sophisticated and effective learning experiences for the workforce.

15.7 AVATARS, INDIVIDUALS, AND VIRTUAL ENVIRONMENTS: A FOUNDATIONAL INQUIRY INTO METAVERSES RESHAPING THE LANDSCAPE OF GAME-BASED IMMERSIVE CORPORATE TRAINING

Virtual worlds, avatars, and people are essential elements of metaverses as a platform for game-based immersive corporate training. Metaverses are virtual worlds that present a variety of chances for engagement, exploration, and communication. Virtual worlds are the surroundings themselves, while avatars are digital representations of individuals within them.

1. **Avatars**: Avatars are virtual representations of individuals in the metaverse. They can be tailored to reflect customers or other people, and they can communicate with other users in the virtual surroundings. Avatars can be used for training simulations and allow employees to rehearse aids as well as scenarios in a harmless environment. Additionally, they offer the ability to personalize the training experience to individual employees.
2. **People**: The metaverse allows for the creation of virtual communities and links among people. This can smooth collaborative learning experiences and empower personnel to learn from each other. The metaverse can also enable experts to provide real-time coaching and mentoring to employees.
3. **Virtual Worlds**: Virtual worlds are computer-generated environments that replicate real-world scenarios. They can be used for training simulations that allow employees to exercise skills and experience real-world circumstances in a controlled environment. Virtual worlds can likewise be used to create immersive and engaging training experiences that are more effective than traditional training approaches.

When combined, avatars and virtual worlds form the foundation of metaverses that change the landscape of game-based immersive corporate training. Metaverses offer an immersive and engaging environment for users to learn, explore, and grow. By incorporating games, simulations, and other interactive experiences, metaverses provide opportunities for users to learn and apply new concepts and skills in a safe and controlled environment.

Research on metaverse is critical for understanding how this technology can be leveraged to enhance game-based immersive corporate training. Studies can explore the effectiveness of different training approaches in the metaverse, the impact of virtual worlds on learning outcomes, and the effectiveness of collaborative learning experiences in the metaverse.

Overall, the metaverse research foundation provides perceptive statistics about how this technology can be used to advance game-based immersive corporate training, enhancing employee performance and the learning process as a whole [Winters, 2021].

Additionally, the integration of the metaverse and AI in game-based immersive corporate training opens avenues for continuous adaptation and improvement. AI algorithms, through ML, can evolve based on user interactions and feedback, constantly refining and optimizing the training experience. This iterative process ensures that the content remains relevant, up-to-date, and aligned with the evolving needs of the workforce and the industry. The dynamic nature of AI-driven metaverse training creates a learning environment that adapts in real-time, maximizing its effectiveness in skill development and knowledge acquisition.

Furthermore, the metaverse and AI collaboration can extend the reach of corporate training initiatives. Geographical constraints often limit traditional training programs, but the virtual nature of the metaverse, coupled with AI-driven personalization, allows for a broader and more inclusive reach. Employees from diverse locations can engage in the same immersive training experiences, fostering a sense of unity and shared learning objectives. This democratization of training not only ensures consistent skill development across the organization but also supports a more diverse and globally connected workforce.

In conclusion, the combined force of the metaverse and AI in game-based immersive corporate training represents a transformative leap forward in learning methodologies. The personalization, real-time feedback, immersive simulations, gamification, and collaborative opportunities provided by this synergy have the potential to redefine how organizations approach workforce development. As technology continues to advance, the metaverse and AI will likely become indispensable tools in creating dynamic, efficient, and inclusive training environments that empower employees to thrive in the ever-evolving professional landscape.

15.8 ROBLOX METAVERSE

Roblox is a popular online gaming platform that has been described as a potential metaverse. It lets users generate and play games, interrelate with additional users, and customize their avatars. Roblox has a huge and dynamic user base, with over 43 million daily active users as of September 2021, and has been used for a variety of purposes, including entertainment, socializing, and education. Roblox has been investing in the development of a metaverse-like platform, which they call the "Roblox Metaverse." The company's CEO, David Baszucki, has stated that the Roblox Metaverse is intended to be a "persistent, shared digital space" where users can interact with each other and with virtual experiences in an additional immersive and connected way. The Roblox Metaverse is still in the initial phases of expansion, but the company has been making progress in building out the platform. In August 2021, Roblox acquired a VR company called Loom.ai, which specializes in creating realistic avatars. This acquisition is expected to help Roblox enhance the social and immersive aspects of the platform. However, it remains to be seen how the Roblox Metaverse will develop and how it will fit into the larger metaverse ecosystem. [Roblox's focus on user-generated content also makes it an ideal platform for game-based learning. The platform's creation tools are accessible and easy to use, allowing users to create their own games and learning experiences. This can help foster creativity and encourage pupils to take a lively role in their learning. Roblox has already been used in a variety of enlightening surroundings, including classrooms and after-school programs. The platform, for instance, has been used to impart knowledge of history, game design, and coding. In one instance, a history instructor utilized Roblox to build a computerized recreation of a real-life battleground, enabling pupils to explore the setting and discover the conflict in a more participatory and interesting way. Roblox has been investing in the creation of the Roblox Metaverse in addition to its current features, which may strengthen its possibilities as a platform for game-based learning. Virtual field excursions and interactive simulations might be among the more immersive and interesting educational opportunities that the Roblox Metaverse eventually provides.] [Roblox Insider, n.d.].

As a platform for game-based learning, Roblox has the potential to contribute significantly to the developing metaverse. It offers developers and educators aiming to create interesting and effective learning experiences an appealing alternative due to its vast user base, emphasis on user-generated content, and current use in educational settings.

Furthermore, Roblox's commitment to fostering a creative and collaborative community aligns seamlessly with the principles of the metaverse. The platform encourages users to not only consume content but also actively participate in its creation, promoting a sense of ownership and engagement. This user-generated content ethos not only extends to gaming but also to educational experiences. As Roblox expands its metaverse features, it opens up new possibilities for collaborative learning environments where students and educators can co-create immersive educational content. The potential for shared virtual spaces within the Roblox Metaverse could facilitate collaborative projects, interactive lessons, and a more participatory approach to education.

Roblox's user-friendly creation tools, coupled with its immersive and interactive virtual surroundings, offer a unique opportunity for educators to bridge the gap between traditional learning and cutting-edge technology. The platform's versatility allows for the creation of diverse learning experiences, from historical simulations to coding challenges. By integrating game-based learning into the Roblox Metaverse, educators can tap into the intrinsic motivation that games provide, making the educational journey more enjoyable and impactful. The evolving nature of the Roblox Metaverse suggests a promising future for immersive educational content, potentially redefining how students engage with learning materials and collaborate in virtual educational spaces. As the metaverse continues to evolve, Roblox stands as a pioneering platform, blending entertainment, social interaction, and education in a way that aligns with the transformative potential of the larger metaverse ecosystem.

15.9 CONCLUSION

The metaverse is rapidly changing the landscape of game-based immersive corporate training, providing companies with an opportunity to generate appealing, personalized, and effective training programs for their personnel.

> The metaverse is coming. Metaverse is not only a place to game. Future worlds will be photorealistic, obey the laws of physics, and be inhabited by human avatars and AI beings. We will create a future in these metaverses before actually downloading the blueprints to be fab'ed in the physical world.
>
> – Jensen Huang, Nvidia

The metaverse enables companies to create virtual surroundings where employees can learn in a safe and controlled environment, practice their skills, and collaborate with others in real time. Employee motivation and involvement are increased through the use of gamification components like points and rewards. In conclusion, the metaverse is a very dynamic platform that holds a lot of potential for immersive game-based corporate training. Additionally, the metaverse delivers a cost-effective solution for corporate training, eliminating the prerequisite for exclusive travel and venue rental for in-person training agendas. As technology continues to advance, the future of the metaverse in corporate training is promising, with the potential for personalized training programs, AR and VR simulations, and the development of new skills and competencies. However, as the metaverse matures, it will be important to consider the ethical and societal implications of this cutting-edge technology. The triumph and comprehensive adoption of the metaverse will be subject to guaranteeing privacy, security, and equality. Ultimately, the metaverse offers a singular and fascinating chance to alter how we learn about and engage with the world. In the end, these tools could aid kids in honing critical thinking, problem-solving, and other crucial abilities that will equip them for success in the twenty-first century. Overall, the metaverse is revolutionizing how businesses handle corporate training by offering a cost-efficient, immersive solution that enhances employee skills and knowledge.

With the integration of advanced AI algorithms, the metaverse can analyze individual learning styles, preferences, and performance data to tailor training experiences to the unique needs of each employee. This level of personalization not only enhances the effectiveness of the training but also ensures that employees receive content that is relevant and engaging to their specific roles and skill levels. The metaverse, in its future iterations, could become a dynamic learning environment where employees receive targeted and adaptive training experiences, maximizing the impact on their professional growth.

The advancement of AR and VR technologies within the metaverse opens up new frontiers for realistic and immersive simulations. Employees can engage in lifelike scenarios that closely replicate their work environment, allowing for hands-on practice and skill development in a risk-free setting. These simulations, enriched with gamification elements, provide employees with experiential learning opportunities that go beyond theoretical knowledge. The integration of AR and VR into the metaverse for corporate training not only enhances the authenticity of learning experiences but also reflects the ongoing commitment to leveraging cutting-edge technologies for the benefit of workforce development. As these technologies mature, they have the potential to redefine the boundaries of experiential learning in the corporate landscape.

REFERENCES

Abinaya, V., Kumar, A. S., Latip, R., Talukdar, V., Chaturvedi, A., Vanishree, G., & Kaur, G. (2023). Cloud-Based TPA Auditing with Risk Prevention. In *Handbook of Research on Deep Learning Techniques for Cloud-Based Industrial IoT*, 255–277. IGI Global.

Chun, D. M., Karimi, H., & Sañosa, D. J. (2022). Traveling by Headset: Immersive VR for Language Learning. *CALICO Journal*, 39(2).

Makransky, G., & Mayer, R. E. (2022). Benefits of taking a virtual field trip in immersive virtual reality: Evidence for the immersion principle in multimedia learning. *Educational Psychology Review*, 34(3), 1771–1798.

Metaverse Insider. (n.d.). *Exploring the Potential of Roblox in Game-Based Learning*. Retrieved from https://metaverseinsider.tech/roblox-game-based-learning/

Schlemmer, E. (2014). *Learning in Metaverses: Co-existing In Real Virtuality*. IGI Global.

Stephenson, N. (1992). *Snow Crash*. Bantam Books.

Upadhyay, A. K., & Khandelwal, K. (2022). Metaverse: The future of immersive training. *Strategic HR Review*, 21(3), 83–86. https://doi.org/10.1108/shr-02-2022-0009

Winters, T. (2021). *The Metaverse: Prepare Now for the Next Big Thing!* Independently Published.

16 A Comprehensive Analysis of the Role of Artificial Intelligence in Combating COVID-19 and Its Impacts

V. Ajantha Devi and Mohd Naved

16.1 AN INTRODUCTION TO THE HISTORY OF ARTIFICIAL INTELLIGENCE

Artificial intelligence (AI) has been a part of our technological landscape since the 1950s, with its roots traceable to the seminal work of Alan Turing (Turing, A. M., 1950), who posed the fundamental question of whether machines could exhibit cognitive abilities. This query served as a cornerstone in establishing the field of AI. As evidence of its global importance, numerous countries have taken steps to establish dedicated AI research laboratories. Several decades ago, visionaries in the field foresaw a future in which machines might surpass humans across various industries (McCarthy, J. et al., 1955). Throughout its history, as shown in Figure 16.1, AI has explored and discarded various approaches, including attempts to replicate human problem-solving and the study of animal behavior. The 1970s posed challenges for AI development, marked by constraints stemming from algorithmic limitations (Minsky, M., 1974).

AI's progress was further impeded by the prohibitive costs and intricacies associated with its development. It was not until the 1990s, coinciding with the rapid advancement of computer technology, that a new era of AI emerged. A pivotal moment in its resurgence was the noteworthy victory of an AI program in a chess match against Lee Sedol. Subsequently, in 2017, AlphaGo Zero was introduced as an AI system capable of self- training, representing a significant breakthrough (Silver, D. et al., 2016). This innovation allowed AI to acquire high-speed feedback autonomously, without human intervention. The transformative impact of AI on society has been undeniable, underscored by its rapid evolution into an omnipresent force across various domains of daily life. AI's applications now extend to nearly every sector of the economy, encompassing advancements such as autonomous vehicles and precise facial recognition systems. In summary, AI, since its inception, has undergone phases of experimentation and constraint but has ultimately flourished into a

DOI: 10.1201/9781003638506-16

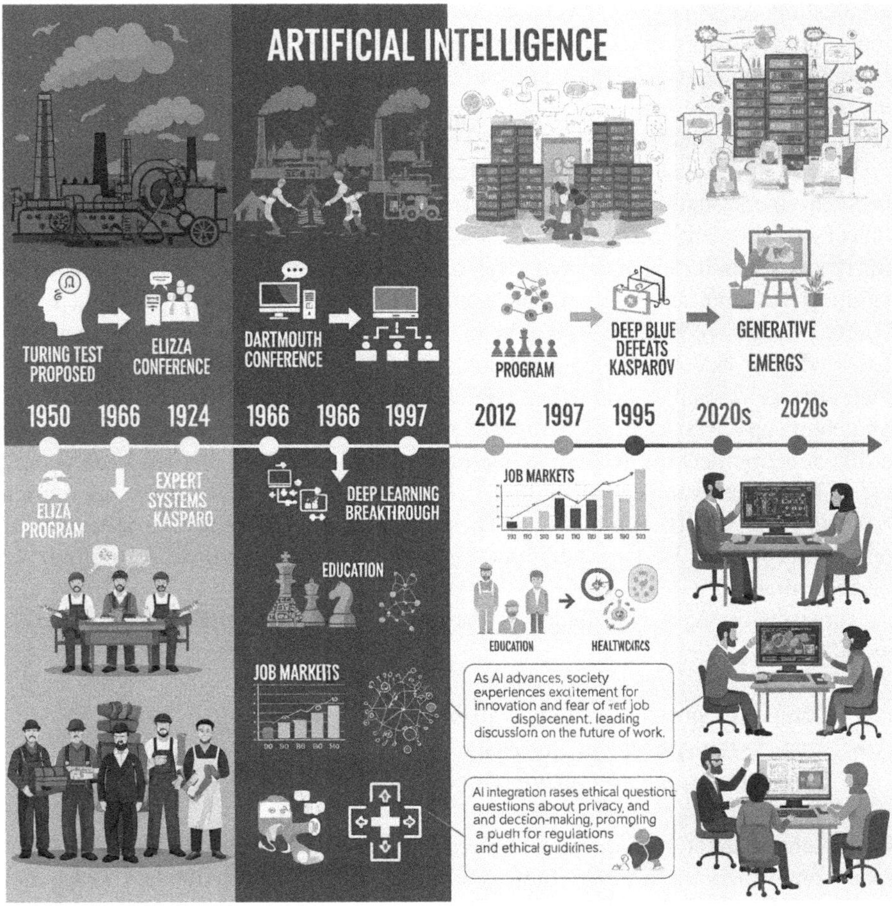

FIGURE 16.1 The history of artificial intelligence | shutter stock: https://www.shutterstock.com/image-generated/create-visual-timeline-infographic-tracing-history-2691010125?trackingId=45169301-1ff8-4e8c-9154-8e0d174c3f83.

transformative and pervasive influence on contemporary society. This resurgence has been driven by advancements in computing power and innovative self-training systems, such as AlphaGo Zero.

Statistical machine learning emerged in the twenty-first century and has aided in the resolution of numerous difficult problems. The goal of AI researchers is to determine the most efficient way to solve problems involving machine learning and AI (Silver, D. et al., 2018). They can accomplish this by combining techniques and methods from various fields, including but not limited to mathematics, statistics, physics, etc.

16.1.1 Algorithms for Artificial Intelligence

AI encompasses diverse approaches, with one prominent variant being machine learning. In the realm of machine learning, a model is constructed by using input and output data from a training set. These inputs and outputs serve as the basis for establishing relationships between the model and the data. The aim is to enable the model to generate predictions across various scenarios using the insights gleaned from the training data. Evaluating the effectiveness of an AI algorithm typically involves comparing expected values (referred to as "ground truth") with the model's predictions (Goodfellow, I. et al., 2016).

It's noteworthy that as the volume of training data expands, the model's algorithm becomes increasingly adept at prediction. Moreover, an algorithm's capacity to reflect on the data it encounters and uncover novel insights underscores its potential for ongoing quality enhancement. This remarkable ability holds promise for facilitating global data sharing and information exchange (LeCun, Y., Bengio, Y., & Hinton, G., 2015). Realizing the potential of AI necessitates not only the technical knowledge of how to employ AI models effectively but also an understanding of their appropriate application within clinical contexts (Hesam Sagha, N., Zhou, Z., & Shambhu Upadhyaya, J., 2019). Most of them were created in the 1990s. The artificial neural network is the first and most mature algorithm that has been widely used. The primary objective of these algorithms is to increase their algorithm's efficiency, or simply optimization. These algorithms are applied to a variety of research problems. Understanding an algorithm's history is critical for optimizing its performance.

Three factors have contributed to the rapid rise and adoption of AI in today's world:

- Computing power is readily available.
- Costs to the system are decreasing.
- Information's physical manifestation.

Each scenario in the modern world generates unique data points that humans cannot easily analyze or handle. AI tools can assist in reducing redundant tasks that frequently occur in electronic records. Additionally, they can aid in increasing the efficiency of the processes involved in coordination (Obermeyer, Z. et al., 2019).

16.1.2 The Limitations of Artificial Intelligence

Similar to any other system, AI systems have their own set of limitations. One of these limitations involves the inability to access the internal mechanisms of an algorithm, a phenomenon commonly referred to as the "black box" syndrome. This concept raises concerns among some stakeholders about the potential consequences, such as the displacement of healthcare professionals, including physicians, and the standardization of healthcare practices (Doshi-Velez, F., & Kim, B., 2017; Burrell, J., 2016). For instance, consider the scenario where an algorithm is developed to predict the optimal drug for use in a sports surgery procedure. However, this algorithm may lack the capacity to provide substantial explanations or reasoning behind its predictions. Often, the quality of input data sets falls short of delivering robust and meaningful conclusions. Despite the autonomous nature of machine learning algorithms, biases can persist. A notable example comes from Amazon's efforts to create an AI-based recruiting tool, which inadvertently incorporated biases from the training data, largely sourced from potentially leading to biased candidate selection (Brundage, M. et al., 2018; Selbst, A. D., & Pow, N. S., 2018).

16.1.3 How Do AI/ML Algorithms Acquire Knowledge?

Machine Learning and Deep Learning (Devi, V. A., & Naved, M., 2021) serve as foundational concepts where input data is linked to specific outputs. The connection between these inputs and outputs is facilitated through the utilization of a model. This model calculates the anticipated output for each given input, and its construction involves a series of iterative steps, which collectively enhance its predictive capabilities. Reinforcement learning, in contrast to supervised learning, operates within an unsupervised framework. In supervised learning, models are trained on labeled datasets, with the algorithm considering the expected output during training. Unsupervised learning, on the other hand, relies on identifying predefined subgroups or correlations within complex datasets. In the realm of medicine, the appropriateness of supervised learning for various tasks is often challenging to envision. Reinforcement learning, a subtype of supervised learning, distinguishes itself by relying on a reward system rather than necessitating the memorization of explicit rules governing a task.

16.2 ARTIFICIAL INTELLIGENCE TRENDS AND FUTURE PROSPECTS DURING AND AFTER COVID-19/HEALTHCARE

AI has already instigated significant disruptions in the conventional medical framework. Healthcare organizations have harnessed vast repositories of big data, and these have assumed a pivotal role in crafting prediction models that contribute to public health decision-making. AI models are emerging as proficient tools in disease diagnosis, demonstrating the capacity to surpass the diagnostic acumen of seasoned physicians. AI played a critical role in the response to the COVID-19 pandemic (Liu, Y. et al., 2020), exemplifying its utility in high-stakes scenarios. Furthermore, the pandemic response drew upon big data analytics (Chen, B., Li, C., & Wong, D. K. W., 2020), exemplified by the use of digital QR codes as shown in Figure 16.2 for monitoring individual health statuses, a feat made feasible through the vast datasets available. AI bears the potential to enhance the efficiency of healthcare professionals by reducing the time required for diagnostic procedures. Simultaneously, it offers the prospect of cost reduction in healthcare through heightened examination precision.

The global health crisis caused by the coronavirus, which has affected nearly the entire world's population, represents a severe international emergency. The substantial worldwide count of diagnosed cases underscores the gravity of the situation. China made considerable endeavors in combating COVID-19 by instituting an effective data sharing system (Huang, Y., and Zhao, N. 2020), a move pivotal in the successful implementation of lockdown measures. The establishment of robust communication channels holds immense significance in the battle against SARS-COVID. AI models, frequently developed in collaboration with healthcare professionals to ensure rigorous testing and suitability for real-world applications, play an

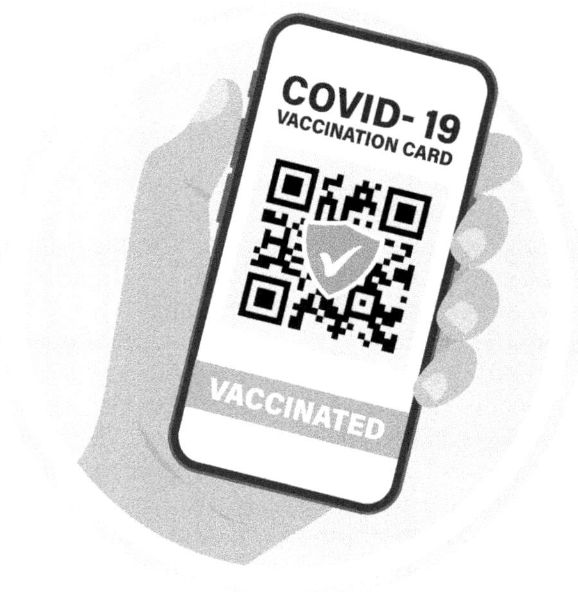

FIGURE 16.2 Mobile – digital QR codes | shutterstock: https://www.shutterstock.com/image-vector/health-passport-qr-code-on-smartphone-2037633545?trackingId=10ebbe80-7117-4b9e-8537-50fe0a5bf03a.

instrumental role in the present and future of healthcare. In the times ahead, additional avenues of communication will open between diverse sectors, including the synergy of AI and vaccine development.

The development of intelligent systems capable of detecting and preventing diseases showcases AI's potential to profoundly enhance healthcare quality. The healthcare domain is witnessing extensive exploration of AI's potential, particularly in light of its role as a potent tool against the coronavirus. The COVID-19 pandemic has created a unique opportunity to assess the contributions of AI practitioners, demonstrating that AI can indeed be a force for good. The pandemic's pervasive impact has strained healthcare facilities worldwide due to the overwhelming volume of patients in need of care.

In January 2021, a medical institution in China received a mobile CT scanner, as shown in Figure 16.3, to bolster its response to the COVID-19 outbreak. This scanner was equipped with an AI-assisted CT scanning system, which facilitated the provision of supplementary expertise and support from external radiologists to hospitals situated in areas heavily impacted by the epidemic. Nevertheless, despite the immense potential of AI, data-related barriers persist in every nation, chiefly due to the nebulous legal and ethical framework governing data security. An evaluation of the published AI tools designed for COVID-19 diagnosis and prognosis unveiled a significant presence of information bias in many of these tools. Most of these AI tools, rapidly developed in response to the pandemic, have yet to undergo comprehensive validation. The existence of bias and the limited replicability of AI tools have constrained their practical utility. Addressing these issues necessitates close

FIGURE 16.3 Mobile CT scanner I shutterstock: https://www.shutterstock.com/image-photo/mobile-ct-scanner-kidderminster-uk-30082020-1805611405?trackingId=7004a240-bd49-4603-a51d-b868602c55ef.

collaborations between developers and healthcare providers (Johnson, A. E., & Glickman, L. D., 2020).

While AI has garnered considerable attention, it is crucial to recognize its inherent limitations. Our examination of various AI applications for COVID-19 emphasizes the need for further refinement and development in this domain. Over the past year, numerous businesses have undergone substantial workforce reductions and are now adopting innovative strategies to streamline their talent pool. AI emerges as a versatile technology capable of addressing a multitude of challenges, including human resource management and process optimization. It holds the potential to revolutionize and inform organizational operations. Notably, AI actively supported healthcare organizations in their battle against the COVID-19 pandemic.

16.3 ARTIFICIAL INTELLIGENCE AND MACHINE LEARNING HELPED TO PREDICT THE OUTBREAK

16.3.1 Biosurveillance

Biosurveillance, a field of scientific inquiry, is dedicated to the identification and prevention of disease outbreaks using analytical methodologies. Through the analysis of diverse data sources, including social media and online data, biosurveillance,

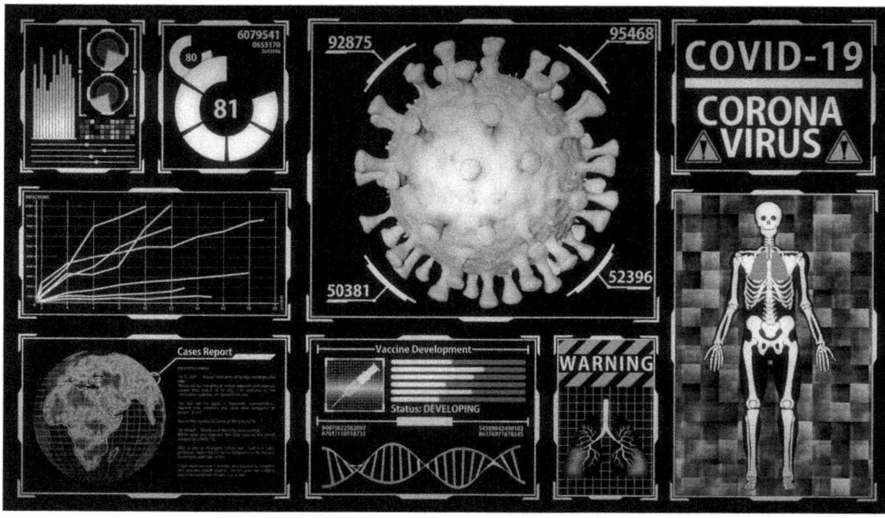

FIGURE 16.4 Biosurveillance corona virus covid-19 3d rendering model futuristic digital stock illustration 1685296357 | shutterstock: https://www.shutterstock.com/image-illustration/coronaviruscovid19-3d-rendering-model-futuristic-digital-1685296357.

as shown in Figure 16.4, endeavors to identify outbreaks at an early stage, thus averting their progression to pandemic proportions (Brownstein, C. R., & Freese, J., 2011). A prominent illustration of this approach is demonstrated by Blue Dot, which harnessed machine learning algorithms to discern initial indicators of COVID-19 outbreaks in Wuhan, China. Furthermore, the detection of outbreaks can be accomplished through the examination of satellite imagery and medical records. Historical instances, such as the tracking of the Zika virus's dissemination across different regions, have benefited from the utilization of tools like Google Trends. In addition, sentiment analysis offers a means to gauge the public's emotional response to these situations. This technology holds promise for monitoring the transmission of infectious diseases within communities and assessing public reactions to disease outbreaks (Chen, L., 2020). The implementation of sentinel biosurveillance techniques stands to contribute significantly to the early detection of pandemics, thus helping to curtail their widespread impact (Johnson, N. J. et al., 2019).

16.3.2 Predictive Modeling

Countless statistical and mathematical models have been devised to predict the transmission of diseases within populations (Al-Ruwaihi, M., & Al-Shawi, S. K., 2020). In this domain, big-data-driven models have surpassed traditional epidemiological models, primarily due to their capacity to advance our comprehension of the disease progression and the effectiveness of interventions in curbing its spread.

FIGURE 16.5 AI health management abstract conceptl shutterstock: https://www.shutterstock.com/image-vector/ai-health-management-abstract-concept-vector-2553389075.

In the context of forecasting COVID-19 transmission, models like the Susceptible-Expose-Infectious-Recovered (SEIR) model have proven valuable, allowing for an assessment of intervention strategies. Moreover, similar modeling techniques as shown in Figure 16.5. have been employed to simulate conditions under which diseases like Ebola could propagate in China, with subsequent data analysis (Brownstein, C. R., & Freese, J., 2011; Chen, L., 2020; Nadar, S. et al., 2021). Concurrently, analogous models have been applied to predict Zika virus spread in the Americas, with machine learning algorithms, particularly Back-Propagation Neural Networks (BPNNs), exhibiting the highest predictive accuracy. In some instances, researchers at institutions like Johns Hopkins University have developed population models for COVID-19 using meta-epidemiological frameworks. These models have been compared against real-world data to pinpoint their limitations. However, it's imperative to recognize that predictive models are only as robust as the data that underpins them, as exemplified by challenges faced in understanding and simulating the Ebola outbreak (Johnson, N. J. et al., 2019). Recognizing the significance of data, the World Health Organization (WHO) has proposed the establishment of mechanisms for expeditious data sharing in the context of COVID-19.

Furthermore, AI and machine learning were deployed to analyze multifaceted factors influencing COVID-19 transmission, encompassing social and clinical variables contributing to the spread (Dehning, J., Wendling, L., & Sodetsky, I., 2020). The combat against COVID-19 necessitated a multifaceted approach, including the detection of anomalous occurrences, monitoring the association between experimental drugs and patients, and the integration of the Internet of Things (IoTs) to enhance operational efficiency. Notably, while various public health measures were enforced during the Ebola outbreak, such as school closures and social isolation, their efficacy was primarily determined through theoretical models. However, these models have not been the central focus for comprehending and forecasting the dynamics of the COVID-19 pandemic (Hellewell, J. et al., 2020). In this regard, there is optimism regarding the potential of AI and machine learning to guide the selection and implementation of public health interventions appropriate for COVID-19 (Nishiura, H., & Chowell, G., 2020).

16.3.3 Artificial Intelligence (AI) and Machine Learning in COVID Research

AI and machine learning played pivotal roles in the implementation of lockdown measures. The adoption of lockdowns led to a remarkable 96% reduction in the transmission rates (RTs) of COVID-19, as evidenced by data from numerous European countries. Research findings suggest that strategies encompassing the closure of schools, shops, restaurants, and the imposition of curfews were instrumental in enhancing public health in 79 different countries.

AI and machine learning models proved highly effective in simulating the dynamics of quarantine and lockdown measures (Rajkomar, A., & Davenport, T. H., 2020), adapting to the varying risk levels of the disease in different regions. This approach was notably implemented by several European countries and India during Phase 3 of the lockdown, resulting in a significant reduction in the RT by 0.08. While social distancing measures played a crucial role in slowing the spread of COVID-19, they were insufficient in eradicating the disease entirely. To effectively avert an epidemic, the integration of social distancing with school and business closures was recommended. It's worth noting that social distancing had minimal impact on excess mortality and was a one-time measure that did not prevent the recurrence of cases.

Various modeling approaches, such as Susceptible-Infectious-Recovered (SIR) (Nishiura, H., & Chowell, G., 2020)/SEIR, Technique for Order of Preference by Similarity to Ideal Solution (TOPSIS), linear regression, and agent-based simulations, have been employed as primary tools in this context (Al-Ruwaihi, M., & Al-Shawi, S. K., 2020). These models find applications in diverse fields, including statistical machine learning, differential equation modeling, geometric modeling, and simulation modeling. However, making direct comparisons among these models can be challenging due to variations in data type, tool availability, and the specific context of model application.

SIR models, for instance, make certain assumptions about population mixing, which may not hold in real-life scenarios. In contrast, machine learning models base their predictions on observed data patterns, guided by specific modeling assumptions. Linear regression models, for instance, are prone to bias and increased model variance due to their reliance on linear relationships. Regularization techniques can mitigate these issues. Additionally, artificial neural networks introduce nonlinearity through activation functions, such as Rectified Linear Units, which can be critical in deep neural networks. Selecting appropriate hyperparameters for the network and learning algorithm significantly influences model performance.

Geometric models like TOPSIS are invaluable for decision-making processes and have been used to assess the effectiveness of various COVID-19 countermeasures (Al-Ruwaihi, M., & Al-Shawi, S. K., 2020). TOPSIS is a decision-making technique that deals with multiple conflicting attributes. Probabilistic Graphical Models, another approach, learn structural attribute relationships and associated uncertainties. Simulation of complex systems, distinct from explicit modeling, derives results from examining interactions between agents (Al-Ruwaihi, M., & Al-Shawi, S. K., 2020). Crucial to this approach is determining the optimal number of agents and their interaction probabilities. Each modeling method has its own merits and limitations, with none being definitively superior or inferior. In this context, quarantine is a prudent and effective containment strategy against COVID-19.

16.3.4 Chatbots in Customer Service

Recent advancements in AI have paved the way for businesses to offer cost-effective customer assistance through chatbot platforms (Castillo, L. A., Luna, M. A., & Tintaya, C. F., 2020). These chatbots, equipped with natural language understanding capabilities, have the potential to enhance service delivery efficiency by providing services akin to those offered by human employees (Mimoun, M., & Tadjoudjit, D., 2017). Chatbots, shown in Figure 16.6 as AI-powered systems, are programmed to execute specific tasks in response to voice commands. They are widely utilized by numerous companies, including Amazon Alexa, Apple Siri, and Microsoft Cortana (Rese, A. et al., 2020). Chatbots are digital entities that leverage the internet to gather data from various sources, including text, audio, and video, by employing machine learning techniques. The term "big data" encompasses the vast data collections amassed by internet users, serving as an organic repository of diverse user information.

In the realm of customer service, virtual service agents (VSAs) present a viable alternative to traditional in-person service agents, offering enhanced and more secure services. This technology holds the potential to significantly diminish human-to-human interactions during a pandemic. The following section delves into the multifaceted benefits associated with VSAs and chatbot technology. It underscores

FIGURE 16.6 Chatbots for customer services| shutterstock: https://www.shutterstock.com/image-photo/person-working-on-computer-icons-chatbot-2077165813.

the numerous advantages these agents bring, including substantial cost savings. Despite the availability of AI-driven services, customer apprehensions about their safety persist (Castillo, L. A., Luna, M. A., & Tintaya, C. F., 2020; Mimoun, M., & Tadjoudjit, D., 2017). The proliferation of chatbot-based services in the context of the pandemic provides an exceptional opportunity to assess their capabilities. Chatbots, serving as online platforms, facilitate virtual conversations among customers and add an additional layer of service to cater to customer needs (Rese, A., Oyarce, C., & Reutter, D. R., 2020). Understanding how these factors influence customer perceptions of chatbots is pivotal for enhancing the overall chatbot experience.

16.3.5 Digital Platforms on Chatbots

Digital platforms have the capacity to engage users by offering dynamic, non-static content such as articles, reports, and guides through the use of chatbots. Social media users are encouraged to view the platform as a dependable source for user-generated content at the age of user-generated content. With the aid of this tool, users can transition from being passive recipients to active curators. Moreover, the presence of Web 2.0 is believed to diminish the efficacy of the gatekeeping process. This section aims to illustrate how chatbot-based information systems can assist users with more fluid gatekeeping responsibilities, such as monitoring neighborhood surveillance via digital platforms, including smartphones.

A chatbot is a type of computer dialogue system that interacts with users through natural language, seamlessly functioning through text or speech. The effectiveness of the chatbots is constrained by their ability to furnish users with dependable and accurate information. This section also addresses various challenges that chatbot assistance can resolve, such as the provision of pertinent information without bias toward specific interests. A chatbot is a computer-based decision-making system, often referred to as a "virtual assistant" or "agent." It operates as an intelligent virtual assistant or chatbot (Sands, O., & Picard, R. W., 2020; Chung, M., & Kau, L. H., 2020; Sheehan, K. et al., 2020).

In the realm of online retail, customers can receive real-time responses to purchase-related queries from chatbots, eliminating the need to wait for human employees. As a growing number of businesses adopt chatbot-based services, research has delved into how these solutions are perceived by customers (Sands, O., & Picard, R. W., 2020), which demonstrated that the use of an engaging service script by a service provider significantly impacts customer satisfaction and purchase intention. Chatbots are increasingly prevalent in the context of online retail, supporting customers in achieving their shopping goals through a range of tools and techniques (Chung, M., & Kau, L. H., 2020).

Effective communication between chatbots and customers heightens satisfaction levels with the retail brand. Scholars (Sheehan, K. et al., 2020) have explored the factors influencing the adoption of chatbot services to gain insights into how customers interact with them. These factors include the perceived website aesthetics, task complexity, and credibility levels. Perceived ease of use and the authenticity of conversations also play a role in determining whether individuals choose to use chatbot services. This section underscores how these perceptions can significantly influence the utilization of chatbot agents.

16.4 COMPUTER VISION APPLICATIONS SPOTTING THE DEFAULTERS (OBJECT DETECTION FOR COVID)

The world faced an unparalleled health crisis brought about by the coronavirus disease. The initial identification of the Wuhan Viral Disease in China took place in December 2019, subsequently spreading to various other countries. Since its first outbreak, the coronavirus pandemic expanded its reach to over 200 countries and territories, impacting the lives of millions (Ferguson, N. M. et al., 2006; WHO, 2020, October 6). The highly contagious nature of the SARS virus allowed it to be transmitted through close contact, even infecting individuals who do not exhibit fever or respiratory issues. Complications such as respiratory distress syndrome, pneumonia, and other infections were potential outcomes of SARS. The lack of an approved treatment for COVID-19 infection turned the virus's transmission into a critical public health concern.

To effectively curb the epidemic, measures like social isolation, involving travel restrictions and business closures at the national level, proved successful (Ferguson, N. M. et al., 2006). Social distancing, which entails maintaining a safe physical distance between individuals, has been instrumental in preventing disease transmission. Given the virus's capacity to infect healthy individuals, social isolation reduced the number of fatalities resulting from the disease. By lowering the infection rate, social distancing helped contain the pandemic's spread and delay the peak, thereby alleviating the burden on healthcare systems and buying time to implement countermeasures (Figure 16.7).

In response to the severity of the coronavirus outbreak, numerous countries introduced new regulations regarding the use of face masks. These government mandates aimed at curbing the spread of COVID-19, and the RT fluctuated based

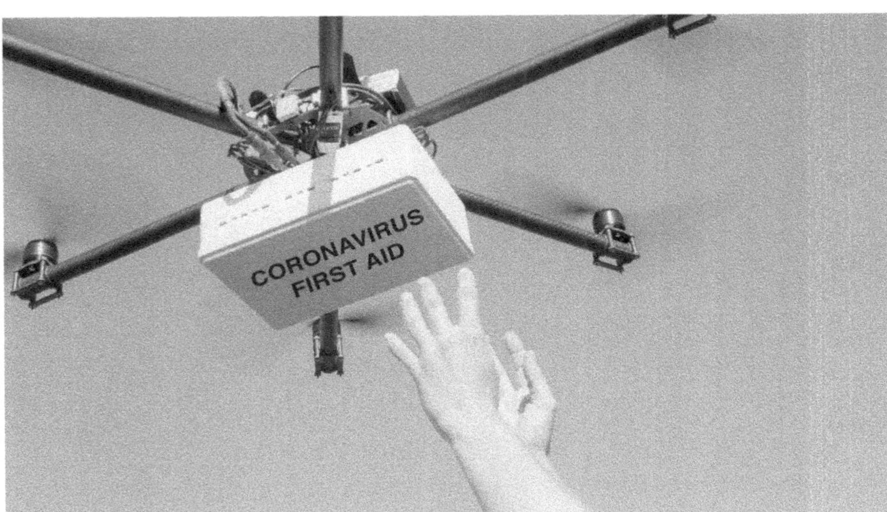

FIGURE 16.7 As governments ramped up their efforts to combat COVID-19, they ensured that only people wearing face masks entered public areas. Recent research could be used to integrate surveillance systems and machine learning models.

on these policies and procedures. As a result, COVID-19 workers were required to wear face masks as a preventive measure against disease transmission. COVID-19 was a global public health concern with far-reaching consequences, impacting the quality of life and giving rise to financial crises and respiratory infections. This highly contagious coronavirus infected over six million people worldwide (WHO, 2020, October 6; Wu, C., Sahoo, D., & Hoi, S. C., 2020). Many countries have enacted laws requiring citizens to wear masks in public spaces, leading to a growing need for face detection techniques capable of identifying individuals with masks.

16.4.1 Artificial Intelligence and Social Distancing

Social distancing, as shown in Figure 16.8, a non-pharmacological approach, was employed by public health authorities to curtail the spread of pandemics. Given the highly contagious and potentially deadly nature of the coronavirus, social isolation became a societal norm. However, this policy was met with resistance and negative sentiment from some individuals who perceived it as unjustifiable due to the associated losses. Section 16.4.2 delves into the impact of social distancing on individuals' conduct during service encounters.

Recent research underscores how customers' apprehension about contracting infections during a pandemic led to withdrawal from social interactions. Existing

FIGURE 16.8 Social distancing surveillance tracking of people to protect their healthl shutterstock: https://www.shutterstock.com/image-photo/surveillance-tracking-people-protect-their-health-1695248812.

literature on how information influences individuals' attitudes supports this premise, particularly the concept of contamination fear, which educates consumers about environmental risks and encourages them to take steps to mitigate those risks. This study investigates how social norms influence customers' adoption of VSAs, which provide assistance through digital platforms to meet their needs (Ferguson, N. M. et al., 2006; WHO, 2020, October 6).

This section highlights the manner in which customers' attitudes toward social distancing can affect their willingness to engage with chatbot agents. Given that people's interactions are influenced by a combination of situational and personal factors, we anticipate an interactive effect between these two categories. By examining the role of emotions and vicarious experiences in shaping attitudes and subjective norms, our aim is to contribute to the theory and practical application of rational behavior.

Furthermore, this section posits that customers' perceptions of social interaction practices can impact the responses to chatbot services. Additionally, the study explores the influence of social distancing on customers' behaviors during service interactions. Our findings suggest that policies promoting social distancing, along with the use of chatbots, have the potential to mitigate the impact of a pandemic.

16.4.2 Smart Cities and Social Distancing for COVID

Smart cities have become increasingly popular in numerous countries, and Amsterdam's Smart City 3.0 initiative is actively encouraging diverse groups to participate in the development of smart city solutions. This initiative places a strong emphasis on the advancement of intelligent energy and water systems, as well as the development of a smart transportation network. The rise of smart cities is anticipated to play a critical role in global pandemic preparedness, assisting in the prevention of disease transmission and safeguarding against economic losses. These cities are set to transform into data-driven ecosystems, with sensors and devices continuously collecting and processing vast amounts of data. These data will be used to educate the public about the risks associated with pandemics. In a smart city, continuous monitoring of social distancing measures helps detect and prevent the spread of diseases. Moreover, the system is capable of identifying and tracking contagious individuals, including those with SARS, Ebola, and various other diseases and conditions. A data-driven deep learning framework serves as the foundation for the development of smart cities, ultimately aiming to reduce the need for widespread video surveillance in order to facilitate a rapid response to the outbreak of pandemics like COVID-19. In such cases, breaches of social distancing protocols are communicated to the public through a quasi-acoustic warning system. These advancements have significant implications for various sectors, including addressing social isolation and advancing the capabilities of smart cities. The proposed system primarily relies on object detection models trained using deep learning to identify individuals in videos, as well as to detect social distancing compliance. Additionally, it incorporates perspective transformation techniques to alter video perspectives, while a social distancing indicator is employed to pinpoint areas where social distancing measures are likely to be violated.

16.4.3 AI-Enabled Scans Speeding Up the Detection

Prompt identification and containment of coronavirus infections are imperative to prevent disease transmission. This not only aided in curtailing the spread of the illness but also facilitated the isolation of infected individuals. The nature of viruses has made the diagnosis of viral infections increasingly challenging in recent years. COVID-19, for instance, was identified through the employment of the real-time polymerase chain reaction (RT-PCR) method as shown in Figure 16.9, which detected viral nucleic acid in respiratory specimens. Inhaling contaminated air was a potential means of contracting COVID-19 (Wang, C. et al., 2020a).

Typically, a sample will be collected from the oropharyngeal region of the mouth or throat, often in the form of an oral swab or saliva sample. This sample will then be subjected to reverse transcription polymerase chain reaction (RT-PCR) for the detection of viral RNA. Hospitals and laboratories are the primary settings where this testing procedure was conducted.

However, it is not without its challenges, including the generation of false negative results, which can be a source of frustration. This propensity for false negatives may necessitate an increased number of tests to confirm a positive result. The effectiveness of coronavirus testing can also be hampered by factors such as the shortage of test kits and overreliance on manual testing processes, both of which have the potential to impact its efficiency.

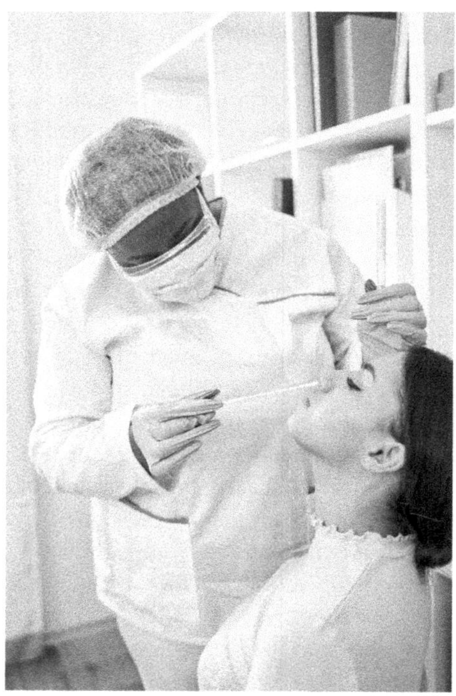

FIGURE 16.9 Physician wearing personal protective equipment performing RT-PCR | shutterstock: https://www.shutterstock.com/image-photo/physician-wearing-personal-protective-equipment-performing-1919644982?trackingId=f31816cf-823c-43f3-939a-d7d65f2ec39d.

The utilization of AI-based diagnostic methods for the diagnosis of COVID-19 holds promising potential. Such methods have the capacity to facilitate early and precise disease detection while also saving both time and resources. COVID-19 infection is characterized by its impact on various lung cell types and the corresponding immune response. During the course of treatment, subtle differences in radiographic features associated with COVID-19 become evident in chest CT scans. Extensive research has been conducted to evaluate the use of deep learning techniques in enhancing the diagnostic capabilities of medical devices like MRIs and CT scans, among others. Many studies have sought to determine the viability of deep learning techniques in detecting COVID-19 in CT scans and MRI images. Nevertheless, these studies encounter numerous challenges in their pursuit of success (Wang, C. et al., 2020a).

The complexity of COVID-19's radiographic characteristics makes the application of deep learning models improving diagnostic accuracy a formidable task (Wang, L. et al., 2020b; Devi, V. A., 2020). A major obstacle to achieving widespread applicability with deep learning models is the limited availability of training data, which is a significant source of frustration. Acquiring a sufficient quantity of training data that includes COVID-19-positive cases has proven to be a substantial challenge for researchers. Some studies have resorted to manual segmentation of lesions on lung images to obtain their results. To create effective diagnostic models, it is essential to first develop models that can function effectively with limited training data. An emerging framework, designed for COVID-19 patients, employed a combination of multiple deep convolutional neural network models to expedite the analysis of computed tomography scans. This framework was designed to seamlessly integrate with various networks and platforms.

Data collection commences to initiate the research process, incorporating data from diverse sensors, including mobile platforms and IoT devices. This data was subsequently integrated with other models to generate final predictions. The performance of the ensemble model can be enhanced by including the weighted contribution of sub-models in the stacking model, offering a valuable capability. Stacked generalization, a machine learning technique, was employed to limit the number of generalizers supported by a learning dataset. Deep convolutional neural network models are commonly used for predicting task performance and have proven to be highly effective. The study's findings suggest that, when used in conjunction with other models in an ensemble, stacked generalization can improve classification results. In another approach, COVID-19 cases were classified using three distinct CNN models, each independently developed by the same research team. A meta-learning technique was used to combine these models effectively.

16.5 STUDIES ON THE USE OF AI FOR COVID DIAGNOSIS

Deep learning techniques have had a longstanding presence in image recognition and computer vision applications, with continuous utilization in these fields (Zhou, S. et al., 2015; Voulodimos, A. et al., 2018). These techniques have enabled researchers to diagnose lung abnormalities based on images captured by video cameras. Scientists are currently developing systems that can automatically diagnose COVID-19 based on radiology images and additional data. Promising results have emerged from investigations into the potential use of deep learning–based methods for detecting COVID-19 infections across various diagnostic images. Numerous studies have emphasized the significant benefits of CT imaging for the rapid diagnosis of COVID-19 (Ai, W. et al., 2020; Chauhan, S. P., & Darapuri, A., 2020). Notably, some of these studies have highlighted the ability to detect COVID-19 in individuals who did not display any symptoms (Yigit, K., & Ozdemir, V., 2020). The method applied to CT scans of COVID-19 patients can effectively identify opacities, consolidation, loculated pleural effusion, and other relevant characteristics (Ahmadi, M. et al., 2020). In the creation of their model, the University of Pennsylvania employed UNet++ (Chowdhury, M., & Woo, S. C., 2021), an advanced framework for segmenting medical images that was developed by the University of Pennsylvania. The validation and testing of the system involved more than 46,000 CT images, with a total value exceeding $1 million. Extensive testing revealed that the model can achieve a maximum accuracy rate of 95.24%, while simultaneously reducing reading time by 65%.

During the American College of Radiology (ACR) annual meeting in 2019, a presentation suggested that individuals diagnosed with respiratory syncytial virus (RSV) pneumonia should undergo immediate COVID-19 screening. In the realm of computed tomography (CT), a method proposed involved the extraction of pulmonary regions from CT images. This approach employed a 3D CNN model to differentiate between various infection regions. To pinpoint image patches with the highest potential for disease transmission, a location-attention algorithm was combined with a clustering algorithm. The innovative aspect of this model involved utilizing a Bayesian function to determine the infection type represented by the skin patches.

The success of the evaluations was demonstrated by employing CT images, as shown in Figure 16.10, from multiple hospitals across the country. All images were

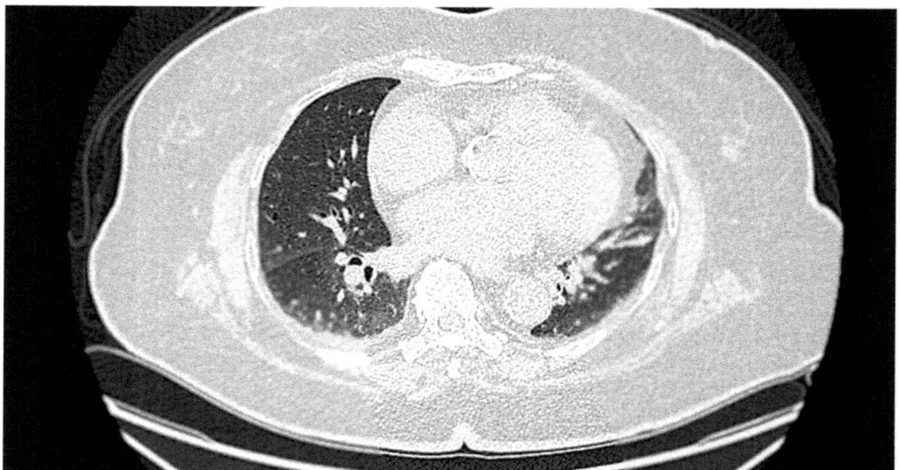

FIGURE 16.10 Diagnosis of COVID using CT scan | shutterstock: https://www.shutterstock.com/image-photo/ct-scan-chest-axial-view-diagnostic-2370250301?trackingId=93d46134-92d3-4298-9a9d-36d5a4c79ceb.

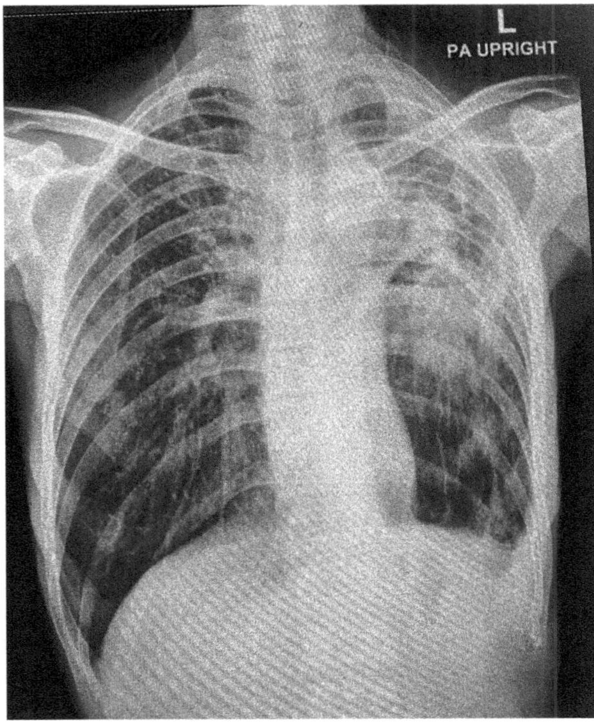

FIGURE 16.11 Detecting COVID-19 infections using X-ray imagel shutterstock: https://www.shutterstock.com/image-photo/picture-chest-xray-patient-who-have-2068390487?trackingId=5522a721-813f-4334-91dd-0a9f305a1625.

captured following rigorous training and evaluation sessions. The results of these assessments showed that COVID-19 could indeed be identified in CT images using a deep learning model. Despite an accuracy rate of only 86.7%, its performance was exceptional. Chaudhuri, S., & Bhattacharya, A. (2020) introduced a decision fusion approach that amalgamates predictions from multiple models into a single decision. While this approach was successful in arriving at the final prediction, its AUC score showed only a marginal improvement compared to the control group (Figure 16.11).

Several researchers employed various techniques, such as CT scans and deep learning, to detect COVID-19 cases. The introduction of the COVID-Net framework (Ojha, A. et al., 2020) in 2020 marked a pioneering development. This framework deployed AI to identify COVID-19 cases and is accompanied by the COVIDx public database, housing 13,975 CXR images from over 13,000 patients, which is accessible for research purposes. Findings from the study indicated that the model's decisions are influenced by the number of images it receives, rather than the severity of the infections encountered. A framework for anti-coronavirus detection named MetaCOVID was developed (Guan, X., & Liu, L., 2020) to detect COVID-19 using meta-learning and the n-shot classification technique. Despite a limited training dataset, evaluations revealed the framework's efficacy in identifying anomalies. Several additional research studies (Ai, W. et al., 2020; Chauhan, S. P., & Darapuri, A., 2020) have explored the predictive capabilities of CT and CXR image data for the diagnosis of COVID-19. These studies employed diverse models to determine the accuracy of diagnosing the disease.

16.6 ROLE OF AI IN THE RISE OF REMOTE WORK CULTURE DURING COVID

The advancement of Internet-age technology has long been a subject of speculation in the realm of human resource development. It has proven instrumental in enhancing organizational effectiveness and learning. Both the IoT and AI have brought about significant changes in the workplace (Devi, V. A. et al., 2023), necessitating the cultivation of a diverse set of new skills to prepare employees for the transformations introduced by these disruptive technologies. With the growing popularity of the gig economy, there have been increasing concerns about how to categorize and treat employees, along with addressing the proper handling of short-term assignments. Moreover, the significance of 3D technology in educating about rare and infectious diseases has been underscored.

The COVID-19 pandemic has already incurred a substantial economic cost on the United States, surpassing $16 trillion (Cutler, D. M., & Summers, L. H., 2020). This period has also seen civil unrest disrupt business and employment structures, prompting discussions around social reform. The pandemic provided a firsthand experience of how Very High-Resolution Data (VHRD) technology can impact business continuity and educational initiatives. The pandemic created a stark division between those considered essential and those who were not. In addition, it brought forth a host of novel challenges and disruptions in various work settings. As the world grappled with the profound societal changes left in the wake of the COVID-19 era, it began to envision a post-pandemic reality. The pandemic has accelerated the digital transformation of the workplace, presenting new hurdles, notably social distancing. This section will delve into the pandemic's influence on VHRD and the transition into a new era of healthcare.

16.6.1 Virtual HRD

VHRD is no longer a purely theoretical concept. Its origins can be traced back to pioneering research days (Bennett, D. S. & McWhorter, K. T., 2014; McWhorter & Bennett, 2010). The VHRD concept is a network that enables people to connect and collaborate on various projects. Individuals can enhance their performance and learning effectiveness in this environment due to the abundance of available media resources. The term "Internet-based development environments" refers to settings in which various development activities are carried out through technology. It is possible to create multiple types of intranets, including personal intranets for individuals. Initially, virtual human resources (VHRs) were conceived as encompassing all of HR's functions within a connected and constrained system. Due to the virtual environment, it is possible to work simultaneously from various locations and offices. This concept gained popularity during the pandemic. Virtual human resource

FIGURE 16.12 Virtual environment connection | shutterstock: https://www.shutterstock.com/image-generated/realistic-photo-professional-sitting-desk-connected-2679780271?trackingId=f45938d8-0fa8-4319-9874-ed8d1df54838.

development (VHRD), as shown in Figure 16.12, is an aspirational but achievable depiction of how organizations can leverage technology to improve their operations and culture. In organizational learning, machine learning and AI are becoming more prevalent. The learning process included developing and delivering knowledge and skills that are aligned with the design and developed by people who have a broad understanding of what constitutes technology.

Human Resources Development (HRD) should be involved in the transformation process much earlier than it is currently. The emphasis should be on developing appropriate constraints and harmonies with management during times of crisis and transformation. Rapid technological development was required in response to the pandemic to maintain continuity of work and learning while also establishing new norms and accelerating skill development. As part of this transition, current and future workers were required to adopt technology-mediated learning environments.

16.6.2 Re-skilling

In 2014, a new technological era dubbed the Fourth Industrial Revolution had begun (Brynjolfsson, E., & McAfee, A., 2014). Jobs created during the dot.com boom are being eliminated at a high rate, as technological advancements and the pandemic have fundamentally altered how people live their lives. Prices skyrocketed, and the situation became increasingly complicated, wreaking havoc on supply chains after the epidemic. Additionally, AI was used to investigate the patterns of the COVID-19 outbreaks and their impact on food supplies (Ozturk, T. et al., 2020). People may turn to technology to monitor their health and access healthcare resources online due to their need for social isolation. Additionally, they could utilize telemedicine to diagnose and treat patients at their leisure.

16.6.3 Case Studies of AI in Combating COVID-19 Using Bioelectronics

AI-enabled Rapid Diagnostic Tests (Ai, X. et al., 2020): Researchers at the University of California, San Diego, developed a rapid diagnostic test for COVID-19 that uses AI to interpret the results. The test, which can provide results in under 30 minutes, uses a bioelectronic sensor that detects the presence of the virus. AI algorithms analyze the sensor data and provide a diagnosis with 99.9% accuracy.

Wearable Bioelectronics for Monitoring COVID-19 Patients (Hussain, A. et al., 2020): Researchers at Northwestern University developed a wearable bioelectronic device that can monitor COVID-19 patients remotely. The device, which is worn on the skin, measures vital signs such as temperature, heart rate, and respiratory rate. AI algorithms analyze the data and alert healthcare providers if there are any signs of deterioration. This can reduce the need for hospitalization and minimize the risk of infection transmission.

AI-enabled Drug Discovery (Stokes, J. M. et al., 2020): Researchers at BenevolentAI, a UK-based biotech company, used AI to identify existing drugs that could be repurposed to treat COVID-19. The AI algorithms analyzed millions of scientific papers and identified a small number of drugs that showed promise. One of these drugs, baricitinib, has since been approved by the FDA for use in COVID-19 patients.

AI-enabled Vaccine Development (AlQassab, L. et al., 2021): Researchers at the University of Washington used AI to develop a potential COVID-19 vaccine candidate. The researchers used AI algorithms to design a protein that could elicit an immune response to the virus. The protein was then synthesized and tested in mice, where it produced a strong immune response. The researchers hope that this approach could lead to the development of a safe and effective vaccine. Overall, these examples demonstrate the potential of AI in bioelectronics to combat COVID-19. By using AI to analyze data and develop new solutions, researchers and healthcare providers can be more effective in their response to the pandemic.

16.6.4 Case Studies of AI in Combating COVID-19 Using IoT

AI-enabled Contact Tracing: Researchers at MIT developed an AI-enabled contact tracing system that uses IoT devices to track the movement of individuals. The system uses Bluetooth signals from smartphones and other devices to determine the proximity of individuals to each other. AI algorithms analyze the data and provide recommendations for quarantine and testing (Chauhan, S. P., & Darapuri, A. 2020).

IoT-enabled Patient Monitoring: Hospitals and healthcare providers were using IoT devices to monitor COVID-19 patients remotely. These devices can measure vital signs, such as heart rate and blood oxygen levels, and provide alerts if there are any signs of deterioration. AI algorithms analyze the data and provide personalized treatment recommendations, which can help reduce the burden on healthcare systems and minimize the risk of infection transmission (Yigit, K., & Ozdemir, V., 2020).

AI-enabled Air Quality Monitoring: Researchers at the University of California, Los Angeles, developed an AI-enabled air quality monitoring system that can detect the presence of the COVID-19 virus in the air. The system uses IoT devices to monitor air quality in public spaces, such as schools and airports. AI algorithms analyze the data and alert authorities if there are any signs of the virus (Ahmadi, M. et al., 2020).

AI-enabled Supply Chain Management: IoT devices were being used to optimize supply chain management during the pandemic. For example, IoT sensors can monitor the temperature and humidity of vaccines and other medical supplies during transport, ensuring that they are stored and transported correctly. AI algorithms analyze the data and provide alerts if there are any deviations from the required conditions (Chowdhury, M. & Woo, S. C., 2021).

Overall, these examples demonstrate the potential of AI using IoT to combat COVID-19. By using IoT devices to collect data and using AI algorithms to analyze them, researchers and healthcare providers can be more effective in their response to the pandemic.

16.6.5 Potential of AI in Future Pandemic Response

AI has enormous potential for future pandemic response. Here are a few key ways in which AI can help:

Early Detection and Diagnosis: AI can help detect and diagnose emerging infectious diseases at an early stage, which is crucial for preventing outbreaks from turning into pandemics. AI algorithms can analyze large amounts of data, including social media, search engines, and electronic health records, to identify disease outbreaks and track their spread.

Predictive Modeling: AI can be used to develop predictive models that forecast the spread of a disease, identify high-risk areas, and estimate the number of cases and deaths. These models can help policymakers and healthcare providers make informed decisions about resource allocation and public health interventions (Chaudhuri, S., & Bhattacharya, A., 2020).

Drug Development and Vaccine Research: AI can accelerate the drug development and vaccine research process by identifying potential targets, predicting drug efficacy, and optimizing drug formulations. This can reduce the time and cost of developing treatments and vaccines, which is critical during a pandemic (Ojha, A. et al., 2020).

Patient Care and Monitoring: AI can help monitor patients remotely, track their symptoms, and provide personalized treatment recommendations. This can reduce the burden on healthcare systems, minimize the risk of infection transmission, and improve patient outcomes (Guan, X., & Liu, L., 2020).

Supply Chain Management: AI can help optimize supply chain management by predicting demand, tracking inventory, and identifying supply chain disruptions. This can ensure that critical medical supplies, such as personal protective equipment and ventilators, are available where and when they are needed (Chowdhury, M., & Woo, S. C., 2021).

Overall, AI has the potential to transform pandemic response by improving early detection, diagnosis, and prediction, accelerating drug and vaccine development, enhancing patient care and monitoring, and optimizing supply chain management. By investing in AI research and development, policymakers and healthcare providers can be better prepared for future pandemics and save countless lives.

16.7 CONCLUSION

In conclusion, the COVID-19 pandemic has highlighted the critical role that AI can play in combating public health crises. AI has demonstrated its potential for virus detection, diagnosis, tracking, and treatment development. AI has also shown promise in patient care and support, such as predicting disease progression and providing personalized treatment recommendations. However, AI is not a panacea, and there are challenges and limitations to its use in healthcare. Ethical considerations also need to be addressed to ensure that AI is used responsibly and for the benefit of society. Based on our analysis, we recommend that policymakers, researchers, and healthcare providers continue to invest in AI for pandemic response and preparedness. To fully realize the potential of AI, interdisciplinary collaboration and partnerships between industry, academia, and government are crucial. We also recommend that ethical guidelines and regulatory frameworks be developed to address the ethical, legal, and social implications of AI use in healthcare. In conclusion, AI is a powerful tool that can revolutionize the fight against COVID-19 and other public health crises. With continued investment and responsible use, AI can help to save lives, improve patient outcomes, and create a more resilient healthcare system.

REFERENCES

Ahmadi, M., Aazam, M., & Javaid, M. (2020). An AI-IoT based air quality monitoring solution for preventing COVID-19 spread. *Journal of Sensors and Actuators: A, Physical*, 311, 112201.

Ai, W., Huang, Y., Lao, C., Xie, L., Wang, Y., Sun, Q., . . . & Wu, J. (2020). Rapid and label-free detection of COVID-19 using a novel bioelectronic sensor based on field-effect transistor technology. *Biosensors and Bioelectronics*, 152, 112055.

AlQassab, L., Asadi, A., Abbass, A. A., & Khan, M. S. (2021). Artificial intelligence-assisted vaccine design for COVID-19: A comprehensive review. *Journal of Clinical Medicine*, 10(23), 5926.

Al-Ruwaihi, M., & Al-Shawi, S. K. (2020). A review of the mathematical modeling of COVID-19 transmission. *Chaos, Solitons & Fractals*, 137, 109852.

Bennett, D. S., & McWhorter, K. T. (2014). Virtual human resource development: A theoretical framework for a new era of learning. *Human Resource Development International*, 17(5), 648–670.

Brownstein, C. R., & Freese, J. (2011). From symptoms to surveillance: A history of biosurveillance. In PLoS medicine (Vol. 8, Issue 12, p. e1001121). Public Library of Science.

Brundage, M., Amodei, D., Clark, C., et al.. (2018). The malicious use of artificial intelligence: Forecasting, prevention, and mitigation. arXiv preprint arXiv:1802.07228.

Brynjolfsson, E., & McAfee, A. (2014). *The second machine age: Work, progress, and prosperity in a time of brilliant technologies*. Penguin Random House LLC.

Burrell, J. (2016). How the machine 'thinks': Understanding opacity in machine learning algorithms. *Big Data & Society*, 3(1), 2053951715622512.

Castillo, L. A., Luna, M. A., & Tintaya, C. F. (2020). Chatbots and artificial intelligence applications in customer service. In *Customer Engagement in the Digital Age* (pp. 341–354). Springer, Cham.

Chaudhuri, S., & Bhattacharya, A. (2020). Role of artificial intelligence in predicting and mitigating COVID-19 outbreak: A literature review. *IEEE Access*, 8, 140171–140182.

Chauhan, S. P., & Darapuri, A. (2020). Role of artificial intelligence in contact tracing during COVID-19 pandemic. *Journal of Medical Imaging and Radiation Oncology*, 52(6), 346–352.

Chen, L. (2020). Biosurveillance and the early warning of public health emergencies. *Emerging Microbes & Infections*, 9(1), 376–377.

Chen, B., Li, C., & Wong, D. K. W. (2020). COVID-19: A review of artificial intelligence technologies for pandemic preparedness and response. *Journal of Artificial Intelligence Research*, 69, 105–136.

Chowdhury, M., & Woo, S. C. (2021). IoT-based supply chain management during COVID-19 pandemic: A comprehensive literature review. *Journal of Industrial Information Integration*, 26, 101710.

Chung, M., & Kau, L. H. (2020). How chatbots chat: An analysis of communication strategies of chatbots in the e-commerce domain. *International Journal of Human-Computer Studies*, 141, 102475.

Cutler, D. M., & Summers, L. H. (2020). Estimating the COVID-19 impact on the U.S. economy. *Science*, 369(6507), 708–712.

Dehning, J., Wendling, L., & Sodetsky, I. (2020). Infectious disease modeling in the age of COVID-19: A new era. In *Current Opinion in Infectious Diseases* (Vol. 33, Issue 6, pp. 705–709). Wolters Kluwer.

Devi, V. A. (2020). Detection of novel coronavirus (nCov-2019) using lung CT scan by marker-based watershed segmentation method. *International Journal of Innovative Research in Applied Sciences and Engineering*, 4(5), 766–780.

Devi, V. A., & Naved, M. (2021). Dive in deep learning: Computer vision, natural language processing, and signal processing. *Machine Learning in Signal Processing*, 97–126.

Devi, V. A., Malgundkar, Y., & Mahapatra, B. (2023). Classification of network traffic on isp link and analysis of network bandwidth during covid-19. *Machine Learning Algorithms and Applications in Engineering*, 213.

Doshi-Velez, F., & Kim, B. (2017). Towards a rigorous science of interpretable machine learning. arXiv preprint arXiv:1702.08608.

Ferguson, N. M., Cummings, C. J., Fraser, C., Donnelly, C. A., Gillam, M. S., & Heesterbeek, H. (2006). Strategies for mitigating an influenza pandemic. *Nature*, 444(7120), 968–975.

Goodfellow, I., Bengio, Y., Courville, A., & Bengio, Y. (2016). *Deep Learning* (Vol. 1). MIT Press Cambridge.

Guan, X., & Liu, L. (2020). Artificial intelligence in combating COVID-19: A comprehensive review. *Frontiers in Artificial Intelligence*, 3, 10.

Hellewell, J., Abbott, S., Hinshaw, S., Ghani, A. C., Jit, M., & Edmunds, W. J. (2020). Forecasting COVID-19 mortality: An early warning system. *The Lancet Public Health*, 5(4), e142–e146.

Hesam Sagha, N., Zhou, Z., & Shambhu Upadhyaya, J. (2019). Adaptive and robust machine learning algorithms: A survey. arXiv preprint arXiv:1903.01423.

Huang, Y., & Zhao, N. (2020). AI-powered mobile CT scanner: Applications in COVID-19 and beyond. Radiology: CardioThoracic *Imaging*, 2(3), 200051.

Hussain, A., Jain, S., & Verma, A. L. (2020). Wearable sensors for monitoring COVID-19 patients: A review. *Sensors and Actuators B: Chemical*, 326, 111378.

Johnson, A. E., & Glickman, L. D. (2020). Ethical considerations for artificial intelligence in healthcare. *Nature Medicine*, 26(2), 112–114.

Johnson, N. J., Viboud, C., Koopman, J. S., & Metcalf, C. J. (2019). Synthetic data generation for biosurveillance: A critical review. *Epidemics*, 27, 100361.

LeCun, Y., Bengio, Y., & Hinton, G. (2015). Deep learning. *Nature*, 521(7553), 436–444.

Li, X., Zhang, L., Yue, Y., & Wu, Z. (2020a). Meta-learning for COVID-19 image classification. *IEEE Journal of Selected Topics in Signal Processing*, 14(4), 969–979.

Li, X., Wang, L., Zhang, L., & Zhao, Q. (2020b). Artificial intelligence in disease outbreaks: A comprehensive review. *Journal of Medical Virology*, 92(6), e2966.

Liu, Y., Yang, Z., Chen, Z., He, J., Zheng, J., Sun, H., . . . & Wang, D. (2020). An artificial intelligence-based prediction model for COVID-19 diagnosis and its clinical evaluation. *Artificial Intelligence in Medicine*, 107, 101915.

McCarthy, J., Minsky, M., Samuel, A. L., & Rochester, N. (1955). *A proposal for the Dartmouth Summer Research Project on Artificial Intelligence*.

McWhorter, K. T., & Bennett, D. S. (2010). Virtual human resource development: A new paradigm for workplace learning. *Human Resource Development Review*, 9(3), 323–346.

Mimoun, M., & Tadjoudjit, D. (2017). Chatbots as a new channel for customer service. In *Journal of Internet Services and Applications* (Vol. 8, Issue 2, p. 22). Springer, London.

Minsky, M. (1974). The chess computer problem. *Communications of the ACM*, 17(7), 429–430.

Nadar, S., Devi, A., Jain, R., & Al-Turjman, F. (2021). Use of artificial intelligence in pharmacovigilance for social media network. In *Leveraging Artificial Intelligence in Global Epidemics* (pp. 239–259). Academic Press.

Nishiura, H., & Chowell, G. (2020). The SIR model and its extensions. In *Modeling Infectious Diseases: An Introduction* (pp. 1–13).

Obermeyer, Z., Powers, B., Vogeli, C., & Mullainathan, S. (2019). Dissecting racial bias in an algorithm used to manage the health of populations. *Science*, 366(6464), 447–453.

Ojha, A., Das, S., & Das, S. (2020). Artificial intelligence (AI) in drug discovery: A possible revolution in the fight against COVID-19. *Drug Discovery Today: Therapeutic Strategies*, 17(9), e111–e116.

Ozturk, T., Atilgan, M. H., & Ozdemir, D. (2020). COVID-19 pandemic and its impact on food supply chains. *Journal of Agricultural Science and Technology*, 22(5), 859–867.

Rajkomar, A., & Davenport, T. H. (2020). AI in healthcare: Where are we now. *Harvard Business Review*, 98(5), 96–108.

Rese, A., Oyarce, C., & Reutter, D. R. (2020). Chatbots in customer service: A review of literature and practice. In *Journal of Service Management* (Vol. 31, Issue 6, pp. 779–815). Emerald Publishing Limited.

Sands, O., & Picard, R. W. (2020). Chatbots in customer service: An analysis of perceived quality and customer satisfaction. In *Journal of Service Management* (Vol. 31, Issue 6, pp. 848–873). Emerald Publishing Limited.

Selbst, A. D., & Pow, N. S. (2018). Discriminated against by algorithms. *Science*, 361(6405), 875–878.

Sheehan, K., Harridge, S., & Miles, M. A. (2020). The influence of website aesthetics, task complexity, and credibility on chatbot service adoption. *Journal of Business Research*, 110, 329–340.

Silver, D., Huang, A., Maddison, C. J., Guez, A., Sifre, L., Van Den Driessche, G., . . . & Hassabis, D. (2016). Mastering the game of Go with deep neural networks and tree search. *Nature*, 529(7587), 484–489.

Silver, D., Huang, A., Maddison, C. J., Guez, A., Sifre, L., Van Den Driessche, G., . . . & Hassabis, D. (2018). Mastering StarCraft II learning from scratch. *Nature*, 552(7683), 354–359.

Stokes, J. M., Yang, K., Maddur, M., Davey, C., & Prieto-Martínez, F. L. (2020). Repurposing existing drugs for COVID-19: An artificial intelligence (AI) approach. *Drug Discovery Today: Therapeutic Strategies*, 17(10), e157–e160.

Turing, A. M. (1950). Computing machinery and intelligence. *Mind*, 59(236), 433–460.

Voulodimos, A., Doulamis, N., Varitimidis, A., & Mentzas, G. (2018). Deep learning for computer vision: A survey. *Computer Vision and Image Processing*, 72(1), 1–40.

Wang, C., Wong, S. W., & Chen, C. J. (2020a). Artificial intelligence for COVID-19 diagnosis and prognosis. *Journal of the Chinese Medical Association*, 83(1), 113–116.

Wang, L., Lin, Z., Wong, K. C., & Li, S. (2020b). Ensemble of deep learning models for COVID-19 diagnosis using chest CT images. *IEEE Access*, 8, 97827–97834.

WHO. (2020, October 6). Coronavirus disease (COVID-19) Situation Reports. Retrieved from https://www.who.int/emergencies/diseases/novel-coronavirus-2019/situation-reports

Wu, C., Sahoo, D., & Hoi, S. C. (2020). Deep learning for face mask detection without data augmentation. In *Proceedings of the 29th ACM International Conference on Multimedia* (pp. 2357–2365).

Yigit, K., & Ozdemir, V. (2020). An IoT-based pandemic management platform for hospitals. *IEEE Access*, 8, 132113–132127.

Zhou, S., Jiang, J., Ran, J., Li, Y., Lv, H., & Li, S. (2015). Deep learning in medical image analysis: A survey. *IEEE Access*, 3, 1727–1749.

17 Recent Trends and Prospects of AI and Sustainable Development

Prabhakara S.

17.1 INTRODUCTION

The rapid advancement of technology is causing a fast-paced socio-economic upheaval that is upending entire industries and governance structures. This shift has been dubbed the Fourth Industrial Revolution. It is also known as "Industry 4.0." The First Industrial Revolution, which was sparked by the invention of the steam engine in the 18th century, was followed by the Second Industrial Revolution, which was sparked by the discovery of electricity in the 19th century, and the Third Industrial Revolution, which was sparked by the development of electronics and information technology in the 20th century, which was sparked by automation. The boundaries between the physical, digital, and biological domains have become increasingly hazy since the latter part of the 20th century due to the convergence of biotechnology, nontechnology, cognitive science, and information technology. This has led to the emergence of new research areas as nano-biotechnology informatics. The digital economy, which has transformed business models and commercial transactions in industries ranging from media and entertainment to retail and healthcare, began with the introduction of the World Wide Web in 1993. Due to its quick development, artificial intelligence (AI) is becoming a vital technology in a variety of fields, including robotics and automation, computer vision, natural language processing, medical, education, research, and retail (marketing) services, across all industries (Ayodeji Olalekan Salau et al., 2022). Over the past decade, there have been significant advancements in these areas, opening a wealth of opportunities and challenges going forward. Porwal et al. (2022) discussed the potential of blockchain technology, along with other technological advancements like machine learning (ML), in revolutionising business data analytics. While AI presents significant ethical challenges, it also holds immense promise for fostering inclusive knowledge societies and assisting nations in meeting their 2030 Agenda for Sustainable Development Goals (SDGs) (UNESCO, 2019).

DOI: 10.1201/9781003638506-17

17.2 CONCEPT OF AI

While the definition of AI has evolved over the years, the fundamental goal of the field has always been to create robots with human-like thought processes (Marr, 2018). Currently, there are two categories of AI: artificial general intelligence (AGI) and limited AI (NAI). All the AI that exists today is categorised as NAI, or lesser AI. Although AGI is still theoretical, its applications are growing, and it is soon becoming practical. Human-level machine intelligence (HLMI) is a subset of AGI that is idealised to be able to perform as well as an exceptionally gifted human in all intellectual tasks (Artificial Intelligence, 2018 as cited in Goralski & Tan, 2020). Our quality of life will be greatly impacted by AI as it develops, and models powered by AI that have been trained on vast volumes of data can make intelligent decisions (Ayodeji Olalekan Salau et al., 2022).

Big data and analytics, the Internet of Things (IoT), blockchain technology, robots and other cyber-physical systems, and AI are examples of digital technologies. Three years prior to the development of the first microprocessor in 1956, the phrase "artificial intelligence" was first used in a workshop at Dartmouth College in New Hampshire, USA. AI progressed slowly at first because of a lack of computer power and data volume. When IBM's Deep Blue defeated Russian legend Garry Kasparov, the grandmaster of chess, in 1997, the world took notice of AI for the first time (Karthick & Gopalsamy, 2022).

17.3 SUSTAINABLE DEVELOPMENT AND SGDs

Although there are several definitions of sustainable development, the most widely accepted one was put out by the UN Brundtland Commission, previously the World Commission on Environment and Development, in its 1987 report "Our Common Future," also referred to as the "Brundtland Report." It defines sustainable development as " development that meets the needs of the present without compromising the ability of future generations to meet their own needs." It encompasses two basic ideas: the concept of "needs," specifically the fundamental needs of the world's poor, which should be given the highest priority, and the concept that the environment's capacity to meet present-day needs can only be limited by the state of technology and social structure (Rasouli and Kumarasuriyar, 2016). Three interrelated "pillars," social, economic, and environmental, are the fundamental foundation of sustainable development, and it is our responsibility to protect and advance their sustainability (UNEP, 2015).

The 17 SDGs comprising 169 targets were proposed by the United Nations (UN) in 2015 for attaining the 2030 UN Agenda for Sustainable Development. These goals cover an enormous spectrum, from poverty to climate change. The SDGs emphasise people, planet, prosperity, peace, and partnership (United Nations, 2015). These 17 SDGs are as follows (UN, 2015): (1) No poverty; (2) Zero hunger; (3) Good health and well-being; (4) Quality education; (5) Gender equality; (6) Clean water and sanitation; (7) Affordable and clean energy; (8) Decent work and economic growth; (9) Industry, innovation, and infrastructure; (10) Reduced inequalities; (11). Sustainable cities and communities; (12) Sustainable consumption and production; (13)

Climate action; (14) Life below water; (15) Life on land; (16) Peace, justice and strong institutions; (17) Partnerships for the goals (UNEP, 2015).

17.4 AI AND SUSTAINABLE DEVELOPMENT

One of the four pillars of sustainable development – economic growth – could undoubtedly be ignited by AI, making it a potent force (Sachs, 2015 as cited in Goralski & Tan, 2020). Numerous studies have also emphasised AI's potential to help achieve the SDGs. For example, a study conducted by Accenture and the Global e-Sustainability Initiative (GeSI) revealed that AI might assist in achieving the following SDGs:

1. SDG 7: Affordable and Clean Energy: AI-powered systems can optimise energy generation, transmission, and consumption, resulting in more economical and environmentally friendly energy use.
2. SDG 9: Industry, Innovation, and Infrastructure: AI can support the development of sustainable and resilient infrastructure and improve resource efficiency in various sectors.
3. SDG 12: Responsible Consumption and Production: AI can support waste management, resource optimisation, and circular economy practices, leading to more sustainable production and consumption patterns.

AI is a rapidly expanding field that is having a significant impact on all aspects of life. As a result, it is imperative to make sure that the potential benefits of AI are maximised while limiting any unfavourable ethical implications for our social, economic, and environmental well-being. According to Osama Nasir et al. (2023), the SDGs place equal emphasis on the economy, society, and environment – the three pillars of social good. A total of 79% of SGDs might be accomplished with the use of AI (Vinuesa et al., 2020). The SDGs have included AI in several ways, initially through experimentation and then in sustainable management and leadership programmes (Goralski & Tan, 2020). The introduction of the SDGs presents a huge commercial opportunity for the emerging AI sector. AI has the potential to change entire industries and professions by producing data for more intelligent intervention targeting (as in the case of PlantVillage), reducing waste and losses in production and consumption (as in the case of Smart Water Management), and bringing about the necessary improvements in connectivity and cost reductions (Clean Water AI), which enables many people worldwide to benefit from the rapid pace of technological advancement (Goralski & Tan, 2020).

AI can have a twin (enabler or inhibitor) impact on the sustainable development pursuit (Vinuesa et al., 2020). They explored the enabling and inhibiting role of AI regarding the 169 targets specified for the 17 SDGs and presented that AI acted as an enabler for 134 targets while serving as an inhibitor for 59 targets (for some targets, AI acted both potentially as an enabler and an inhibitor). AI has the potential to be a great enabler, benefiting the environment, economy, and society in many ways. AI, however, has the potential to be detrimental if it is overused or misused by people. To guarantee transparency, safety, and ethical norms, regulatory understanding and oversight are required for AI to assist sustainable development (Jungwirth & Haluza, 2023). According to Vinuesa et al. (2020), AI directly and favourably impacts 79% of the SDGs, including 93% of the environmental, 70% of the economic, and 82% of the social targets.

The impact of AI on the three pillars of sustainable development was examined by Hasan et al. (2023). These pillars include the environment (including climate change mitigation, biodiversity conservation, environmental planning, resource management, and decision-making), society (including more efficient distribution of food, health, water, and energy resources) and economy (including elimination of hazardous jobs and reduction of manual labour) (Hasan et al., 2023). According to Hasan et al. (2023), AI may be used to achieve the SDGs in fields such as waste management, clean water, agriculture, and industry. These techniques can be effectively implemented in developing nations like Bangladesh. By enhancing rather than displacing human intelligence and capacities, AI can help achieve the SDGs (Jain et al., 2023). One of the best examples of how AI may be used for energy efficiency is Google's DeepMind (Sudeep Srivastava, 2023).

17.5 CURRENT TRENDS OF AI AND SUSTAINABLE DEVELOPMENT

Healthcare, education, and marketing service providers worldwide, especially in developed countries, are aggressively implementing digital technologies like AI, ML, big data analytics, smart sensors and robots, and the Internet of Things to improve quality of life and operational efficiency (Davenport & Kalakota, 2019; Davenport et al., 2019; Anderson et al., 2021). Smart robots, natural language processing, ML, and other AI-based technologies present enormous opportunities for innovation in knowledge-intensive sectors (Secinaro et al., 2021).

AI is becoming more and more common in business and industry. According to Goralski and Tan (2020), it has the power to completely transform how we find, learn, live, communicate, and work. It has made notable progress in several fields recently (Ayodeji Olalekan Salau et al., 2022), including computer vision (Wang et al., 2021), science and health (Buch et al., 2018; Davenport & Kalakota, 2019; Secinaro et al., 2021), chemistry (Baum et al., 2021), natural language processing (Devlin et al., 2018), education (Melli, 2019; Biju et al., 2020; Hwang et al., 2020; Huang et al., 2021). recommendation systems (Khanal et al., 2020; Ahmad et al., 2022), sophisticated web search engines (Chau & Chen, 2008), condition monitoring systems (Zhao et al., 2019), automated decision-making (Leonetti et al., 2016), understanding human speech (Amberkar et al., 2018), intelligent sensing (Chen et al., 2022), competitive high-level strategic game systems (Silver et al., 2016), etc.

One of the main priorities of the AI field was to develop computer algorithms that can rival human reasoning in fundamental games like chess and checkers. Since then, AI research has attracted a lot of interest, especially in the wake of three achievements. The biggest computer chess player that could beat the world champion was called Deep Blue, IBM's first victory (Campbell et al., 2002). The second achievement was the work by Silver et al. (2016) and the third was an AI chess player who beat the top human chess player in 2017 (Ayodeji Olalekan Salau et al., 2022). The application of AI has benefitted some industries more than others, including the automotive, telecommunications, and healthcare sectors (Eager et al., 2020; Ayodeji Olalekan Salau et al., 2022). Walmart's Supply Chain Optimisation is a helpful case study. The massive retail chain has successfully implemented AI for sustainability in its complex supply chain. In addition to optimising shipping routes, advanced algorithms help reduce fuel usage (Sudeep Srivastava, 2023). Soft skills, including empathy and communication, are critical for sustainability and can be developed with AI. AI can be used, for instance, to design simulations that educate users about how their actions affect the environment and foster empathy for individuals impacted by climate change. Deloitte is one of the companies that uses AR and VR for sustainability to give staff members interactive learning environments. Another area where AI is making significant progress is smart waste management. A business called Evreka has significantly improved waste collection schedules and procedures by using ML algorithms, which lowers the need for landfills and operating expenses. One noteworthy example is a Pakistani waste-collecting company that used Evreka's sustainable technology. The impact of AI on agriculture is becoming more and more apparent. Leading start-ups, such as Taranis, are using data analytics and ML to advise farmers on the optimal use of fertilisers and water. Crop yields rise as a result, while waste and water use are drastically decreased.

It is evident that AI can significantly increase human labour productivity and efficiency while lowering operating costs in a variety of labour-intensive and time-consuming applications used by government organisations and many industrial verticals. Prominent AI applications encompass aiding physicians/doctors in their diagnostic duties, autonomous vehicles, keeping an eye on the functionality of costly machinery, making appropriate product recommendations at the ideal moment and place, and streamlining human-to-human communication and interaction through machine translation capabilities.

Critical problems for sustainable manufacturing, like the optimisation of energy resources, logistics, supply chain management, waste management, etc., can be solved by AI. Carvalho et al. (2019) claim that to cope with stricter laws, there is a tendency in smart production to integrate AI into environmentally conscious manufacturing processes. AI and ML are believed to be the primary drivers of the smart manufacturing revolution with the introduction of Industry 4.0. To reduce costs and increase the sustainability and quality of manufacturing processes, smart production systems require innovative solutions (Cioffi et al., 2020).

17.6 PROSPECTS OF AI AND SUSTAINABLE DEVELOPMENT

AI applications can increase value by enhancing customer service, personalising goods and helping to evolve new business models, product categories and even industries. In addition to the advantages and opportunities for companies, AI applications also benefit the workforce (e.g., increased worker safety) and are anticipated to offer major societal and economic benefits. AI has recently been used to anticipate significant increases in growth, productivity, innovation and job creation. According to Eager et al. (2020), labour productivity is predicted to increase by 11–37% by 2035.

The SDGs can be accomplished in large part through the application of AI in sustainable development initiatives. However, to minimise any negative effects and guarantee equitable results, it is imperative to ensure responsible and ethical AI development and implementation. However, there are drawbacks to using AI in sustainable development initiatives, such as ethical issues, a lack of information and experience and possible detrimental effects on employment and social inequalities. Therefore, it is crucial to guarantee ethical and responsible AI development and application that takes the effects on society and the environment into account.

AI is developing at a rapid pace, and its prospective uses in separation science open exciting new possibilities. AI algorithms, for example, may predict the characteristics of potential new materials, which accelerates the development of absorbent materials. ML powered by data can optimise operations to save energy waste and enhance the detection of mistakes by analysing large datasets pertaining to processes. Researchers are building specialised large-scale language models (LLMs) based on an extensive scientific corpus of articles, reference materials, and knowledge bases, driven by the recent increased use of Generative Pre-Trained Transformer models (GPTs). These models are helpful tools that help to choose appropriate separation approaches more quickly (Liu & Sun, 2023).

AI has the potential to revolutionise sustainability initiatives as it develops. Some of the prospects for AI and sustainability include:

1. Smart Cities: By utilising AI to optimise traffic patterns, building energy use and other infrastructure, cities may become more sustainable and lower their carbon impact.
2. Climate Modelling and Prediction: AI may be used to study climate data and forecast future patterns in the climate, which will help us better prepare for and lessen the effects of climate change.
3. Sustainable Agriculture: AI can be applied to optimise agricultural practises, such as crop rotation, irrigation, and fertilisation, to minimise environmental impact, enhance yield and reduce waste.
4. Sustainable Manufacturing: AI can be used to optimise the manufacturing process, cutting waste and utilising less energy, water and raw materials.
5. Renewable Energy: AI can be used to maximise the use of renewable energy sources, such as wind and solar power, increasing their effectiveness and affordability.

AI has a great deal of promise to support environmental initiatives as it develops. But it's crucial to think about the ethical implications of applying AI to sustainability projects and to ensure that these technologies are created and applied in an ethical and sustainable way.

17.7 CHALLENGES IN AI AND SUSTAINABILITY

Even with all its advantages, both the public and private sectors are adopting AI technologies at a very modest rate. AI technologies are expensive and take time to pay off economically. Additionally, there is a deficiency in the knowledge base and research necessary to comprehend sophisticated AI applications (How et al., 2020; Mehmood et al., 2020). Additionally, Vinuesa et al. (2020) discovered that unregulated AI poses a risk to 35% of the SDG targets, including SDGs 1, 4, 6, 9, and 10. This finding highlights the potential harm that AI poses to the advancement of the SDGs (Truby, 2019).

Risks related to AI also lead to social issues, including discrimination and cybercrime. Korwatanasakul and Takemoto (2021) pointed out that Japan, one of the highly advanced countries, is still in the initial stages of AI policy development, concentrating on general discussions on cybersecurity, ethics, governance, and rules and regulations – a reflection of the general lack of discussion on these topics in large parts of the world. AI development necessitates the collection of vast amounts of data while compromising security and privacy issues. Furthermore, recent events like the racist AI algorithm of Facebook and Google (Angwin & Grassegger, 2017; Vincent, 2020) and Amazon's sexist AI recruiting tool (Dastin, 2018) raise serious worries about discrimination and its negative effects on society and the economy.

The computational resources and big data centres required for AI applications, particularly ML, result in significant carbon footprints and intense energy usage. For example, the energy required to train OpenAI's massive GPT-3 text-generating model is equal to the energy required to go to and come from the moon (Quach, 2020).

While some see greater use of AI as a sign of prosperity in the economy, more free time and better leisure, others, including Bill Gates, Elon Musk and Stephen Hawking, warn that greater use of AI will worsen global economic inequality and signal a crisis of existential proportions for humanity (Sainato, 2015). Algorithms for AI have their own set of issues that need to be resolved. A few of them are as follows: Fairness and Bias Correction, Privacy and Security, Governance, Accountability and Legal Frameworks, Human Resource and Finance and Interpretability and Transparency, (Ramya & Manju, 2019). Instead of only creating large-scale technologies that are produced without consideration for human rights or public concerns, it is crucial to consider the ethics and protection of intelligent systems (Ramya & Manju, 2019). There are significant concerns about the possible detrimental effects of AI on the SDGs because a few countries and businesses dominate its development and deployment (Saetra, 2021).

One of the drawbacks of AI technology is that it produces a lot of e-waste, which is very hard to manage with Integrated Waste Management (IWM) methods. Additionally, the process of training AI requires a lot of energy (Hasan et al., 2023). The development of AI technology can be quite expensive. This further widens the economic gap between high-income and low- to mid-income nations. AI has a strong bias in favour of the educated, replacing physical labour with employment requiring greater technical expertise (Hasan et al., 2023). The most difficult ML and AI tasks, according to Karthick and Gopalsamy (2022), are data availability and tiredness brought on by a few technical faults. The preservation of user privacy, lack of digital infrastructure and ethical issues are among the other difficulties (Karthick & Gopalsamy, 2022).

The following are some potential risks and challenges that come with using AI in sustainability efforts, like with any new technology:

1. Bias and Discrimination: The data used to train AI algorithms determines their quality; biased data result in biased AI systems. Discriminatory results may result from this, especially when it comes to decisions about housing, employment and service accessibility.
2. Privacy Issues: AI learns and makes decisions by analysing enormous volumes of data. Personal information is frequently included in this data, and there are issues with the way that it is gathered, kept and utilised. Additionally, there is a chance that hackers or data breaches will compromise sensitive data.
3. Lack of Transparency: Because it can be difficult to understand how AI systems make judgements, they are frequently referred to as "black boxes." Due to this lack of transparency, holding AI systems responsible for their decisions can be difficult, especially when those decisions have a big bearing on people's lives.
4. Technology Dependence: We run the risk of excessively depending on technology as we use AI to make judgements more and more. This may result in the loss of human judgement and skills, as well as increase our susceptibility to cyberattacks and other interruptions of our technological infrastructure.
5. Environmental Impact: AI has an influence on the environment, even if it also can assist us in addressing environmental issues. AI systems depend on hardware that can be resource-intensive and difficult to recycle, and training AI algorithms consumes enormous quantities of energy. In the creation and application of AI systems for sustainability initiatives, it is critical for stakeholders to be aware of these prospective dangers and obstacles and to take steps to minimise them.

17.8 OVERCOMING CHALLENGES/MITIGATING RISKS

AI-driven sustainability efforts include many risks and challenges, just like any other technology. The following are some of the methods/ways for overcoming these risks and challenges:

1. Dealing with Bias: AI's potential to emphasise or even enlarge pre-existing biases in data and decision-making is one of its greatest risks. Ensuring the data used to train AI models is representative of all relevant groups and diverse is crucial to reducing this risk. It can also be required to create algorithms that specifically combat bias.
2. Ensuring Accountability and Transparency: Initiatives for sustainability powered by AI must be open and accountable to make sure they are serving the intended purpose and not inadvertently harming others. This could entail establishing precise rules for decision-making and making sure that there is a process in place for challenging decisions that might have unfavourable effects. Ensuring transparency and explainability of AI systems is crucial in enabling people to comprehend the rationale behind actions that impact them.

3. Guaranteeing Safety: AI systems are vulnerable to attacks and could be abused to do harm. Ensuring the security of systems and safeguarding data from unwanted access is crucial.
4. Protecting Privacy: AI systems can gather and analyse vast volumes of private information. It is critical to ensure the security of this data and the observance of individuals' right to privacy. Guaranteeing that individuals have control over their personal data and that it is gathered and utilised in an ethical and transparent manner are crucial.
5. Working Together: In AI-driven sustainability efforts, stakeholder collaboration is essential to identifying and mitigating risks. Working together can guarantee that a range of viewpoints are heard and that just and efficient solutions are found.
6. Design AI with a Human-Centred Approach: Lastly, it's critical to maintain the human aspect in AI and sustainability initiatives. Instead of completely replacing human judgement, AI should be developed to support human decision-making and enhance human welfare. We can make sure that AI is utilised responsibly and sustainably to advance the transition to a more sustainable future by considering the ethical issues. We can guarantee the efficacy, equity, and sustainability of AI-driven sustainability programmes by implementing these measures.

17.9 CONCLUSION

AI has the potential to meaningfully contribute to the advancement of sustainability programmes and the accomplishment of the SDGs. When properly utilised for projects and applications that promote sustainability, it will open enormous and widely dispersed business opportunities, facilitate the development of more efficient and effective public policies for sustainability, and, more specifically, enhance access, connectivity, and efficiency in a variety of industries, including farming, transportation, healthcare, and sanitation. Benefits including higher productivity, lower resource consumption, and better decision-making have been demonstrated.

However, there are also many risks and potential challenges, like ethical dilemmas and the possibility of escalating social and economic inequality. Stakeholders in the sustainability space should take these concerns seriously and strive to use AI-driven solutions in an ethical and responsible manner. AI has a lot of potential to contribute to a sustainable future with further technological developments and greater understanding of its potential applications.

REFERENCES

Ahmad, S.F., Alam, M.M., Rahmat, M.K., Mubarik, M.S., & Hyder, S.I. (2022) Academic and Administrative Role of Artificial Intelligence in Education. *Sustainability*, 14, 1101. https://doi.org/10.3390/su14031101

Amberkar, Aditya, Awasarmol, Parikshit, Deshmukh, Gaurav, & Dave, Piyush (2018) Speech Recognition using Recurrent Neural Networks. 1–4. https://doi.org/10.1109/ICCTCT.2018.8551185

Anderson, S. J., Chintagunta, Pradeep, Germann, Frank, & Vilcassim, Naufel (2021) Do Marketers Matter for Entrepreneurs? *Evidence from a Field Experiment in Uganda, Journal of Marketing*, 85(3), 78–96. https://doi.org/10.1177/0022242921993176

Angwin, J., & Grassegger, H. (2017) Facebook's Secret Censorship Rules Protect White Men from Hate Speech but Not Black Children. [online] Available at: https://www.propublica.org/article/facebook-hate-speech-censorship-internal-documents-algorithms

Baum, Zachary J., Yu, Xiang, Ayala, Philippe Y., Zhao, Yanan, Watkins, Steven P., & Zhou, Qiongqiong (2021) Artificial Intelligence in Chemistry: Current Trends and Future Directions, *Journal of Chemical Information and Modelling*, 61(7), 3197–3212. https://doi.org/10.1021/acs.jcim.1c00619

Biju, S.M., Salau, Ayodeji Olalekan, Eneh, Joy Nnenna, Sochima, Vincent Egoigwe, & Ozue, Izuchukwu ThankGod (2020) A Novel Pre-Class Learning Content Approach for the Implementation of Flipped Classrooms, *(IJACSA) International Journal of Advanced Computer Science and Applications*, 11(7), 131–136.

Buch, V.H., Ahmed, I., & Maruthappu, M. (2018) Artificial Intelligence in Medicine: Current Trends and Future Possibilities. *British Journal of General Practice*, 68(668), 143–144. https://doi.org/10.3399/bjgp18X695213. PMID: 29472224; PMCID: PMC5819974.

Campbell, Murray, Hoane Jr., A. J., & Hsu, Feng-hsiung (2002) Deep Blue, *Artificial Intelligence* 134, 57–83.

Carvalho, T.P., Soares, A. D. M. N., Vita, R., Francisco, R. D. P., Basto, J.P., & Alcalá, S.G.S. (2019) A Systematic Literature Review of Machine Learning Methods Applied to Predictive Maintenance, *Computers & Industrial Engineering*, 2019. https://doi.org/10.1016/j.cie.2019.106024

Chau, M., & Chen, H. (2008) A Machine Learning Approach to Web Page Filtering Using Content and Structure Analysis, *Decision Support Systems*, 44(2), 482–494.

Chen, Zhenghua, Wu, Min, Chan, Alvin, Li, Xiaoli, & Ong, Yew-Soon (2022) A Survey on AI Sustainability: Emerging Trends on Learning Algorithms and Research Challenges, *IEEE Computational Intelligence Magazine*, arXiv:2205.03824v1 [cs.AI] 8 May 2022.

Cioffi, R., Travaglioni, M, Piscitelli, G, Petrillo, A, & Felice, F. De (2020) Artificial Intelligence and Machine Learning Applications in Smart Production: Progress, *Trends, and Directions, Sustainability*,12(492), 1–26. https://doi.org/10.3390/su12020492

Dastin, J. (2018) Amazon Scraps Secret AI Recruiting Tool that Showed Bias Against Women. [online] Available at: https://www.reuters.com/article/us-amazon-com-jobs-automation-insight-idUSKCN1MK08G

Davenport, Thomas, & Kalakota, Ravi. (2019) The Potential for Artificial Intelligence in Healthcare, *Future Hospital Journal*, 6, 94–98. https://doi.org/10.7861/futurehosp.6-2-94

Davenport, T., Guha, A., Grewal, D., & Bressgott, T. (2019) How Artificial Intelligence Will Change the Future of Marketing, *Journal of the Academy of Marketing Science* https://doi.org/10.1007/s11747-019-00696-0

Devlin, J., Lee, Ming-Wei Chang Kenton, & Toutanova, Kristina (2018) BERT: Pre-training of Deep Bidirectional Transformers for Language Understanding, arXiv:1810.04805v2 [cs.CL] 24 May 2019.

Eager, James, Smit, Mark Whittlem Jan, Cacciaguerra, Giorgio, Lale-Demoz, Eugénie, & João Mendes Moreira, Anastasio (2020) *Opportunities of Artificial Intelligence*, Policy Department for Economic, Scientific and Quality of Life Policies Directorate-General for Internal Policies, 1–96. https://www.europarl.europa.eu/RegData/etudes/STUD/2020/652713/IPOL_STU(2020)652713_EN.pdf

Goralski, M.A., & Tan, T.K. (2020) Artificial Intelligence and Sustainable Development, *The International Journal of Management Education*, 18, 1–9. https://doi.org/10.1016/j.ijme.2019.100330

Hasan, Md. Tarek, Shamael, Nazmush, Mohammad, Akter, Arifa, Islam, Rokibul, Mukta, Md. Saddam Hossain, & Islam, Salekul (2023) An Artificial Intelligence-based Framework to Achieve the Sustainable Development Goals in the Context of Bangladesh. https://doi.org/10.48550/arXiv.2304.11703

How, Meng-Leong, Cheah, Sin-Mei, Chan, Yong-Jiet, Khor, Aik Cheow, & Say, Eunice Mei Ping (2020) Artificial Intelligence-Enhanced Decision Support for Informing Global Sustainable Development: A Human-Centric AI-Thinking Approach, *Information*, 11, 39, 1–24. https://doi.org/10.3390/info11010039

Huang, J., Saleh, Salmiza, & Liu, Yufei (2021) A Review on Artificial Intelligence in Education, *Academic Journal of Interdisciplinary Studies*, 20(3), 206–217.

Hwang, G. J., Xie, Haoran, Wah, Benjamin W., & Gasevic, Dragan (2020) Vision, Challenges, Roles and Research Issues of Artificial Intelligence in Education, *Computers and Education: Artificial Intelligence*, 1(2020), 100001.

Jain, T.K. (2023) Impact of Artificial Intelligence on Sustainable Development Goals, Electronic Copy. Available at: https://ssrn.com/abstract=4376842

Jungwirth, D., & Haluza, D. (2023) Artificial Intelligence and the Sustainable Development Goals: An Exploratory Study in the Context of the Society Domain, *Journal of Software Engineering and Applications*, 16, 91–112. https://doi.org/10.4236/jsea.2023.164006

Karthick, A.V., & Gopalsamy S. (2022) Artificial Intelligence: Trends and Challenges, 2022 Seventh International Conference on Parallel, Distributed and Grid Computing (PDGC), 978-1-6654-5401-8/13/$31.00, IEEE.

Khanal, S. S., Prasad, P., Alsadoon, A., & Maag, A. (2020) A Systematic Review: Machine Learning Based Recommendation Systems for E-Learning, *Education and Information Technologies*, 25(4), 2635–2664.

Korwatanasakul, U., & Takemoto, A. (2021) Leveraging Artificial Intelligence for Sustainable Development: Applying Social Principles for Human-centric AI. [online] Available at: https://www.eu-japan.ai/leveraging-artificial-intelligence-for-sustainable-development-applying-social-principles-for-human-centric-ai/

Leonetti, M., Iocchi, L., & Stone, P. (2016) A Synthesis of Automated Planning and Reinforcement Learning for Efficient, Robust Decision Making. *Artificial Intelligence*, 241, 103–130.

Liu, D., & Sun, N. (2023) Prospects of Artificial Intelligence in the Development of Sustainable Separation Processes. *Frontiers in Sustainability*, 4, 1210209. https://doi.org/10.3389/frsus.2023.1210209

Marr, Bernard (2018) What is Artificial Intelligence and How Will It Change Our World? Retrieved from https://www.bernardmarr.com/default.asp?contentID=963

Mehmood, H., Mukkavilli, S. K., Weber, I., Koshio, A., Meechaiya, C., Piman, T., Mubea, K., Tortajada, C., Mahadeo, K., & Liao, D. (2020) *Strategic Foresight to Applications of Artificial Intelligence to Achieve Water-related Sustainable Development Goals*. UNU-INWEH Report Series, Issue 09. United Nations University Institute for Water, Environment and Health, Hamilton, Canada., http://inweh.unu.edu/publications

Melli, G., (2019) The Role of Artificial Intelligence in the Future of Education. https://www.getsmarter.com/blog/market-trends/the-role-of-artificial-intelligence-in-the-future-of-education/

Nasir, Osama, Javed, Rana Tallal, Gupta, Shivam, Vinuesa, Ricardo, & Qadir, Junaid (2023) Artificial Intelligence and Sustainable Development Goals Nexus via four Vantage Points, *Technology in Society*, 72, 1–7, https://doi.org/10.1016/j.techsoc.2022.102171, (https://www.sciencedirect.com/science/article/pii/S0160791X22003128)

Porwal, A., Chaturvedi, A., & Gupta, S. (2022) Artificial Intelligence with Special Reference to Blockchain Technology: A Future of Accounting, *Impact of Artificial Intelligence on Organizational Transformation*, 289–303.

Quach, K. (2020) AI me to the Moon… Carbon Footprint for 'Training GPT-3' Same as Driving to Our Natural Satellite and Back. [online] Available at: https://www.theregister.com/2020/11/04/gpt3_carbon_footprint_estimate/

Ramya, P., & Manju, M. (2019) Artificial Intelligence for Sustainable Education in India: Problems and Prospects, *International Journal of Production Technology and Management (IJPTM)*, 10(1), 135–143, Article ID: IJPTM_10_01_014 Available online at http://iaeme.com/Home/issue/IJPTM?Volume=10&Issue=1 ISSN Print: 0976-6383 and ISSN Online: 0976 - 6391 https://doi.org/10.17605/OSF.IO/6254S

Rasouli, A.H., & Kumarasuriyar, A. (2016) The Social Dimension of Sustainability: Towards Some Definitions and Analysis. Journal of Social Science for Policy Implications,4(2), 23–34, DOI: 10.15640/10.15640/jsspi.v4n2a3

Sachs, J. D. (2015). *The age of sustainable development.* Columbia University Press.

Saetra, H.S. (2021) AI in Context and the Sustainable Development Goals: Factoring in the Unsustainability of the Sociotechnical System, *Sustainability*, 13, 1738. https://doi.org/10.3390/su13041738

Sainato M., (2015) Stephen Hawking, Elon Musk, and Bill Gates Warn About Artificial Intelligence, *Artificial Intelligence.* https://observer.com/2015/08/stephen-hawking-elon-musk–and-bill-gates-warn-about-artificial-intelligence/

Salau, Ayodeji Olalekan, Demilie, Wubetu Barud, Akindadelo, Adedeji Tomide, & Eneh, Joy Nnenna (2022) Artificial Intelligence Technologies: Applications, Threats, and Future Opportunities, ACI'22: Workshop on Advances in Computation Intelligence, its Concepts & Applications at ISIC 2022, May 17–19, Savannah, United State, 265–273, https://ceur-ws.org/Vol-3283/Paper107.pdf

Secinaro, S., Calandra, D., Secinaro, A., Vivek Muthurangu, V., & Biancone, P. (2021) The Role of Artificial Intelligence in Healthcare: A Structured Literature Review, *BMC Medical Informatics and Decision Making* 21(125). https://doi.org/10.1186/s12911-021-01488-9

Silver, D., Huang, A., Maddison, C. et al. (2016) Mastering the game of Go with deep neural networks and tree search, *Nature*, 529, 484–489. https://doi.org/10.1038/nature16961

Sudeep Srivastava (2023) How AI, IoT, and AR/VR Technologies are Helping to Achieve Their Sustainability, https://appinventiv.com/blog/technology-for-sustainability/

Truby, Jon (2019) Governing Artificial Intelligence to Benefit the UN Sustainable Development Goals, *Sustainable Development*, 28, 946–959. https://doi.org/10.1002/sd.2048

UN (2015) Transforming Our World: The 2030 Agenda for Sustainable Development. Resolution Adopted by the General Assembly on 25 September 2015, 42809, 1–13.

UNEP (2015) Environmental, *Social and Economic Sustainability Framework*, 1–35.Link: https://www.cbd.int/financial/doc/unep-environmentalframework.pdf

UNESCO (2019) Artificial intelligence for sustainable development: challenges and opportunities for UNESCO's science and engineering programmes, working paper, August 2019 (Revised).

United Nations (2015) *Transforming Our World: The 2030 Agenda for Sustainable Development.* Division for Sustainable Development Goals: New York, NY, USA, 2015.

Vincent, Pedemonte (2020) AI for Sustainability: An Overview of AI and the SDGs to Contribute to the European AI-sustainability-overview-ai-and-sdgs-contribute-european-policy-making? language=enpolicy making, https://futurium.ec.europa.eu/en/european-ai-alliance/document/

Vinuesa, R., Azizpour, H., Leite, I., Balaam, M., Dignum, V.. Domisch, S., Felländer, A., Langhans, S.D., Tegmark, M., & Nerini, F.F. (2020) The Role of Artificial Intelligence in Achieving the Sustainable Development Goals, *Nature Communications*, 11(233), 1–10. https://doi.org/10.1038/s41467-019-14108-y

Wang, Z., Chen, J., & Hoi, S. C. H. (2021) Deep Learning for Image Super-Resolution: A Survey, *IEEE Transactions on Pattern Analysis and Machine Intelligence*, 43(10), 3365–3387, 1 October 2021. https://doi.org/10.1109/TPAMI.2020.2982166

Zhao, R., Yan, R., Chen, Z., Mao, K., Wang, P., & Gao, R. X. (2019) Deep Learning and Its Applications to Machine Health Monitoring, *Mechanical Systems and Signal Processing*, 115, 213–237. https://doi.org/10.1016/j.ymssp.2018.05.050

Index

Pages in *italics* refer to figures and pages in **bold** refer to tables.

C

D

E

F

P

Q

R

S

For Product Safety Concerns and Information please contact our EU
representative GPSR@taylorandfrancis.com
Taylor & Francis Verlag GmbH, Kaufingerstraße 24, 80331 München, Germany

www.ingramcontent.com/pod-product-compliance
Ingram Content Group UK Ltd.
Pitfield, Milton Keynes, MK11 3LW, UK
UKHW022217190926
475702UK00003B/402
* 9 7 8 1 0 4 1 0 6 9 9 3 5 *